Proceedings of the 8ᵗʰ European Congress of Neurosurgery

Barcelona, September 6–11, 1987

Edited by

F. Isamat, A. Jefferson, F. Loew, L. Symon

Volume 1

Intraoperative and Posttraumatic Monitoring and Brain Protection
Cerebro-vascular Lesions
Intracranial Tumours
Benign Intracranial Cystic Lesions, Hydrocephalus, CSF-Volumes
Central Pain Syndromes

Acta Neurochirurgica
Supplementum 42

Springer-Verlag Wien New York

Professor Dr. Fabian Isamat
Servicio de Neurocirugía, Hospital de Bellvitge, University of Barcelona, Spain

Dr. Antony Jefferson
Newport, Dyfed, United Kingdom

Professor Dr. Friedrich Loew
Neurochirurgische Universitätsklinik, Homburg/Saar, Federal Republic of Germany

Professor Lindsay Symon, TD, FRCS
Department of Neurological Surgery, Institute of Neurology, The National Hospital, London, U.K.

With 100 Figures

Library of Congress Cataloging-in-Publication Data. European Congress of Neurosurgery (8th: 1987: Barcelona, Spain) Proceedings of the 8th European Congress of Neurosurgery: Barcelona, September 6–11, 1987/edited by F. Isamat ... [et al.]. IX, 252 p. 21 × 27.7 cm.— (Acta neurochirurgica. Supplementum, ISSN 0065-1419; 42). 1. Nervous system—Surgery—Congresses. I. Isamat, F. II. Title. III. Series. [DNLM: 1. Neurosurgery—congresses. W1 AC8661 no. 42/WL 368 E89p 1987]. RD593.E95 1987. 617'.48-dc19. 88-20167

ISSN 0065-1419

ISBN-13:978-3-7091-8977-1   e-ISBN-13:978-3-7091-8975-7
DOI: 10.1007/978-3-7091-8975-7

Preface

The 8th European Congress of Neurosurgery which took place in Barcelona from September 6 to 11, 1987, was an unforgettable experience. Many factors contributed to its success: the splendid ambience of Barcelona, the generous hospitality and warm friendship of the hosts, and the marvellous organization and high scientific standard of papers, workshops and discussions.

For financial reasons it would not have been possible to publish all the papers presented during the congress. In order to preserve as much as possible of the scientific results, the Programme Committee of the European Association of Neurosurgical Societies had already selected before the congress a certain number of abstracts to be published as full papers. These are compiled in two supplement volumes of Acta Neurochirurgica and deal with the following main topics:

Volume 1: 1. Intraoperative and posttraumatic monitoring and brain protection
2. Cerebro-vascular lesions
3. Intracranial tumours
4. Benign intracranial cystic lesions, hydrocephalus, CSF-volumes
5. Central pain syndromes

Volume 2: 6. Spinal cord and spine pathologies
7. Basic research in neurosurgery.

The Editors
F. Isamat, A. Jefferson, F. Loew, L. Symon

Contents

I. Intraoperative and Posttraumatic Monitoring and Brain Protection

II. Cerebro-vascular Lesions

III. Intracranial Tumours

IV. Benign Intracranial Cystic Lesions, Hydrocephalus, CSF-Volumes

V. Central Pain Syndromes

I. Intraoperative and Posttraumatic Monitoring and Brain Protection

Acta Neurochirurgica, Suppl. 42, 3–7 (1988)

Assessment of Reversible Cerebral Ischaemia in Man: Intraoperative Monitoring of the Somatosensory Evoked Response

L. Symon, F. Momma, and **T. Murota**

Gough-Cooper Department of Neurological Surgery, Institute of Neurology, The National Hospital, London, U.K.

Summary

Central conduction time (CCT) has been monitored in 37 patients undergoing temporary arterial occlusion in aneurysm surgery. 17 patients had internal carotid, 17 had middle cerebral, and 4 had basilar artery occlusion. Internal carotid or middle cerebral artery occlusion lasting less than 12 minutes has not been associated with postoperative morbidity, in any case without appreciable change in CCT after occlusion. Prompt prolongation of CCT was warned the surgeon, but CCT prolongation up to 10 ms could occur without permanent neurological deficit, except in one Grade 4 patient. 10 of 18 patients who lost the N 20 cortical potential showed postoperative neurological deficit, which was promptly recoverable in 7 patients. The speed of loss or recovery of N 20 enabled a patient's prognosis to be predicted. Irrecoverable postoperative deficit is unlikely if the N 20 takes longer than 4 minutes to disappear, to reappears within 20 minutes after recirculation.

Keywords: Somatosensory evoked potentials; temporary clipping; cerebral ischaemia.

Introduction

A threshold relationship between the maintenance of the somatosensory evoked response and local cortical blood flow has been established in anaesthetized primates[2]. Another lower flow threshold for the maintenance of ionic homeostasis has been described, disruption of which would eventually lead to cellular death[1]. Electrical silence can, thus, be sustained without irrecoverable damage to the cell membrane, or permanently disturbed neuronal function. Clinical correlates of primate thresholds have been established in man and monitoring of the somatosensory evoked responses has been evaluated in the actue phase of ischaemia for protection of the brain[4].

Clinical Materials and Methods

We have monitored somatosensory evoked responses during aneurysm surgery in approaching 150 cases and the value of monitoring CCT has been reported previously[5]. 37 patients underwent temporary arterial occlusion (Tables 1, 2, and 3), and changes in CCT have been related to postoperative neurological status.

Patient's Neurological Status

The unmodified Hunt and Hess scale[3] was used. Seven patients were in Grade 0, ten in Grade 1, ten in Grade 2, eight in Grade 3, and two in Grade 4 preoperatively.

Operation

All operations were undertaken by one surgeon (L.S.) under a standard anaesthetic technique described elsewhere[6]. Scoville clips were used in all cases for temporary occlusion.

Monitoring Technique

Our standard technique has been previously described[5]. A small frontotemporal skin incision enabled the area accessible to this group of aneurysms to be free of electrode sites. The contralateral median nerve was stimulated with a repetition rate of 3–4 Hz and 128 responses were averaged using a Digitimer D 200 system, allowing rapid measurement of CCT, traces were then stored on floppy discs. Postoperative measurements were made in the recovery room and on return to the ward.

Results and Discussion

1. Preservation of Conduction Time

We have reported false-negative monitoring of CCT during bilateral anterior cerebral occlusions resulting in bifrontal infarction, reflecting ischaemic areas not involved in the neural pathways monitored[5]. With internal carotid or middle cerebral artery occlusion up to 12 minutes in duration, no case with preserved conduction has been associated with postoperative morbidity. Monitoring of CCT has correctly assessed the safety of these brief occlusions.

2. Prolongation of Conduction Time

The conduction time was considerably increased in 6 patients (4 internal carotid, 2 middle cerebral artery

Table 1. *ICA Occlusion with SEP Recording—17 Patients*

Patient (Grade)	Time of occlusion	Change in CCT Operated side	Change in CCT Opposite side	Time to disappear- ance of N 20	Recovery time of N 20	Post- operative morbidity
OPS-7 (0)	10 min	7.0–10.0	6.8–7.0		9 min	recoverable
OPS-14 (2)	3 min 35 sec	6.4–6.4	6.4—6.4			no
OPS-19 (1)	1 min 20 sec	9.2-flat	8.8–8.8	1 min	8 min	recoverable
OPS-29 (1)	26 sec	6.4–6.6	6.0–6.0			no
OPS-41 (1)	27 min 44 sec	8.2-flat	8.0–8.0	1 min	7 min	recoverable
OPS-46 (2)	26 min	7.6-flat	7.6–7.8	5 min	19 min	no
OPS-49 (0)	26 min	7.6-flat	7.6–7.6	1 min	40 min	permanent
OPS-56 (1)	1 min 25 sec	5.4–10.0	5.4–6.2		12 min 50 sec	no
OPS-60 (3)	2 min	6.8–8.0	6.6–6.6		18 min	no
OPS-63 (2)	3 min	6.8–7.0	6.8–6.8			no
OPS-73 (2)	8 min	7.2-flat	7.2–7.2	6 min	1 min	no
OPS-79 (4)	3 min 50 sec	6.4–7.4	6.8–7.0		20 min	permanent
OPS-82 (3)	2 min	5.6–5.6	5.6–5.6			no
OPS-90 (0)	38 min	8.2-flat	7.6–7.6	2 min	3 min	no
OPS-95 (0)	69 min	7.8-flat	7.8–7.8	30 sec	4 hours	permanent
OPS-114 (0)	12 min	6.8-flat	6.6–6.8	30 sec	7 min	no
OPS-129 (1)	5 min 10 sec	6.6 8.8	6.6—6.8		3 min	no

occlusions), resulting in an interhemispheric difference longer than 0.6 ms. Concerning occlusions less than 10 minutes in duration, delay of conduction up to 10 ms could occur without permanent neurological deficit except in one Grade 4 patients, but rapid prolongation of CCT to such an extent should be regarded as a warning signal.

3. Disappearance of N 20

Eighteen patients lost the cortical potential in this series (Twice in one patient with middle cerebral artery occlusion). Postoperative neurological deficit was seen in 10 of 18 patients, but recovered in 7. The speed of loss or recovery of N 20 has appeared of value to predict patient's prognosis. Prompt loss of N 20 within 1 minute has occurred in 6 patients, 5 of whom showed postoperative deficit (Fig. 1). Cerebral bloow flow presumably reduced rapidly to levels below threshold in these cases reflecting poor collateral circulation, as demonstrated in cortical ischaemic lesions in primates[2]. No case in which the N 20 took longer than 4 minutes (Fig. 1) or reappeared within 20 minutes after release has been associated with irrecoverable neurological deficit (Fig. 2).

Table 2. *MCA occlusion with SEP recording—17 patients*

Patient (Grade)	Time of occlusion	Change in CCT		Time to disappear-ance of N 20	Recovery time of N 20	Post-operative morbidity
		Operated side	Opposite side			
OPS-8 (2)	8 min	7.0–7.4	7.0–7.0		5 min	no
OPS-10 (3)	4 min 57 sec	6.8–6.8	6.6–6.6			no
OPS-12 (2)	1 min	7.8–7.8	7.6–7.6			no
OPS-16 (2)	2 min 45 sec	6.0–6.0	5.8–5.8			no
OPS-18 (3)	12 min	8.0–8.0	8.0–8.0			no
OPS-39 (3)	2 min	7.2–7.2	7.2–7.2			no
OPS-60 (3)	6 min	6.8–8.0	6.6–6.6		18 min	no
OPS-61 (3)	5 min 15 sec	7.0–8.6	6.6–6.6		5 min	no
OPS-65 (0)	6 min 18 sec	9.0-flat	8.8–8.8	3 min 45 sec	45 min	recoverable
OPS-68 (3)	6 min 40 sec	9.4-flat	9.4–9.4	1 min 35 sec	14 min	no
OPS-77 (2)	10 min 45 sec	6.4–6.4	6.2–6.2			no
OPS-86 (0)	6 min 45 sec	7.4–7.4	7.4–7.4			no
OPS-101 (1)	7 min 48 sec	7.0-flat	6.8–7.0	2 min	3 min 25 sec	recoverable
OPS-110 (3)	15 min 50 sec	6.4-flat	6.4–6.4	30 sec	90 min	recoverable
OPS-118 (1)	8 min 13 sec	6.8-flat	6.6–6.6	9 min	4 min 47 sec	no
OPS-119 (2)	6 min	6.4-flat	6.4–6.4	4 min	18 min	recoverable
OPS-123 (2)	12 min 4 min	7.6-flat 8.4-flat	7.4–7.4 7.4–7.4	6 min 2 min	2 min 1 min	recoverable

4. Change in N 20 Amplitude

The ratio between the N 20 and N 14 amplitude disclosed no extra information except during occlusion[5]. In one case of posterior communicating aneurysm for example, the proximal internal carotid artery was occluded for 5 minutes and 10 seconds without any postoperative neurological deficit. It was not until 6 minutes after occlusion that the conduction delayed considerably from 6.6 to 8.8 ms. Although the N 20 amplitude gradually diminished after occlusion, this could alert the surgeon to developing ischaemia. Temporary occlusion enabled the aneurysm to be obliterated satisfactorily and quickly without losing the cortical potential (Fig. 3).

Conclusion

Intraoperative monitoring of central conduction time has thus a proven value in internal carotid, middle cerebral, or basilar artery occlusion in terms of protection of the brain.

1. Preservation of the conduction, even if it is in-

Table 3. *BA Occlusion with SEP Recording—4 Patients*

Patient (Grade)	Time of occlusion	Change in CCT		Time to disappear-ance of N 20	Recovery time of N 20	Post-operative morbidity
		Operated side	Opposite side			
OPS-71 (4)	15 min 50 sec	8.0-flat	7.0-flat	5 min	26 min	recoverable
OPS-81 (1)	13 min 30 sec	8.2-flat	7.4-flat	3 min	12 min	no
OPS-106	12 min	7.4-flat	7.4-flat	N 20 disappeared 40 minutes before occlusion on both sides, and took 3 hours to recover on the affected side and 1 hour on the other.		recoverable
OPS-128 (1)	3 min 40 sec	8.6–9.0				
	6 min 25 sec	9.0–9.2				
	9 min 30 sec	8.8-flat		9 min	10 min	no

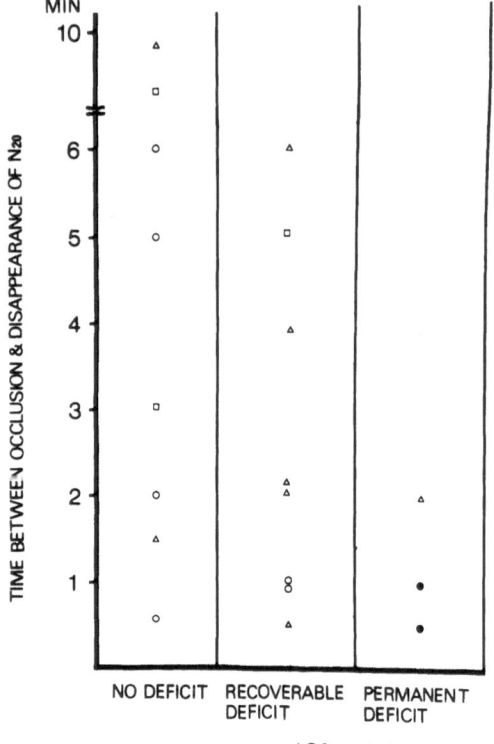

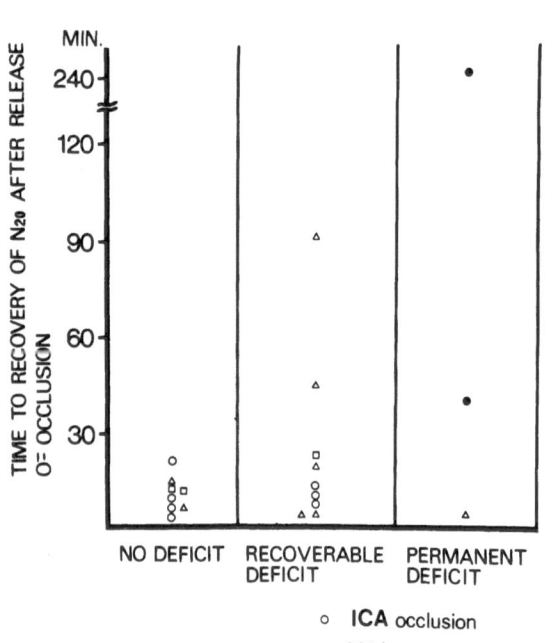

Fig. 1. The relationship between the speed of loss of N 20 and postoperative morbidity is shown. Black circles indicate fatal outcome. No case in which the N 20 potential took longer than 4 minutes to disappear has been associated with postoperative irrecoverable neurological deficit. Rapid disappearance of N 20 should be regarded as a danger signal

Fig. 2. The speed of recovery of N 20 appeared of use in predicting patient's prognosis. Postoperative irrecoverable deficit is unlikely when the cortical potential reappears within 20 minutes after recirculation. There was one exception in which the N 20 recovering 9 minutes after release of MCA occlusion resulted in an incompletely recoverable dysphasia

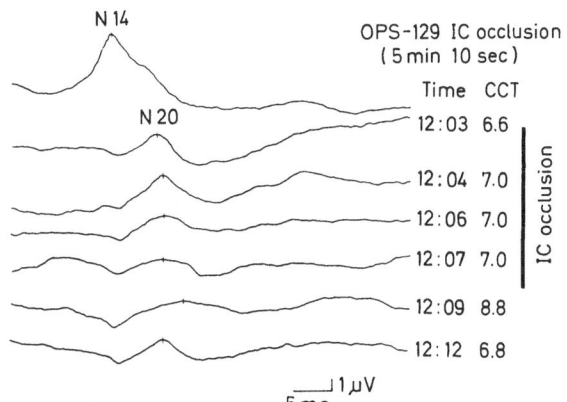

Fig. 3. Changes in the conduction time and N 20 amplitude after ICA occlusion are shown. Changes in CCT were preceded by those in N 20 amplitude which alerted the surgeon to developing ischaemia in this case. The posterior communicating aneurysm was quickly obliterated and temporary occlusion was released just before losing the cortical potential

creased to the level of 10 ms, may guarantee the safety of temporary occlusion for prolonged periods.

2. The speed of loss of N 20 following can be of use to recognize the degree of ischaemia, prompt disappearance of the N 20 is obviously a danger signal. Slow disappearance over 4 minutes after occlusion may allow temporary occlusion to be carried on for another 10 minutes.

3. The speed of recovery of N 20, which seems independent of the duration of occlusion, is likely to predict patient's prognosis after surgery.

4. Permissible time for temporary arterial occlusion may be assessed with intraoperative use of the somatosensory evoked response.

References

1. Astrup J, Symon L, Branston NM, Lassen NA (1977) Cortical evoked potential and extracellular K^+ and H^+ at critical level of brain ischaemia. Stroke 8: 51–57
2. Branston NM, Symon L, Crockard HA, Pasztor E (1974) Relationship between the cortical evoked potential and local cortical blood flow following acute middle cerebral artery occlusion in the baboon. Exp Neurol 45: 195–208
3. Hunt WE, Hess RM (1968) Surgical risk as related to time of intervention in the repair of intracranial aneurysms. J Neurosurg 28: 14–20
4. Momma F, Wang AD, Symon L (1987) Effects of temporary arterial occlusion on somatosensory evoked responses in aneurysm surgery. Surg Neurol 27: 343–352
5. Symon L, Wang AD, Silva IEC, Gentili F (1984) Perioperative use of somatosensory evoked responses in aneurysm surgery. J Neurosurg 60: 269–275
6. Wang AD, Silva IEC, Symon L, Jewkes D (1985) The effects of halothane on somatosensory and flash visual evoked potentials during operations. Neurol Res 7: 58–62

Address for correspondence: Lindsay Symon TD, FRCS (Ed), FRCS, Professor of Neurosurgery, The Gough-Copper Department of Neurological Surgery, Institute of Neurology, The National Hospital, Queen Square, London WC1N 3BG, U.K.

Acta Neurochirurgica, Suppl. 42, 8–13 (1988)

Monitoring of Brain Function by Means of Evoked Potentials in Cerebral Aneurysm Surgery

A. Ducati, A. Landi, M. Cenzato, E. Fava[1]**, P. Rampini, M. Giovanelli,** and **R. Villani**

Institute of Neurosurgery, University of Milano and [1] CNR Centre for Muscle Physiology, Milano, Italy

Summary

Deliberate arterial hypotension is currently used to operate upon cerebral aneurysms. However, it is not ascertained whether this practice is really safe for all patients, especially those presenting with preoperative vasospasm. 50 patients, requiring surgical treatment for cerebral aneurysm, have been submitted, during surgery, to the recording of Somatosensory Evoked Potentials (SEPs) on median nerve stimulation. This technique allows the functional evaluation of neural pathways mediating the somatosensory stimuli and of primary somatosensory cortex; it is known that a decrease of cerebral perfusion may affect the SEP waveforms in terms of reduced subcortical conduction velocity (i.e., increased central conduction time, CCT) and of reduced cortical response amplitude. These changes may be apparent before a permanent neurological damage is produced. Preoperative SEP recording demonstrated a prolonged CCT, possibly related to vasospasm, in 9 patients, a normal clinical evaluation notwithstanding (grade I and II).

During intraoperative deliberate hypotension, a SEP change has always been produced. No postoperative damage has been observed, however, as long as the CCT did not exceed 9 msec for 10 minutes (maximum normal CCT value is 6.7 msec) and as the cortical response had been visible throughout the whole surgical procedure. The critical value of CCT has been reached at a mean arterial pressure (MAP) lower than 60 Torr in patients with a normal preoperative SEP recording; at the opposite, in patients presenting with a prolonged preoperative CCT, the value of 9 msec was arrived at with a MAP value that is generally accepted as safe for all patients (75 Torr).

In conclusion, SEP monitoring has proved to be a useful technique to identify the minimum safe MAP value for the individual patient undergoing surgery for cerebral aneurysm.

Introduction

Intraoperative monitoring of Central Nervous System function by means of Evoked Potential (EP) recording has become a common practice during neurosurgical procedures. Several reports deal with EP changes during posterior fossa, spine, spinal cord, and extracranial carotid artery operations (Grundy 1982, Grundy *et al.* 1982, Markand *et al.* 1984, Ducati *et al.* 1988, Jones 1988, Landi *et al.* 1988).

Only a few papers discuss the application of this technique to aneurysm surgery (Symon *et al.* 1979, Symon and Wang 1986, Kidooka *et al.* 1987, Koht *et al.* 1987). Either global or focal changes in cerebral perfusion commonly take place in aneurysm surgery, due to the use of deliberate arterial hypotension, brain retraction, temporary clipping of the parent vessel or inadvertent closure of terminal arteries. When these changes exceed the subject threshold for neural ischemia, severe or even fatal consequences complicate the postoperative course. There is now a sound evidence that Somatosensory Evoked Potentials (SEPs) can detect, in the individual patients, inadequate perfusion of both cortical and subcortical areas (Grundy 1985). This is the rationale for using SEP monitoring during aneurysm surgery. Intraoperative SEP alterations may also depend upon non pathological factors, such as anesthetic agents and hypothermia (Grundy 1982). There is therefore the need for a precise identification of the tolerance limits of intraoperative SEP modifications, i.e., the changes related to cerebral ischemia and possibly associated with a complicated postoperative course. The literature data are conflicting with this respect. Symon and Wang (1986) report that a Central Conduction Time (CCT) up to 10 msec is without significance, while Kidooka *et al.* (1987) observe a patient who died after surgery whose CCT never exceeded 8.1 msec. Moreover, CCT measurements study the subcortical conduction alone, disregarding all the cortical responses taking place after the early negativity (N 20). It is well known that in case of reduced middle cerebral artery (MCA) perfusion the earliest and more severe ischemia is to be expected in the peripheral areas of

MCA territory, where the SEP late waves are generated. We observed that, during extracranial carotid surgery, the first sign of cortical hypoperfusion is the disappearance of late SEP components (Ducati *et al.* 1987 a).

The aim of the present study is to contribute information concerning the value of SEP cortical component analysis and the tolerance limits of CCT during aneurysm surgery, related to anesthesiological and surgical management, in order to increase the individual patient's safety. Furthermore, we studied the preoperative SEPs in order to identify reliable parameters that could modify the intraoperative management and make it tailored for the individual patient's needs.

A preliminary report of this experience has been published elsewhere (Ducati *et al.* 1987 b).

Patients and Methods

50 patients underwent SEP recording during operation for direct approach to intracranial aneurysm over a 3 years period. The cases were non consecutive, but no selection criteria have been employed. In 15 cases the aneurysm was located in the ACA-ACoA complex; in 19 cases on the ICA-PCoA complex; in 12 on the MCA; in 4 cases in the VB circulation.

Preoperative neurological evaluation was conducted according to the classical Hunt and Hess scale (1968). 5 patients were grade I; 21 were grade II; 15 were grade III; 9 were grade IV. 38 were operated upon within the third post-bleeding day.

SEP recording was carried out at least once before surgery, using the Amplaid MK 15 apparatus (Amplaid – Italy). The median nerve at the wrist was stimulated with surface electrodes, cathode 25 mm proximal to the anode, at an intensity 3 times threshold for a twitch in the opponens pollicis muscle. Short latency responses were collected on a 30 msec time basis and the stimulation frequency was 4 pulses per second. Active analog filters were employed (12 dB/octave rolloff) with a bandpass of 10–2,500 Hz. Subcutaneous platinum needle electrodes were placed on the parietal regions (C 3' and C 4') and over the C-2 spinous process. A 3 channels bipolar recording was carried out, referring the active electrodes (C-2 spinal – C 3' and C 4') to a midfrontal electrode (Fpz). 250 traces were averaged. The cervical N 14 response and the N 20 cortical response (on the hemisphere contralateral to the stimulated nerve) were analyzed, measuring their amplitudes and their peak latencies. The central conduction time (CCT) was computed, as the difference between N 20 and N 14 latency (Hume and Cant 1978). The normal value of CCT was derived from a control population age and sex matched. It appeared to be $5.7 \pm - 0.4$ msec; maximum normal CCT value was 6.7 msec (mean + 2.5 SD). Cortical SEP responses were collected on a 200 msec time basis, using a stimulation frequency of 1 pulse per second. The recording was carried out between C 3'/C 4' and Fpz, using a filtering bandpass of 2–100 Hz. The amplitude of the N 20–P 25 primary response and amplitude/symmetry of the later waves were considered. As stated before, there is good evidence that the generators of the N 20–P 25 component and of the later waves are different, being the former in the primary somatosensory cortex, the latter in the posterior parietal cortex (Ducati *et al.* 1984).

2 of the 4 VB aneurysms were approached through the posterior fossa. In these cases, brainstem acoustic evoked potentials (BAEPs) were collected in addition to SEPs. 2,000 click stimuli of 100 dB p.e. sound pressure level were delivered, at a frequency of 11 Hz, through a shielded earphone to the right ear and averaged on a 12 msec time basis. Filtering bandpass was 100–2,000 Hz. Active electrode was on Cz; reference electrode was on the mastoid of the stimulated ear. The interpeak latency between wave I (acoustic nerve activity) and wave V (contralateral upper pontine generators) was computed; in normals, its mean value is 4.02 msec, with a maximum value of 4.5 msec (mean + 2.5 SD).

Anesthesia was induced with a short acting barbiturate (Thiopental 5–7 mg/kg) and Succinylcholine chloride (1 mg/kg); maintenance was achieved by Nitrous Oxide in Oxygen (2 : 1) and Enflurane (0.8–1%). Deliberate hypotension was produced by continuous infusion of Nitroglycerin (4 microgram per milliliter). Body temperature was maintained between 35 and 36 °C. Mean arterial pressure (MAP) and Electrocardiogram were continuously monitored; blood gases and serum electrolytes were checked every 30 minutes.

Results

Preoperative SEP measurements demonstrated that the mean value of CCT had some relation with the clinical grading. Grade I patients presented with a CCT of 5.5 ± 0.9 msec, grade II with 6.2 ± 0.8, grade III with 6.7 ± 0.5, grade IV with 8.5 ± 0.6. Grade I and grade II mean CCT, although at the opposite limits of the normal range, are not statistically different. Grade III values are different from grade I but not from grade II. Grade IV CCT values are highly different from both grade I and II. The variability of values was higher among low grade patients; we observed one grade I patient with an abnormal bilateral CCT (patient A.A., ACoA aneurysm; right median CCT of 6.7 msec and left median CCT of 6.9 msec). In this particular patient, transcranial doppler sonography demonstrated a bilateral MCA flow speed exceeding 150 cm/sec, as compared with a normal value of 80. Furthermore, angiography showed the presence of spasm on both MCAs. The correspondence among CCT, flow speed measured with transcranial doppler sonography and angiography has been confirmed in other 4 patients belonging to grade III and IV.

After induction of anaesthesia CCT became always longer by 0.6–0.8 msec on both sides, as compared to preoperative measurements.

When hypotension was induced, a further bilateral prolongation of CCT took place. MAP level was kept around 95 Torr during the initial phase of surgery; it was decreased to around 75 Torr when approaching the aneurysm; and further decreased to 60 Torr during dissection and clipping of the aneurysm. The duration of low-grade hypotension (60 Torr) varied between 12

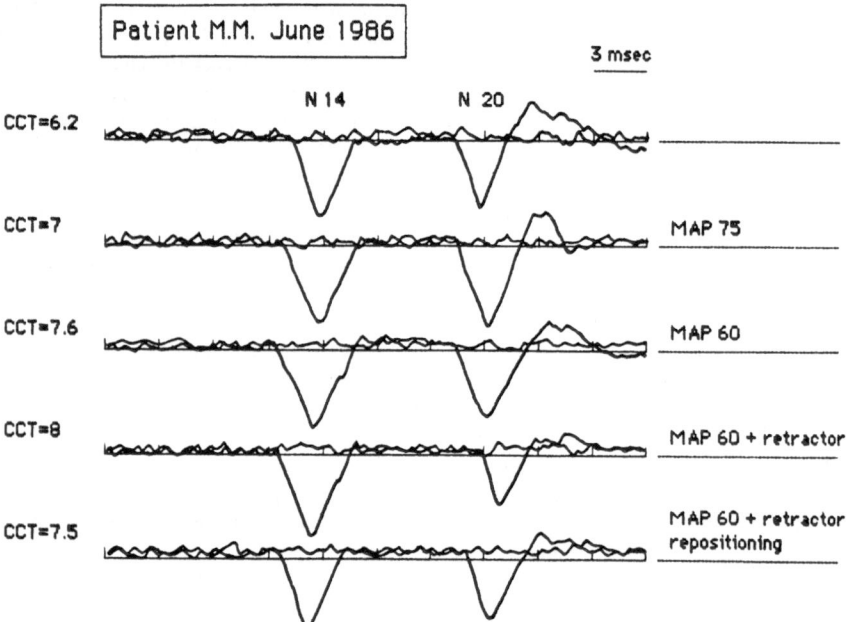

Fig. 1. The figure shows a sequence of s.l. SEPs traces collected during a surgical procedure for an ACoA aneurysm. Traces recorded in SC 2-Fpz and C 4-Fpz are superimposed to demonstrate CCT value. Note the effect of retractor positioning and of hypotension on CCT value, CCT duration is shown on the left side of the picture. In this and in the following figures the active electrode negativity is shown as a downward deflection

and 38 minutes (median 21 minutes). Temporary clipping was never employed.

The CCT prolongation was different at the two MAP values, being in mean 0.5 msec and 1.1 msec for MAP of 75 and 60 Torr respectively, with reference to the basal surgical recording. On 3 occasions, the prolongation of CCT was significantly longer (i.e., > 0.5 msec) on the side of the surgical approach as compared to the opposite hemisphere. It was judged that the brain retractors could play a role in this asymmetry: in fact, repositioning of the spatula to decrease the pressure on the temporal veins obtained a restoration of the homolateral CCT (Fig. 1). When considering the maximum absolute value reached by the CCT, not just the prolongation of it, we observed that four patients presented with an intraoperative CCT duration in excess of 9 msec at least in one side (2 between 12 and 13 msec, one around 10 msec and one at 9.5 msec).

As concerns the cortical SEP responses, the traces obtained before surgery and in the initial part of the operation showed the presence of a primary component (N 20–P 25 peak), followed by at least one positive-negative deflection between 50 and 150 msec (late component). During hypotension at 60 Torr, the late components disappeared in 4 patients; in two of them they did not recover when the MAP was increased again.

One patient died 4 days after surgery. Clinical evaluation carried out one week after surgery revealed that

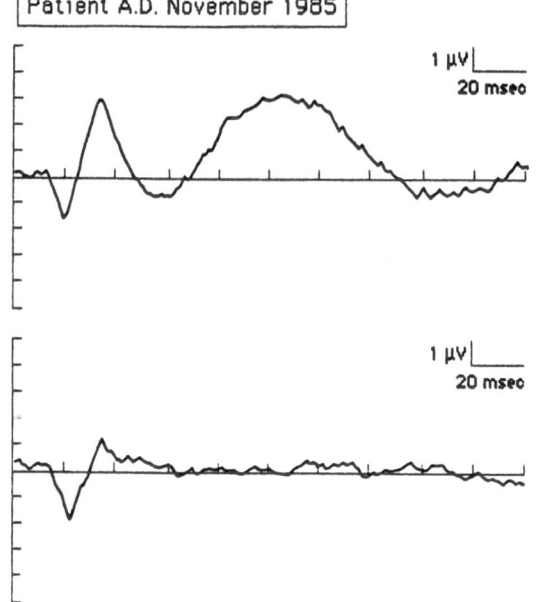

Fig. 2. Cortical SEPs recorded during surgery: Top: right hemisphere response to median nerve stimulation. The primary response (N 20–P 25) is followed by a large late component. Bottom: left hemisphere response to median nerve stimulation. No significant prolongation is apparent as concerns latency of the early negativity (N 20). The amplitude of the primary response (N 20–P 25) is apparently reduced, but not to a statistically significant degree (not smaller than 50% of the contralateral response). The late component is absent. This change, appeared after clipping of a left MCA aneurysm, persisted till the end of surgery and in the postoperative course. The patient presented with severe speech disorders

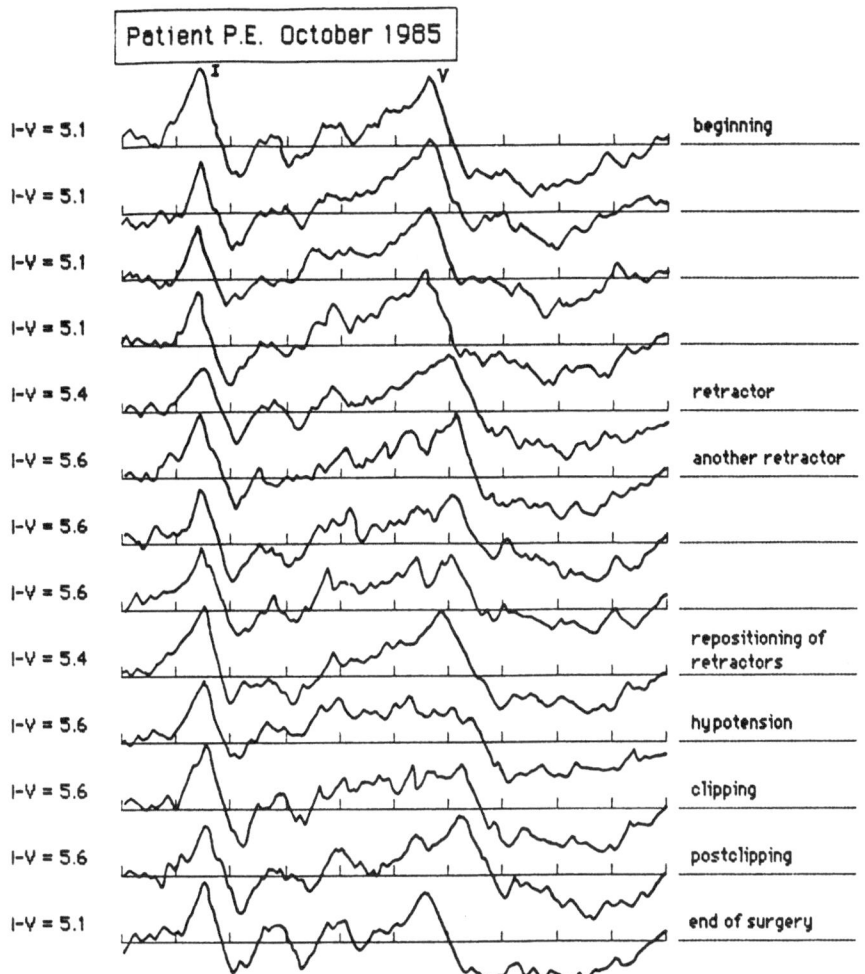

Fig. 3. Sequence of BAEP traces collected during a surgical procedure for a basilar artery aneurysm. Retractor positioning and hypotension cause in crease of the I–V interpeak latency (ipl). A I–V ipl as long as 5.6 msec is still compatible with an uneventful postoperative course

other 3 patients had postoperative complications: two were hemiparetic and one was aphasic. The other patients were either improved or unchanged as compared to preoperative evaluation.

Comparing intraoperative SEP recording with postoperative clinical status, we note that patients with postsurgical complications either showed an intraoperative prolongation of CCT to become longer than 9 msec or presented with persistent absence of SEP late component. In detail, the only fatal outcome had a CCT of 9.5 msec on both sides; the two hemiparetic patients had a unilateral CCT of 10 and 13 msec respectively, in presence of a contralateral CCT shorter than 9 msec; the aphasic patient had a CCT around 8 msec bilaterally, but late waves were selectively absent on the left hemisphere (Fig. 2).

It is worth to note that the prolongation of CCT was of the same order in these patients as it was in the group with a good outcome; what changed, in the two groups, was the absolute value reached after the prolongation, that depended upon the preoperative du-

ration of CCT. The following is an illustrative comparison. Two female patients (Z.C., 58 years, and P.E., 56 years) had identical preoperative score (grade II) for aneurysms in the ACoA. They had different presurgery CCTs (6.7 on both sides the former; 7.1 and 8.6 the latter). The intraoperative prolongation of CCT was of 2 msec for Z.C.; of 1.5 msec for P.E. Z.C. reached the value of 8.7 bilaterally; P.E. reached the value of 10 msec on one side (8.7 on the contralateral hemisphere). Z.C. had an uneventful postoperative course; P.E. presented with a persisting dense hemiparesis on the side of the prolonged CCT.

Only one patient, in the group that was unchanged after surgery, presented with an intraoperative CCT of 12 and absence of late waves; this one was a grade IV patient, with a preoperative left hemiparesis and a preoperative CCT of 10 msec.

Finally, in the two cases in which we recorded the acoustic responses during a posterior fossa approach, we observed a prolongation of the I-V interpeak latency of 0.5 and 0.7 msec respectively, when the retractor was applied to the nervous tissue or to the vessels (Fig. 3). Wave V could always be identified throughout the surgical procedure. No postoperative deficit was apparent.

Discussion

Concerning preoperative recordings, our data are in agreement with the pioneering papers by the group of National Hospital, London GB (Symon *et al.* 1984, Wang *et al.* 1984, Rosenstein *et al.* 1985, Symon and Wang 1986, Symon *et al.* 1986). A significant prolongation of CCT is found only in the group of patients belonging to grade IV of the Hunt and Hess scale. However, the individual patients with low-grade preoperative clinical status can indeed present with a definitely pathological conduction time. We observed that this particular subject is at higher risk, as compared to other patients presenting with a normal CCT. In fact, with the type of anesthesiological management we employed, a CCT prolongation of around 2 msec is to be expected during hypotension; when the critical threshold of 9 msec is reached and exceeded, a postoperative complication has been the rule. Symon and Wang (1986) report that a CCT prolongation up to 10 msec is without significance in their experience. This can be ascribed to the different anesthesiological management, i.e., higher Halothane concentration (up to 2%), no or shorter use of deliberate hypotension, temporary vascular occlusion and possibly different body temperature. The CCT is a very sensitive but completely a-specific measure of central conduction velocity, that is affected by pathological and non-pathological factors in the same manner. Therefore, one has to define his own tolerance limits for neural ischemia, that are dependent upon the anesthesiological approach. Kidooka *et al.* (1987) state that a CCT prolongation exceeding 1.2 msec adversely affect the postoperative conditions; they, however, used halothane in a single patient, and in this one the prolongation was longer than 1.5 msec. We do not agree that the CCT prolongation itself has a definite meaning; at the opposite, our conclusion is that, whatever is the intraoperative CCT prolongation, if the CCT exceeds an absolute threshold limit of 9 msec, a postoperative complication will ensue. This has been illustrated in the two patients with similar clinical condition, different preoperative CCT, similar intraoperative prolongation and different postoperative conditions. The hypotension has always produced a CCT prolongation that was proportional to the MAP reduction. We conclude that hypotension has to be employed with great caution in patients with preop-

erative CCT prolongation, irrespective of the clinical grade.

It has been argued (see Discussion in Kidooka *et al.* 1987) that median nerve SEPs could not be suited for monitoring ACoA aneurysms, since the distal territory of the ACAs are the sensory areas of the foot: the posterior tibial nerve SEPs are suggested instead. This is theoretically correct. However, we never used temporary clipping of the parent arteries to approach and clip aneurysms; therefore, the ischemic risk is mainly due to the use of hypotension and of brain retractors. Under these circumstances, the median nerve SEP monitoring better fits the requirements. This may also explain why we never observed the absence of the cortical N 20 during surgery, that is a rather common finding in other papers (Symon and Wang 1986, Kidooka *et al.* 1987).

The analysis of cortical responses, in particular of the late components, has proven to be useful for exploration of areas outside the primary somatosensory cortex. We suggest to add this recording to the classical CCT measurement, in order to get both an estimate of the subcortical conduction and of the cortical responsiveness.

Finally, when operating in the posterior fossa, the brainstem acoustic evoked potentials (BAEPs) may be successfully used to identify whether the applied retraction to neural or vascular structures is well tolerated. A I–V interpeak latency of 5.6 msec has not been followed by any postoperative complications.

References

1. Ducati A, Pavani M, Cortellazzi P (1984) Somatosensory and acoustic-brainstem evoked responses in ischemic cerebrovascular diseases. Phronesis 4: 275–281
2. Ducati A, Cenzato M, Landi A, Sironi VA, Massei R, Beretta L, Prati R, Bortolani E, Trazzi R (1987a) Intraoperative monitoring of Somatosensory Evoked Potentials and of Electroretinography during vascular surgery of supra-aortic trunks. Chirurgia Cardiovascolare III: 63–67
3. Ducati A, Landi A, Cenzato M, Fava E, Ferrari de Passano C, Villani R (1987b) Monitoraggio intraoperatorio con potenziali evocati nella chirurgia degli aneurismi cerebrali. In: Paoletti P, Knerich R, Urcivoli R, Spairani C (eds) Attualita' in Neurochirurgia. Monduzzi, pp 65–70
4. Ducati A, Cenzato M, Landi A, Fava E, Signoroni G, Giovanelli M, Villani R (1988) BAEPs and SSEPs monitoring during posterior fossa surgery. In: Grundy BL, Villani R (eds) Evoked potentials: intraoperative and ICU monitoring. Springer, Wien New York, in press
5. Grundy BL (1982) Monitoring of somatosensory evoked potentials during neurosurgical operations: methods and applications. Neurosurgery 11-4: 556–575

6. Grundy BL (1985) Intraoperative applications of evoked responses. In: Owen JH, Davis H (eds) Evoked potentials testing. Grune & Stratton, Inc, pp 159–212

7. Grundy BL, Jannetta PJ, Procopio PT, *et al* (1982) Intraoperative monitoring by means of brainstem auditory evoked potentials. J Neurosurg 57: 674–681

8. Jones SJ (1988) Normal and pathologic factor affecting sensory tracts potentials in the human spinal cord during surgery. In: Grundy BL, Villani R (eds) Evoked potentials: intraoperative and ICU monitoring. Springer, Wien New York, in press

9. Hume AL, Cant BR (1978) Conduction time in central somatosensory pathways in man. Electroenceph Clin Neurophysiol 45: 361–375

10. Hunt WE, Hess MR (1968) Surgical risk as related to time of intervention on the repair of intracranial aneurysms. J Neurosurg 28: 14–20

11. Kidooka M, Nakasu Y, Watanabe K, Matsuda M, Handa J (1987) Monitoring of somatosensory evoked potentials during aneurysms surgery. Surg Neurol 27: 69–76

12. Koht A, Schramm J, Watanabe E, Schmidt G (1987) Somatosensory evoked potentials monitoring in cerebral aneurysms surgery. Clin Neurol Neurosurg 89-2/1: 71

13. Landi A, Ducati A, Cenzato M, Fava E, Arvanitakis D, Mulazzi D (1988) Somatosensory evoked potentials monitoring during cervical spine surgery. In: Grundy BL, Villani R (eds) Evoked potentials: intraoperative and ICU monitoring. Springer, Wien, in press

14. Markand ON, Dilley RS, Moorthy SS, Warren C jr (1984) Monitoring of somatosensory evoked potentials responses during carotid endarterectomy. Arch Neurol 41: 375–378

15. Rosenstein J, Wang ADJ, Symon L, Suzuki M (1985) Relationship between hemispheric cerebral blood flow, central conduction time and clinical grade in aneurysmal subarachnoid hemorrhage. J Neurosurg 62: 25–30

16. Symon L, Hagardine J, Zawirski M, *et al* (1979) Central conduction time as an index of ischaemia in subarachnoid hemorrhage. J Neurol Sci 44: 95–103

17. Symon L, Momma F, Schwerdtfeger K, Bentivoglio P, Costa e Silva IE, Wang A (1986) Evoked potential monitoring in neurosurgical practice. In: Symon L *et al* (eds) Advances and technical standards in neurosurgery, vol 14: 25–70. Springer, Wien New York

18. Symon L, Wang ADJ (1986) Somatosensory evoked potentials: their clinical utility in patients with aneurysmal subarchnoid hemorrhage. In: Cracco R, Bodis-Wollner I (eds) Evoked potentials. Alan R Liss, Inc, New York

19. Wang ADJ, Cone J, Symon L, Costa e Silva IE (1984) Somatosensory evoked potentials monitoring during the management of aneurysmal SAH. J Neurosurg 60: 264–268

Address for correspondence: Prof. Dr. A. Ducati, Institute of Neurosurgery, University of Milano, Via F. Sforza 35, I-20122 Milano, Italy.

Acta Neurochirurgica, Suppl. 42, 14–17 (1988)

Safety Limits of Controlled Hypotension in Humans

S. Yamada, F. Brauer[1], D. Knierim, T. Purtzer, T. Fuse, W. Hayward, D. Lobo[1], and L. Dayes

Section of Neurosurgery, Department of Surgery and [1]Department of Anesthesiology, Loma Linda University School of Medicine, Loma Linda, California, U.S.A.

Summary

Controlled hypotension is a safe and convenient means of allowing a surgeon to perform intracranial aneurysm, arteriovenous malformation and vascular tumor surgery. The mean arterial pressure between 40 and 60 mmHg induces diminished pulsatile arterial pressure, thus preventing rupture of these abnormal vasculatures. It is still possible to maintain cerebral metabolism in the functional level within this 40–60 mmHg blood pressure range. This statement is based on our experience of the physical and neurological outcome of patients after surgery, and on analyses of somatosensory evoked potential and redox of cytochrome a, a_3 in the mitochondria of the cerebral cortical cells.

Keywords: Controlled hypotension; somatosensory evoked potential; oxidative metabolism; cerebrovascular surgery.

Introduction

Controlled hypotension has two advantages during neurosurgical procedures[3, 10, 15]. First, it minimizes blood loss during surgery such as for intracranial aneurysms, arteriovenous malformations (AVMs) or vascular tumors. The other advantage is that it diminishes pulsatile arterial pressure which is a major factor in prevention of rupture during surgery for an aneurysm or AVM.

Controlled hypotension is extremely valuable during dissection of an AVM located in the functional area[15] for the following reasons. Resection of an AVM in the functional area requires technical precision to preserve brain tissue and maintain micro-circulation to the area. It is also imperative that bleeding be controlled to preclude the possibility of damage to the surrounding brain from bleeding of even small arterial or venous channels of the AVM, which occurs under normal arterial pressure.

A complication of hypotension is that it may bring about decreases in oxygen delivery to the brain with consequences to cell survival. We present here studies of evoked potentials and oxidative metabolism during hypotensive anesthesia in neurosurgical patients with a goal of defining more precisely the safety limits beyond which hypotension threatens brain damage.

Material and Method

Twenty patients were subjected to simultaneous studies of evoked potentials and oxidative metabolism; 16 patients had arteriovenous malformations, 2 patients had aneurysms and 2 patients had parsagittal meningiomas. Ages ranged from 18 to 57 years among 11 male and 9 female patients.

The patient was intubated intratracheally with intravenous administration of thiopental sodium, and then inhaled with combination of nitrous oxide and oxygen and 1% isofluorane. The fraction of inspired oxygen (FiO_2) was maintained at 0.3. After craniotomy, the dura mater was opened and the cerebral cortex was exposed.

Simultaneous measurements of the following were made during anesthesia and surgical procedures:

1. Somatosensory Evoked Potentials (SSEP)

A needle electrode was inserted through the scalp so that its tip contacted the skull in the parietal area. Another needle electrode was inserted at the glabella. When the sensory area was visible in the dural opening, a disc electrode was directly placed on the sensory cortex. Stimulation of the median or ulnar nerve was carried out

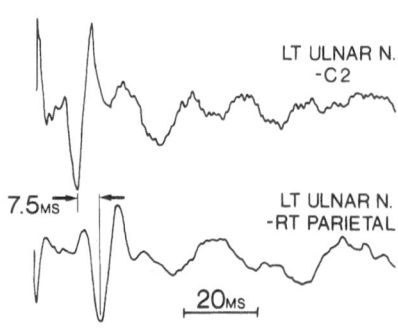

Fig. 1. Measurement of central conduction time from C2-level

through a bipolar electrode with a 30 second train of square wave pulses. Each pulse was 0.3 millisecond in duration, and 4.1 Hz, and 3–5 milliampere in amplitude. Bilateral recordings were taken in the sensory cortex on the side opposite to nerve stimulation to identify symmetrical potentials. Another electrode was also inserted into the second cervical spinous process, and evoked potentials were recorded from the cervical cord. This allowed calculation of central conduction time (Fig. 1). SSEP recordings were analyzed with Nicolet CA 1000 Clinical Averager and DC 2000 Disc Controller. The body temperature was maintained between 36° and 36.5° to prevent variability in conduction time caused by temperature changes[14].

2. Reflection Spectrophotometry

Oxidative metabolism was evaluated by non-invasive reflection spectrophotometry[9], which recorded changes in the reduction/oxidation (redox) ratio of cytochrome a, a_3 before, during and after controlled hypotension. In this procedure, two beams of light at 605 and 590 nm were projected to the sensory cortex. Reflected light was collected by a microscope and detected with a photomulitplier tube. The procedure is based on the fact that cytochrome a, a_3, the terminal cytochrome in the mitochondrial respiratory chain, absorbs more light at 605 nm when reduced than when oxidized. Analysis of data obtained from redox shifts of cytochromes is based upon previous studies in mitochondria in vitro and brain tissues in vivo[11]. These studies have demonstrated that respiratory chain redox ratios shift toward oxidation when there is requirement to phosphorylate ADP to ATP, and they shift toward increased reduction when oxygen supply is diminished. Since neurons rely on ATP for energy, oxidation of the mitochondrial respiratory chain is essential to maintain cerebral electrical activity and cell survival. Severe and prolonged reduction of cytochromes indicates metabolic derangement, which, if prolonged, may become irreversible.

During recordings of SSEP and mitochondrial redox activity, the patients were first subjected to lowered FiO_2 (0.15) for 3 minutes, as a safe limit[16], since PaO_2 remained above 65 mmHg, and hemoglobin oxygen saturation at 95%.

After FiO_2 was returned to 0.3, controlled hypotension was induced by intravenous administration of nitroprusside for comparison of SSEP and redox activity.

Results

1. Somatosensory Evoked Potentials

SSEPs were unchanged from control during induced hypotension with a mean arterial pressure (MAP) down to between 40 and 50 mmHg. Occasionally transient decrease in amplitude was noted when blood pressure reached this level as shown in Fig. 2 a. At between 30 and 40 mmHg decreases in SSEP amplitude and prolonged latency were noted (Fig. 2 b). These changes became increasingly apparent below 30 mmHg.

2. Redox Activity of Cytochrome a, a_3

When controlled hypotension was induced, cytochrome a, a_3 became increasingly reduced as expected of mitochondria under conditions of declining oxygen-

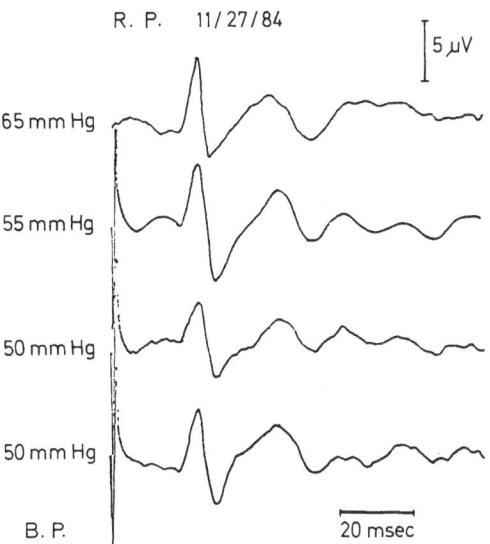

Fig. 2 a. SSEPs before and during controlled hypotension. See transient decrease in amplitude. The amplitude returned to normal within a few minutes

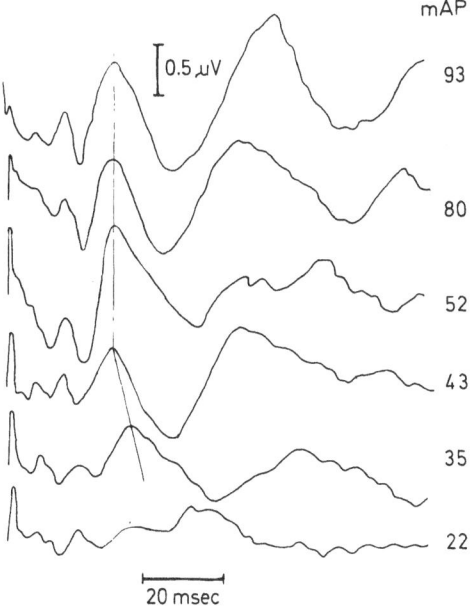

Fig. 2 b. SSEPs before and during controlled hyotension. Decrease in amplitude and delay in latency were noted at MAP 30–40 mmHg. These changes began to rectify when MAP was raised to 40–50 mmHg. Recordings with the pressure between 85 mmHg and 40 mmHg are omitted

ation (Fig. 3 a). Reduction of cytochrome a, a_3 first became apparent when MAP was decreased to 60 mmHg and reduction of cytochrome reached a plateau at approximately 50 mmHg. This corresponded to the reduction shift reached by FiO_2 at 0.15, and was considered to be acceptable as a safety range (Fig. 3 b).

L.B. AVM (CENTRAL AREA)

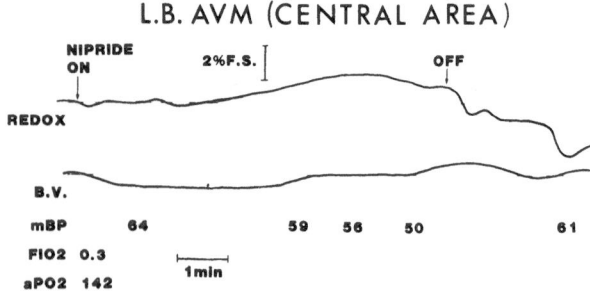

Fig. 3a. Redox of cytochrome a, a_3 during controlled hypotension. The reduction level reached to maximum at 50 mmHg in MAP. This was followed by reoxidation although the MAP was unchanged. Hyperoxidation was noted after MAP returned to the original

L.B. AVM (CENTRAL AREA)

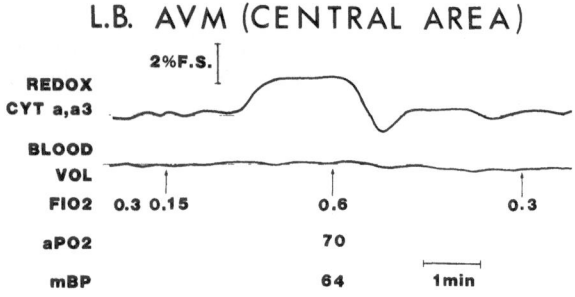

Fig. 3b. The reduction level during low FiO_2 (0.15) was compared to the level reached by controlled hypotension. This is a safe limit of reduction

When MAP was maintained at 50 mmHg, reoxidation of cytochrome a, a_3 was noted (20–40% towards the baseline). This reoxidation appeared to coincide with decreases in reflectance at 590 nm, which are likely to indicate increases in blood volume. These increases in blood volume probably served to compensate for decreased blood oxygenation. Below MAP of 40 mmHg, there was a precipitous increase in the level of reduced cytochrome a, a_3. In some cases SSEP amplitude slightly decreased at MAP of 50 mmHg, but the amplitude returned to control within a few minutes. This is somewhat coincident to reoxidation.

When MAP was elevated above 40 mmHg, cytochrome a, a_3 promptly became reoxidized and re-establishment of normal MAP was accompanied by return of the cytochrome to control redox status. Somatosensory evoked potentials also returned to normal when the blood pressure returned above 40 mmHg.

All 20 patients made good physical and mental recovery including 16 patients with AVMs who were subjected to prolonged hypotension for 6 or more hours of MAP at 40–50 mmHg.

Discussion and Conclusions

Subsequent to the first description of SSEPs by Dawson[2], development of portable averaging computers has facilitated the use of SSEPs for evaluation of neurological disorders[6]. Various authors have attempted to correlate intraoperative SSEP[1, 4, 5, 7, 8, 12, 13] or auditory evoked potential findings[5, 6] with post-operative neurological findings.

In the present study we recorded SSEPs and energy metabolism of the cerebral hemisphere to find if these brain activities are impaired under controlled hypotension and to define the safety limit of controlled hypotension for prevention of irreversible brain damage.

Our studies suggest that there are significant effect upon electron transport in mitochondria but not upon electrophysiology at the range of hypotension required for surgery (40–60 mmHg). Below this MAP, effects of hypotension upon oxidative metabolic activity are greater and effects upon evoked potentials become apparent.

Since MAP of 40–50 mmHg did not preclude subsequent restoration of mitochondrial redox status and since evoked potentials were maintained at this level of hypotension, it is suggested that such hypotension is compatible with surgical safety.

Since evoked potentials were altered at MAP below 40 mmHg, it is suggested that this level of hypotension is below safety limits. However, morphological studies in animal models are required to assure that these in vivo measurements are indicating proper limits of safety.

These data also suggest that the monitoring of evoked potentials and/or mitochondrial activity provide useful tools for establishing effects of hypotension in each patient.

As a first estimate, we consider that the concurrent and parallel deterioration of metabolism and electrophysiology at MAP below 40 mmHg should be avoided to eliminate any possibility of irreversible brain damage.

References

1. Astrup J, Symon L, Branston NM, Lassen NA (1977) Cortical evoked potential and extracellular K^+ and H^+ at critical levels of brain ischemia. Stroke 8: 51–57
2. Dawson GD (1947) Cerebral responses to electrical stimulation of peripheral nerve in man. J Neurol Neurosurg Psychiatry 10: 137–140
3. Drake CG (1979) Cerebral arteriovenous malformations; considerations for and experience with surgical treatment in 166 cases. Clin Neurosurg 28: 145–208

4. Eisenberg HM, Turner JW, Teasdale G, Rowan J, Feinstein R, Grossman RG (1979) Monitoring of cortical excitability during induced hypotension in aneurysm operation. J Neurosurg 50: 595–602

5. Friedman WA, Kaplan BL, Day AL, Sypert GW, Curran MT (1987) Evoked potential monitoring during aneurysm operation: observations after fifty cases. Neurosurgery 5: 678–687

6. Grundy BL (1982) Monitoring of sensory evoked potentials during neurosurgical operations: methods and applications. Neurosurgery 11: 556–575

7. Grundy B, Nelson PB, Lina A, Heros RC (1982) Monitoring of cortical somatosensory evoked potentials to determine the safety of sacrificing the anterior cerebral artery. Neurosurgery 11: 64–67

8. Hargadine JR, Branston NM, Symon L (1980) Central conduction time in primate brain ischemia – a study in baboons. Stroke 11: 637–641

9. Jöbsis FF, Keizer JH, LaManna JC, Rosenthal M (1977) Reflectance spectrophotometry of cytochrome a, a_3 in vivo. J Appl Physiol 43: 858–872

10. Pertuiset B, Sichez JP, Philippon J, Fohanno D, Horn Y (1978) Mortalité et morbidité après exérèse chirurgicale totale de 162 malformations artérioveineuses intracrâniennes. Rev Neurol 135: 319–327

11. Rosenthal M, Martel D, LaManna JC, Jöbsis FF (1976) In situ studies of oxidative energy metabolism during transient cortical ischemia in cats. Exp Neurol 50: 477–494

12. Symon L (1980) The relationship between CBF, evoked potentials and the clinical features in cerebral ischaemia. Acta Neurol Scand [Suppl] 78: 175–190

13. Symon L, Wang AD, Costa e Silva IE, Gentili F (1984) Perioperative use of somatosensory evoked responses in aneurysm surgery. J Neurosurg 60: 269–275

14. Tani S, Hayward W, Yamada S, Days L (1986) Somatosensory evoked potentials in rats. Loma Linda University Surgeon 4: 6–11

15. Yamada S (1982) AVMs in the functional area: surgical approach and regional cerebral blood flow. Neurol Res 4: 283–322

16. Yamada S, Zinke DE, Sanders D (1981) Pathophysiology of "Tethered cord syndrome". J Neurosurg 54: 494–503

Address for correspondence: Dr. S. Yamada, Section of Neurosurgery, Department of Surgery, Loma Linda University School of Medicine, Loma Linda, CA 92350, U.S.A.

Acta Neurochirurgica, Suppl. 42, 18–21 (1988)

Intraoperative Monitoring of the Motor Function: Experimental and Clinical Study

M. Kaneko, A. Fukamachi, H. Sasaki, N. Miyazawa, T. Yagishita, and **H. Nukui**

Department of Neurosurgery, Yamanashi Medical College, Yamanashi, Japan

Summary

Manipulation of the lesions adjacent to the primary motor area or the motor pathway is troublesome for neurosurgeons because they lack an effective method to determine the primary motor area or to monitor motor function in the operative room. It will be of great value to establish a monitoring method of the corticospinal tract under general anaesthesia. We recorded the motor evoked potential (MEP) from direct motor cortex stimulation in cats and showed that it derives almost purely from the corticospinal tract. Then we used this technique during the operation of the resection of tumours near the primary motor area or the motor pathway.

1. Experimental study: Twenty adult cats were used in this study. Recording electrodes were flexible bipolar catheter electrodes inserted into the spinal epidural space. Stimulating electrodes were silver ball electrode on the cortex (anode) and needle electrode in the temporal muscle (cathode). Stimulation of 4–24 V, 5–10 Hz and 0.2 msec in duration were done and evoked potentials signals were averaged 60 to 512 times. MEP with multiple peaks was obtained that had a 112 msec conduction velocity in the spinal cord. We found the same signals from the stimulation of ipsilateral cerebral peduncle. Radiofrequency lesioning of ipsilateral cerebral peduncle produced a loss of MEP. These results show that MEP derives from the corticospinal tract. Significant wave form change, with components of short latency, was noted by the excessively intense stimuli. We supposed that superimposition of the signals from the extrapyramidal pathways, excited in the brain stem, results in this change.

2. Clinical study: MEPs were recorded intraoperatively in ten patients who had tumours, 9 gliomas and 1 metastatic tumour, adjacent to the primary motor area of the motor pathway. Flexible catheter electrodes were inserted percutaneously into the cervical and thoracic spinal epidural space. Following the dural opening, several points on the cerebral cortex were stimulated. MEP was obtained only by stimulation on the precentral gyrus, so the detection of the motor strip was easy and accurate. Resection of the tumours could be done extensively. Decrease of amplitude was noted during tumour resection in four cases but appearance or worsening of muscle weakness did not occur in any case.

Keywords: Primary motor area; identification; corticospinal tract; motor evoked potential.

Introduction

Manipulation of lesions adjacent to the primary motor area or the motor pathway is troublesome for neurosurgeons because they lack an effective method to determine the primary motor area or to monitor the motor function during operation. It would be of great value to establish a method of monitoring the corticospinal tract under general anaesthesia. We recorded the motor evoked potential (MEP) from direct stimulation of the motor cortex in cats and showed that it derives almost purely from the corticospinal tract. Then we used this technique during the resection of tumours near the primary motor area or the corticospinal tract with good results.

Experimental Study

Materials and Methods

Twenty adult cats were used in this study. Animals were anaesthesized with ketamine (10 mg/kg) and atropine (0.02 mg/kg), and femoral arterial and venous lines were placed. Following tracheostomy, the animals were immobilized with pancuronium bromide (0.1 mg/kg) and fixed in a stereotaxic frame for cats under controlled ventilation. The animals were kept on thermostatically controlled heating blankets and rectal temperature was maintained approximately at 38 °C. Anaesthesia was maintained with repeated intravenous injection of ketamine.

Recording electrodes for the MEP were bipolar catheter electrodes (Unique Medical Co., Ltd., Tokyo) inserted into the spinal epidural space via a laminectomy. Negative pole of the electrodes was placed in the cranial side and positive pole in the caudal. Craniotomy was performed over the motor cortex in the frontal region. Stimulation was effected between silver ball electrode on the cortex (anode) and electroencephalogram needle in the ipsilateral temporal muscle (cathode). Evoked potential apparatus was MEB-5100 (Nihon Kohden Inc., Tokyo). Stimulating conditions were 4–40 V in intensity, 2–10 Hz in frequency and 0.1–0.5 ms in duration. The resulting efferent activity was gained with filter settings of 5 Hz and 3,000 Hz, and averaged 60 to 512 times per trial.

Electrical activity or neural noise in the depth of the brain[1] was recorded by bipolar concentric semimicroelectrodes inserted stereotaxically into the cerebral peduncle of the midbrain. Using the same track and the same electrode, stimulation was carried out and the resulting efferent activity was recorded by the same method as for cortical stimulation. Lesioning of the ipsilateral cerebral peduncle or the dorsolateral part of the medulla was done with radiofrequency, and the lesions were verified after the animals were scarified.

Results

MEPs were elicited by the cortical stimulation localized on the orbital, coronal and sigmoid gyri. Threshold for the appearance of a clear signal was usually 4–8 V. MEPs had multiple peaks and the conduction velocity of the first component was 39 ± 5 msec. The travelling wave was registered by the recordings at several different sites of the spinal cord and its conduction velocity in the spinal cord was 101 ± 16 msec. After increasing the stimulation intensity early components of short latency appeared. Threshold for the appearance of these early components was usually 14–20 V; their amplitude increased in proportion to the stimulation intensity, but shortening of the latency was not noted. These results suggest that some tracts in addition to the corticospinal tract were excited at deeper sites in the brain according to the increase of the stimulation intensity. Recording of neural noise in the ipsilateral midbrain could have confirmed the cerebral peduncle. At the most ventral portion of the midbrain corresponding to the cerebral peduncle, typical neural activity of the white matter that is a low neural noise with small positive sporadic spike discharges was recognized. Stimulation of this point created signals which had strong resemblance to the MEPs from the cortical stimulation in their wave forms and latency. Stimulation of more dorsal structures corresponding to the reticular formation or the red nucleus created signals resembling the early components elicited by the intense cortical stimulation.

Radiofrequency lesioning of the ipsilateral cerebral peduncle produced a complete loss of MEP, but the early components were still elicited by intense cortical stimulation. These early components decreased in amplitude after bilateral lesioning of the dorsolateral portion of the medulla including the rubrospinal and reticulospinal tracts (Fig. 1). These results show that MEPs from threshold cortical stimulation are derived from the corticospinal tract passing trough the ipsilateral cerebral peduncle and early components from intense cortical stimulation are elicited by exciting other

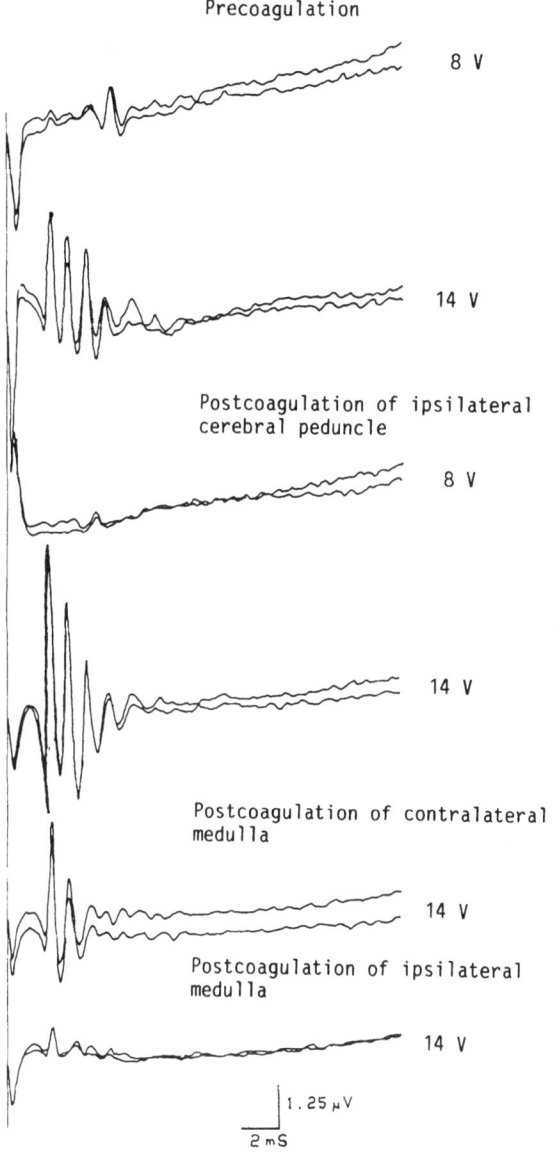

Precoagulation

8 V

14 V

Postcoagulation of ipsilateral cerebral peduncle

8 V

14 V

Postcoagulation of contralateral medulla

14 V

Postcoagulation of ipsilateral medulla

14 V

1.25 μV
2 mS

Fig. 1. Lesioning study in cat 17. Recording site was the upper thoracic epidural space. Coagulation of the ipsilateral cerebral peduncle almost completely abolished the MEP induced by threshold stimulation (8 V in this case), but 14 V stimulation still created early components of the wave with considerable increase of amplitude. After lesioning of the bilateral dorsolateral parts of the medulla these were abolished almost completely

tracts, including the rubrospinal and reticulospinal tracts, in addition to the corticospinal tract in the midbrain or the pons. Therefore, we concluded that threshold stimulation is essential for monitoring the corticospinal tract.

Clinical Study

Materials and Methods

Ten patients who had tumours, nine gliomas and one metastatic tumour, adjacent to the primary motor area or the internal capsule were included in this study. Recording electrodes were bipolar catheter electrodes (0.7 mm in diameter) in the spinal epidural space placed percutaneously through a 17-G epidural needle on the preoperative day or after the induction of anaesthesia. Recording sites were the lower cervical level in most patients and the lower thoracic level or both in some patients. Following the dural opening of craniotomy, silver ball electrodes (anode) were placed on the cortex and electroencephalogram needle (cathode) in the midline of the scalp. Stimulation intensity was set at threshold level for the appearance of a clear signal, usually 12–30 V. The number of stimuli was usually 60–128 times per trial. Several points on the exposed brain were stimulated and the motor strip was identified. Ultrasonography was used to detect the subcortical part of the tumour and define the margin accurately. During the tumour resection, MEPs were monitored by stimulating the motor cortex.

Results

MEPs were constantly obtained with one to four negative peaks at the cervical epidural space. More simple waveforms with one to two peaks were noted at the lower thoracic level. MEPs could be elicited by the stimulation to the arm or leg area of the precentral gyrus. Stimulation to the gyrus adjacent to the precentral one could not elicit MEP. Stimulation to the face area, or lateral portion of the precentral gyrus, also created no response at the spinal cord because efferent activity from this area does not reach to the spinal cord.

All tumours were resected extensively. After the tumour resection, the amplitude of the first component showed no change in six patients and decreased 23–40% in four (Fig. 2). In the former, no alteration of the muscle strength was found postoperatively in any patient. In two patients of the later four, however, very

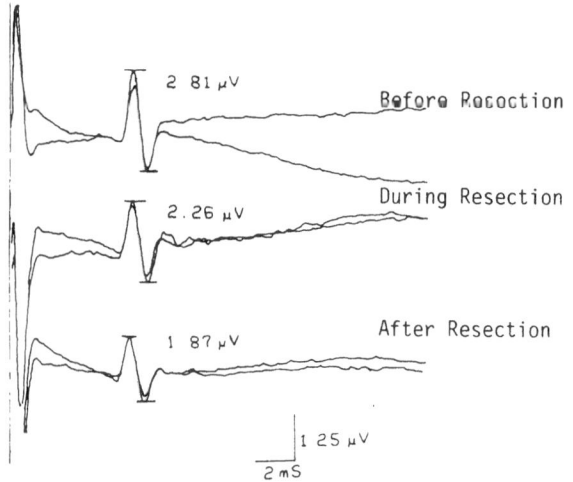

Fig. 2. Intraoperative MEPs in a patient with deep temporal glioma adjacent to the internal capsule. Amplitude of MEPs decreased 33% after tumour resection, but no motor deficits were noted postoperatively

mild muscle weakness of the limbs was noted just after the recovery from anaesthesia, which disappeared completely within 12 hours. Another two patients had no detectable paresis of the limbs. Later components of the MEP could be recorded in six patients. They disappeared during the manipulation of the lesions in three, and two of them showed a transient motor weakness postoperatively. In summary, two patients who showed a decreased first component and absent later components had transient postoperative muscle weakness which resolved in a short time.

Discussion

Evoked potentials elicited by the electrical stimulation of the motor cortex in cats were reported first by Patton and Amassian[5]. They showed that evoked potentials recorded at the spinal cord included D wave, the first component of the wave, and several later waves which were called I waves. D wave was thought to result from direct stimulation of the pyramidal cells, and I waves were from indirect stimulation through interneurone networks. Similar potentials were obtained by the transcranial stimulation of the motor cortex by Levy et al.[4]. These potentials were abolished by the transection of the medullary pyramid or both lateral and ventral columns of the spinal cord, but it is uncertain whether a lesion in the more rostral site of the central nervous system can affect the potentials. We could constantly obtain MEPs that had initial D wave and later I waves, and showed that they passed trough the ipsilateral cerebral peduncle. MEPs obtained by threshold stimulation were completely abolished by radiofrequency lesioning of the ipsilateral cerebral peduncle.

The strength of the stimulation must be considered here. In our experiments, intense stimulation resulted in excitation of the extrapyramidal tracts in the brain stem. This is an important fact because one may confuse these potentials with the MEPs if an inappropriate stimulation intensity is used. Therefore recording with the minimal strength of stimulation is needed in order to monitor the corticospinal tract in the upper brain stem or the supratentorial region. To detect the signals following minimal stimuli the most effective recording methods should be chosen. Spinal epidural recording by bipolar catheter electrodes is sensitive and could minimize artifacts, therefore we think that this method is suitable for the recording of MEPs.

Intraoperative monitoring of the motor function by the MEP has been reported by a few authors[3, 4, 7]. The effectiveness of the MEP for the supratentorial lesions is still obscure. Defining the motor strip is important when manipulating lesions near the motor area. Preoperatively one can presume the relationship of the lesion and the precentral gyrus from the neuroradiological findings[6], but the precentral gyrus may often be displaced by the tumor, so the radiological method is not reliable for the intraoperative delineation of the motor strip. In our experience, MEPs were elicited by the stimulation only to the precentral gyrus so the determination of the motor strip was easy and accurate. In this series amplitude of the D wave decreased in four patients, and I waves disappeared in three. Two of them showed a slight transient motor weakness, but it resolved in a short time in both patients. Decrement of the D wave may be a warning of injury to the pyramidal cells or the motor pathway, but the extent of any decrement which predicts whether or not a major deficit will occur remains unclear. Tsubokawa suggested that it might be 50%[7]. According to Patton, I waves are abolished by cooling of the brain surface. Therefore it is necessary to rule out the cooling of the exposed brain or hypothermia during the operation for evaluating the effect of the MEP. Our results showed a possibility that disappearance of I waves could also be a warning sign of the neuronal damage.

Although further study is needed to substantiate the efficacy of this test, we think that the MEP is very helpful when performing extensive surgery for lesions near the motor area or the motor pathway.

References

1. Fukamachi A, Ohye Ch, Saito Y, Narabayashi H (1977) Estimation of the neural noise within the human thalamus. Acta Neurochir (Wien) [Suppl] 24: 121–136
2. Gorman ALF (1966) Differential patterns of activation of the pyramidal system elicited by surface anodal and cathodal cortical stimulation. J Neurophysiol 29: 547–565
3. Levy WJ (1987) Clinical experience with motor and cerebellar evoked potential monitoring. Neurosurgery 20: 169–187
4. Levy WJ, York DH, McCaffrey M, Tanzer F (1987) Motor evoked potentials from transcranial stimulation of the motor cortex in humans. Neurosurgery 15: 287–302
5. Patton HD, Amassian VE (1954) Single multiple unit analysis of cortical stage of pyramidal tract activation. J Neurophysiol 17: 345–357
6. Ring BA, Waddington M (1967) Angiographic localization of motor strip. J Neurosurg 26: 249–254
7. Tsubokawa T (1986) Clinical significance of the centrofugal evoked potentials to cortical surface or scalp stimulation for evaluation of motor function during the surgery. Proceedings of the 5th annual meeting of the Japanese Congress of Neurological Surgeons. Japan Upjohn, Tokyo, pp 111–122

Address for correspondence: Masami Kaneko, M.D., Department of Neurosurgery, Yamanashi Medical College, 1110 Shimokato, Tamaho, Nakakoma, Yamanashi, 409-38, Japan.

Acta Neurochirurgica, Suppl. 42, 22–26 (1988)

Intraoperative Identification of the Central Sulcus: a Practical Method

T. Aiba and **Y. Seki**

Department of Neurosurgery, Toranomon Hospital, Tokyo, Japan

Summary

A practical method of intraoperative identification of the central sulcus using cortical evoked potentials (EP) was reported. The method herein proposed is unique with stimulation to the trigeminal and tibial nerve in addition to median nerve stimulation, depending on the location of the lesion. Thus, a wide coverage of the rolandic fissure was possible. Eighteen adult patients with peri-rolandic lesions were investigated. Fourteen cases had a malignant tumour but all except 3 showed only a minor neurological deficit preoperatively.

In 13 cases, the central sulcus could be identified through the phenomenon of the phase-reversal of cortical EP across the central fissure. Of the remaining 5 cases, cortical EP were not detected in 3 (all showed hemiplegia and/or aphasia) and EP without phase-reversal were obtained in 2. As a result, in most cases showing only a trivial neurological deficit before surgery, aggressive removal of the tumour was successfully carried out without producing severe motor dysfunction.

Keywords: Central sulcus; cortical evoked potential; brain tumour; postoperative sequelae.

Introduction

In cases with lesions adjacent to the sensorimotor area, recognition of the pre- and postcentral gyri is essential to perform an aggressive neurosurgical attack without producing a functional deficit. Therefore, precise identification of the central sulcus is important as an aid to definition of these critical areas. We report a practical method which utilizes the cortical recording of somatosensory evoked potentials (SEP) to localize the wide range of the central sulcus during craniotomy.

Patients and Methods

Eighteen adult patients were investigated in this study. The cases are summarized in Table 1.

Cortical SEP were recorded monopolarly with custom-made 3×3 or 2×4 electrode arrays which were composed of thin silicone flap. The centers of the electrode were separated by 1 cm. The reference electrode was placed on the dura surrounding the exposed cortex.

Depending on the location of the lesion, contralateral trigeminal, median and tibial nerves were stimulated. Percutaneous needle electrodes were applied immediately adjacent to the median nerve at the wrist and to the tibial nerve at the ankle. The interelectrode distance was approximately 2 cm, with negative electrode placed proximally. For trigeminal stimulation, two clip-shaped electrodes were applied to the inner surface of the upper and lower lip. A rectangular pulse of 0.2 msec duration and the intensity of 5 mA was determined as a standard stimulus, and repeated at a rate of 2.3 per second. But sometimes 10 mA stimuli were given to the tibial nerve. After the induction of the general anesthesia, these electrodes were equipped and muscle contraction (thumb twitch and so on), evoked directly by the stimuli of the standard intensity, was confirmed prior to the injection of neuromuscular blocking agents.

Amplification bandpass was set between 5 Hz and 1.5 kHz. One hundred responses were averaged with Pathfinder II (Nicolet, U.S.A.) which also generated stimulation pulses. Up to 8 channels were recorded simultaneously. The time analysed was 50 msec after stimulus onset for the trigeminal and median nerve stimulation, and 100 msec for tibial nerve stimulation.

At 2.3 sec stimulation rates, one hundred responses were obtained in as little as 50 seconds. Two or three cortical areas were examined every one nerve stimulation. In addition to median nerve, trigeminal or tibial nerve stimulation was added depending on the height of the lesion, so that total time needed to identify the central sulcus was at most 15 minutes.

Results

Cortical evoked potentials by the median nerve stimulation were examined in 17 out of 18 cases. Of these 17 cases, the cortical SEP was obtained with typical phase-reversal across the central sulcus in 13 (case 1 to 12, and 15), undetectable in 3 (case 16, 17, and 18) and of parietal pattern in 1 (case 13). In these 13 cases of which the central fissure could be identified, diphasic negative-positive potential [mean negative peak latency: 22.4 msec (19.5–24.9), positive: 29.1 (22.6–32.9)] was shown in the postcentral gyrus, while a phase-reversed potential (mean positive peak latency: 23.3 msec, negative: 31.1 msec) was observed in the pre-

Table 1

No	Case	Age	Sex	Histological Diagnosis	Side	Site	Height of Convexity	Symptoms & Signs	Trigeminal N	Median N	Tibial N
						Location of Lesion				Intraoperative Response to	
1	M.M	70	M	Metastasis	Lt	F-C	Mid-	Rt facial & upper limb weakness	$N_{16.2}$ $P_{19.8}/P_{18.0}$	$N_{22.5}$ $P_{32.9}/P_{24.6}$	N_{44-45} P_{54-55}
2	S.Y	35	M	Glioblastoma	Rt	F-C	Mid- & Low	Lt upper limb weakness	?	$N_{23.1}$ $P_{31.8}/P_{24.3}$ $N_{30.2}$	
3	T.S	78	F	Meningioma	Lt	F-C	Mid- & Low	Dementia		$N_{21.2}$ $P_{29.0}/P_{23.9}$	
4	F.T	47	F	Metastasis	Rt	F-C	High	Lt upper limb weakness	undetectable	$N_{24.9}$ $P_{28.0}/P_{28.0}$	
5	N.M	48	M	Malignant lymphoma	Lt	F	Mid- & High	Convulsion	$N_{15.6}$ $P_{25.4}/(P)N_{24.4}$	$N_{22.2}$ $P_{31.6}/P_{22.8}$	
6	M.T	70	F	Meningioma	Lt	C	Mid- & Low	Convulsion	?	$N_{23.2}$ $P_{32.0}/P_{23.2}$ $N_{30.8}$	
7	I.M	49	M	Astrocytoma III	Lt	F	High	Convulsion	undetectable	$N_{22.4}$ $P_{27.6}/P_{23.2}$ $N_{34.4}$	
8	I.H	59	M	Glioblastoma	Rt	F	Mid-, High & Low	Convulsion	$N_{16.8}$ $P_{27.5}/P_{19.7}$ $N_{33.5}$	$N_{24.2}$ $P_{31.6}/P_{26.4}$ $N_{38.8}$	
9	O.T	56	M	Glioblastoma	Lt	C	Mid- & Low	Aphasia (mild)	$N_{14.8}$ $P_{22.0}/P_{15.0}$ $N_{21.7}$	$N_{21.9}$ $P_{29.5}/P_{22.0}$ $N_{29.2}$	
10	Y.A	70	F	Meningioma	Rt	F	High (mesial)	Headache		$N_{23.3}$ $P_{28.1}/P_{25.0}$ $N_{36.6}$	$N_{47.8}$ $P_{58.8}/P_{42.8}$ $N_{55.4}$
11	Y.N	29	M	Ateriovenous malformation	Rt	P	Mid-	Convulsion	$N_{13.9}$ $P_{17.1}/P_{13.6}$ $N_{18.8}$	$N_{21.4}$ $P_{24.6}/P_{21.1}$ $N_{26.4}$	
12	O.S	44	M	Astrocytoma III	Lt	F	Mid-	Headache	$N_{12.3}$ $P_{17.1}/P_{12.2}$ $N_{17.2}$	$N_{19.5}$ $P_{22.6}/P_{18.1}$ $N_{22.5}$	
13	S.M	44	F	Metastasis	Rt	P	High	Headache		$N_{21.4}$ $P_{32.7}$	$P_{50.6}$ $N_{67.8}$
14	T.S	49	F	Metastasis	Rt	F	High	Headache			$P_{50.0}$ $N_{68.6}$
15	H.M	59	F	Glioblastoma	Rt	C	Low	Lt facial weakness	undetectable	$N_{21.9}$ $P_{28.4}/P_{20.6}$	
16	O.H	62	F	Glioblastoma	Rt	C	Mid- & Low	Lt hemiplegia	undetectable	undetectable	
17	S.T	66	F	Radiation necrosis	Lt	F-C	Low	Rt hemiplegia & Aphasia	undetectable	undetectable	
18	H.K	37	F	Radiation necrosis	Lt	F	Mid-, High & Low	Rt hemiplegia & Aphasia	undetectable	undetectable	

C: Central, F: Frontal, P: Parietal   /: Phase Reversal, N: Nerve

central sulcus. Through the phenomenon of the phase-reversal, the central sulcus could be determined as long as 4 to 5 cm.

In three cases without cortical responses by the median nerve stimulation (case 16, 17, and 18), a severe neurological deficit had been observed before surgery, and preoperative SEP was abolished in 2 radiation necroses. In the case showing only the diphasic negative-positive potential (case 13), the central sulcus was supposed to lie anterior to the area of the craniotomy, due to the mass effect of the parietal tumour.

Cortical responses by the trigeminal nerve stimulation were examined in 14 cases, most of which had a lesion of the mid- and low convexity. In 6 out of these 14 cases, phase-reversed potential was also obtained, with similarity to the median nerve stimulation. Mean negative peak latency was 14.9 msec and positive 21.5 msec in the postcentral gyrus, while positive 15.7 msec and negative 23.1 msec in the precentral gyrus. Compared to the median nerve stimulation, the range of the central sulcus definable by the trigeminal

nerve stimulation was so short as about 2 cm. However, the central sulcus was possible to be extended continuously from the line drawn by the stimulation of the median nerve stimulation. There were 8 cases without any apparent potential. Two cases (case 13 and 14) were from the location of the lesion (high convexity), and other 2 (case 2 and 6) were presumably due to the application of the electrodes away form the suitable place. In the remaining 2 glioblastomas and 2 radiation necroses (case 15–18), remarkable destruction of the cortex was observed at surgery.

Tibial nerve stimulation was tried in 3 cases with high convexity lesion and 1 case with interhemispheric tumour. In only one patient (case 10), phase-reversed potential was obtained when the electrodes were applied to the upper mesial surface of the hemisphere, while in the other 3 cases, diphasic positive-negative (case 13 and 14) or negative-positive potential (case 1) was observed on the convexity side, irrespective whether the electrode arrays were placed in the frontal or parietal lobe.

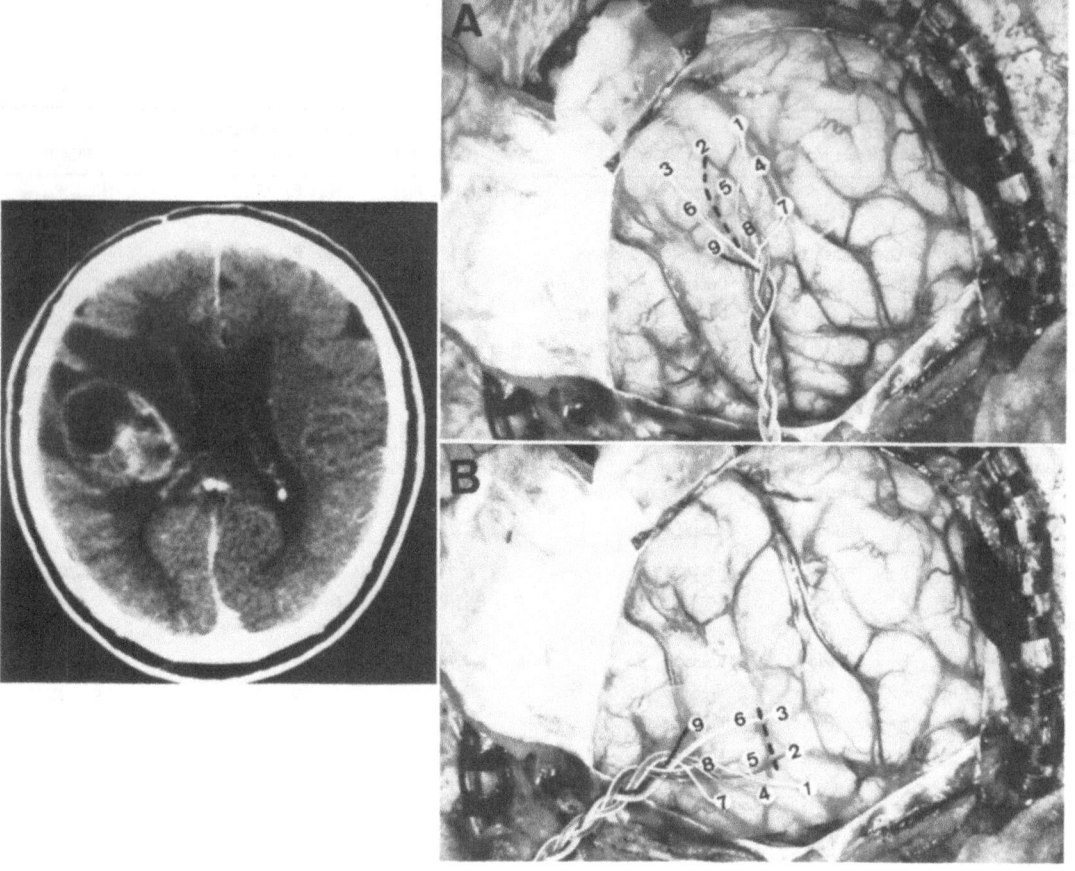

A Median Nerve Stimulation

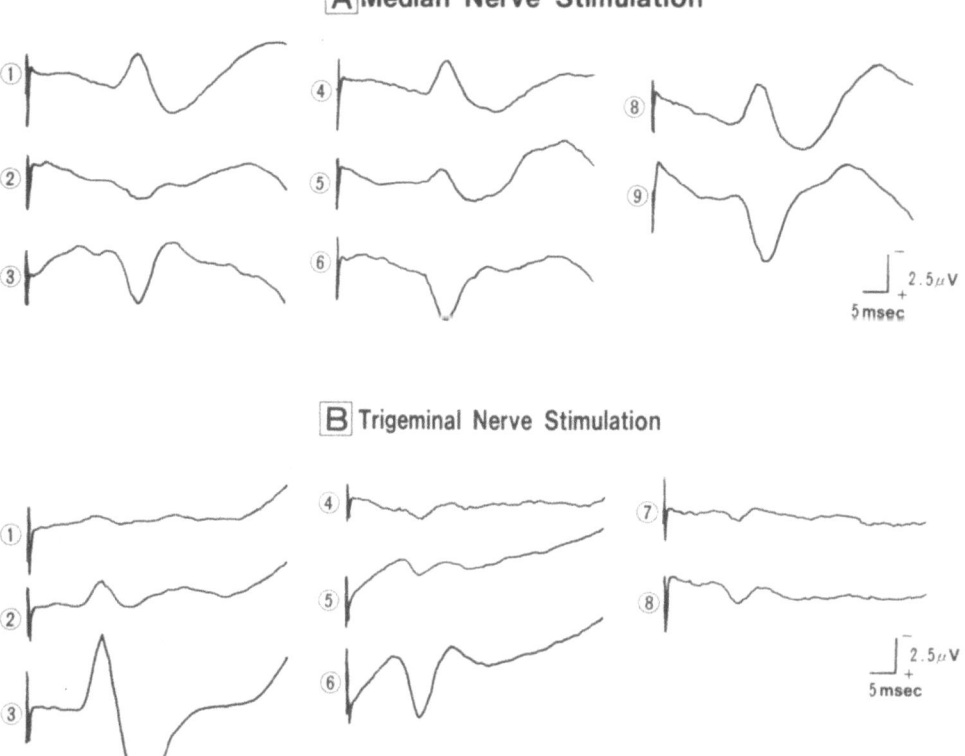

B Trigeminal Nerve Stimulation

Fig. 1

Case 9 is shown in figure as an example of intraoperative identification of the central sulcus for gliomas.

This 56-year-old man had a mild expressive aphasia without apparent motor dysfunction. Enhanced CT scan (upper left of figure) revealed a tumour in the vicinity of the left rolandic area, and the tumour seemed to be partially cystic under the cortex of mid- and low-convexity. At surgery, after the identification of the central sulcus (2 operative photographs and traces of potential in figure), the cortex was incised just posterior to the postcentral gyrus, and then the tumour could be subtotally removed without producing a major neurological deficit.

Discussion

After the introduction of CT scan and MRI, early diagnosis of cerebral lesions has now become possible. In cases with brain tumour, it means the detection of lesions showing only a trivial neurological deficit, such as headache and focal convulsion. Among the major postoperative sequelae, motor disturbance is one of the greatest problems, particularly in the surgery for the lesions adjacent to the sensorimotor area. Therefore, in such lesions, recognition of the pre- and postcentral gyri is essential to perform neurosurgical treatment without producing a functional deficit. When neurosurgical procedures were planned in the rolandic area, the line of "the central sulcus" used to be drawn, on trial, by Ring and Waddington's method[5]. However, this is not adequate in a tumour case, because the line is determined only from the shape of the cranium and the actual sulcus is often displaced by mass effect.

On the other hand, from the recent neurophysiological studies of surgical candidates with intractable epilepsy, the central sulcus can be identified by the use of cortical SEP[2]. However, in most of them, the central sulcus, as well as the epileptic focus, was shown by chronically indwelling subdural electrodes extraoperatively[3]. Few reports have been published so far on the intraoperative identification of the central sulcus in surgery for brain tumours[1].

In our series, among cortical responses to the stimulation of the median nerve, the initial negativity on the postcentral gyrus had the mean peak latency of 22.4 msec. The value was somewhat later than N 1 in Lueders's report[2]. This is probably due to the difference

of pathology and the effect of general anesthesia. In that report, all of the patients were epileptic and potentials were recorded in the awake state. Lueders also reported on the high amplitude P 2 in the postcentral gyrus. P 2 was observed only in a very limited area as shown in his article, whereas N 1 was more widely distributed and never observed anterior to the central sulcus. Therefore, N 1 is particularly suitable as a landmark of the postcentral gyrus. In fact, from our series, the potential of N 1 was shown, accompanied with phase-reversal in the precentral gyrus, in all cases (13 cases) where identification of the central sulcus was necessary.

Trigeminal cortical responses were obtained by stimulation to the lips. By the analogy with the potentials recorded from the median nerve stimulation, the first negativity also showed phase-reversal across the central fissure. The mean peak latency of that potential was 14.9 msec and about 2 msec later than the value of normal subjects recorded on the scalp electrode[6]. The reason for this delay may be the same as in the median nerve stimulation. Compared with responses to median nerve stimulation, the potentials were detected from a relatively small area of the cortex in spite of the wide cortical representation of the lips[4]. This may result from stimulating the lips, instead of the trigeminal nerve trunk itself. However, in 6 out of 8 expected cases, the central sulcus could be defined and extended to the portion identified by median nerve stimulation, thus covering most of the sulcus of the convexity side.

On tibial nerve stimulation, the number of the examinations was too small to draw definite conclusions. However, it seems to be very promising at least for lesions of the interhemispheric rolandic region.

In this report, we have shown an easy and not time-consuming procedure to localize the central sulcus during craniotomy, consisting of multi-channel simultaneous recording of cortical SEP with electrode arrays and stimulation of the appropriately selected peripheral nerve depending on the location of the lesion. In most cases showing only a trivial deficit before surgery, radical removal of the tumour was successfully carried out without producing severe postoperative neurological sequelae.

Fig. 1. An example of intraoperative identification of the central sulcus (case 9). Top: contrast CT scan showing the partially cystic tumour under the cortex in the parieto-central region. Two intraoperative pictures indicating how to apply the electrode arrays to the cortex. Interrupted lines show the central sulcus identified by the cortical responses. Refer to bottom of figure. Upper photograph (A): median nerve stimulation. Lower photograph (B): trigeminal nerve stimulation. (In the actual recording, electrode plate is surrounded with cottonoid to exclude air bubble between plate and cortex.) Bottom: upper traces (A) indicates the responses to the median nerve stimulation. Note the phase-reversal between 1–3, 5–6, and 8–9. Lower traces (B) to the trigeminal nerve stimulation. Phase-reversal is observed between 3–6 and 2–5

Acknowledgement

The authors thank Mr. Y. Shirai and Dr. Y. Ishiyama for their excellent technical advice.

References

1. Gregorie EM, Goldring S (1984) Localization of function in the excision of lesions from the sensorimotor region. J Neurosurg 61: 1047–1054
2. Lueders H, Lesser RP, Hahn J, Dinner DS, Klem G (1983) Cortical somatosensory evoked potentials in response to hand stimulation. J Neurosurg 58: 885–894
3. Morris HH, Lueders H, Hahn JF, Lesser RP, Dinner DS, Estes ML (1986) Neurophysiological techniques as an aid to surgical treatment of primary brain tumours. Ann Neurol 19: 559–567
4. Penfield W, Rasmussen T (1955) Sensorimotor representation of the body. In: The cerebral cortex of man. Macmillan, New York, pp 11–65
5. Ring BA, Waddington MM (1967) Ascending frontal branch of middle cerebral artery. Acta Radiol [Diagn] 6: 209–220
6. Seki Y, Aiba T, Shirai Y, Ishiyama Y (1987) Trigeminal somatosensory evoked potentials: Recording technique and normal responses. No To Shinkei 39: 105–112 (Engl Abst) (Jpn)

Address for correspondence: Yojiro Seki, M.D., Department of Neurosurgery, Toranomon Hospital, 2-2-2 Taranomon, Minato-ku, Tokyo 105, Japan.

Monitoring

A detailed account of the monitoring technique has been given elsewhere[6]. In summary, a 1 mm diameter silver ball recording electrode was positioned transtympanically on the middle ear promontory and referenced to the ipsilateral ear lobe. Click stimuli of varying intensity and polarity were then delivered through a tube held in the external ear canal. Responses were averaged using a Digitimer D 200 signal analyser and stored on floppy disc. Averages of only 128 sweeps were required, and with a click repetition rate of 9 Hz clear waveforms were obtained in approximately 20 seconds, allowing rapid assessment of changes caused by surgical trauma.

Audiological Follow-up

Postoperative assessment included repeat pure-tone audiometry and SD in all patients. Selected patients also underwent repeat fine-needle transtympanic ECochG.

Results and Discussion

Reasons for intra-operative monitoring of evoked potentials include:

i) Recognition of warning signs of potential neurological damage in the pathway monitored and the possibility of reversing the trend before the damage becomes irreversible.

ii) Prediction of postoperative prognosis.

iii) Recognition of surgical manoeuvres causing irreversible neurological deficit which may change surgical practice for future similar operations.

iv) Extending physiological and anatomical knowledge by, for example, establishing intracranial generators of the surface recorded EPs.

We began using intraoperative ECochG recording in the hope of recognising potentially reversible injury to the auditory pathways, particularly the cochlear nerve, at an early stage and then to modify the surgery to preserve function. Our experience has, however, been fairly disappointing in this respect as the following examples illustrate.

ECochG Preserved

Seven patients retained hearing after operation (Fig. 2, Table 1). Although changes were seen in waveform during surgery, these were generally minor and required no changes in surgical technique to reverse them. Of more interest were the two patients who had normal waveforms throughout operation and yet were both deaf. In one case the cochlear nerve was thought to have been divided during tumour excision, yet this had virtually no effect on the waveform. This phenomenon has been noted in animal experiments with monitoring of round window potentials during cochlear nerve section. Only slight changes are seen in the compound

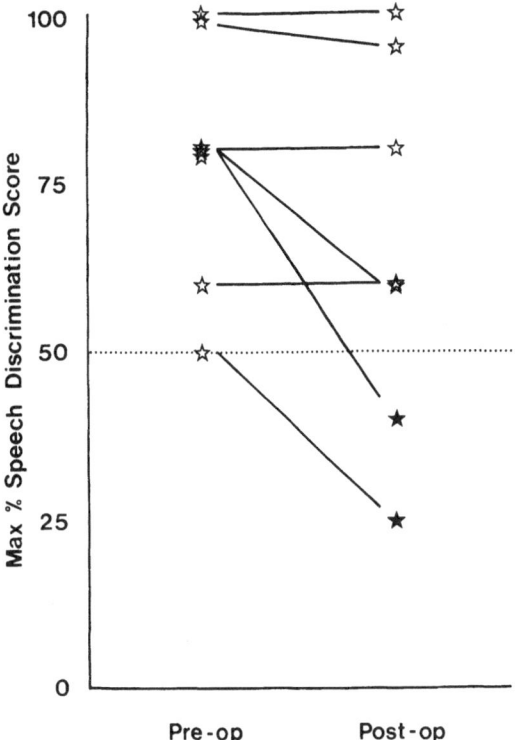

Fig. 2. Comparison of pre- and postoperative maximum speech discrimination scores to demonstrate the quality of hearing that can be preserved. Only two patients (black stars) had hearing of less than 50% postoperatively, and one of these was borderline before operation

Table 1. *Hearing Preservation Related to Tumour Size*

Tumour diameter (sample number)	Postoperative outcome	
	Hearing	Deaf
< 1.5 cm (n = 7)	5	2
1.5–3.0 cm (n = 9)	1	8
> 3.0 cm (n = 8)	1	7

action potential until 48–72 hours after section, when retrograde nerve degeneration causes their loss[1, 5, 9]. The other deaf patient with intraoperatively preserved potentials had a grossly intact cochlear nerve, suggesting possible axonotmesis as a cause of deafness, and it has been shown that the junction of the central and peripheral parts of the nerve is a weak area liable to traction injury[8]. On later repeat ECochG (Fig. 3) she was found to have preserved cochlear potentials but no N_1 or N_2[7], again analogous to chronic nerve section experiments in animals. This lack of ability to detect injury to the cochlear nerve is a serious handicap to the use of ECochG monitoring for hearing preservation.

Acta Neurochirurgica, Suppl. 42, 27–30 (1988)
© by Springer-Verlag 1988

Intraoperative Monitoring of the Electrocochleogram and the Preservation of Hearing During Acoustic Neuroma Excision

L. Symon[1], **H. I. Sabin**[1], **P. Bentivoglio**[1], **A. D. Cheesman**[2], **D. Prasher**[3], and **H. Barratt**[3]

[1] Gough-Cooper Department of Neurological Surgery, Institute of Neurology, The National Hospital, London, U.K., [2] The Royal National Throat, Nose and Ear Hospital, London, U.K., [3] MRC Neuro-Otology Unit, The National Hospital, London, U.K.

Summary

We have monitored the electrocochleogram (ECochG) of 24 patients, using a transtympanic electrode, during acoustic neuroma excision. All patients had unilateral tumours with good preoperative hearing and complete excision was achieved in each case. Of the 24 patients, seven retained some hearing, however, a further two patients had normal ECochG waveforms at the end of operation but were nevertheless deaf. Thus, there is not an invariable correlation between immediate preservation of the ECochG and hearing.

As expected, tumour size was important in hearing preservation. Five of seven patients with tumours less than 1.5 cm in diameter retained some hearing after operation, whereas 15 of 17 patients with tumours greater than 1.5 cm in diameter were deaf.

Keywords: Acoustic neuroma; evoked potentials; hearing preservation.

Introduction

The value of attempting to preserve hearing during acoustic neuroma surgery is controversial. Neurosurgeons usually excise these tumours through the posterior fossa, but it is claimed by otologists that the risk to the facial nerve using this approach outweighs the benefit of preservation of what may be severely impaired hearing[11]. Instead, they generally advocate translabyrinthine surgery, with fewer risks to the facial nerve, but guaranteed postoperative unilateral deafness. Possible ways of increasing the rate and quality of hearing preservation during posterior fossa excision of these tumours have, therefore, been investigated, including monitoring of the brainstem auditory evoked response (BAER)[2] and of the ECochG[3, 6].

Materials and Methods

Patient Selection

All patients with suspected acoustic neuroma underwent neuro-otological investigation including pure-tone audiometry, phoneme speech discrimination (SD), BAER and latterly fine-needle transtympanic ECochG (Fig. 1) to establish a threshold for stimulation. Only those patients with SD scores of greater than 50% at 60 dB nHL were included, and after consent was obtained 24 patients were monitored during a 30 month period.

Surgery

Operations were undertaken by one neurosurgeon (LS) with an otologist (ADC). All tumours were completely excised using a posterior fossa approach with the patient in the park bench position, and all were subsequently confirmed histologically to be neurilemmomas.

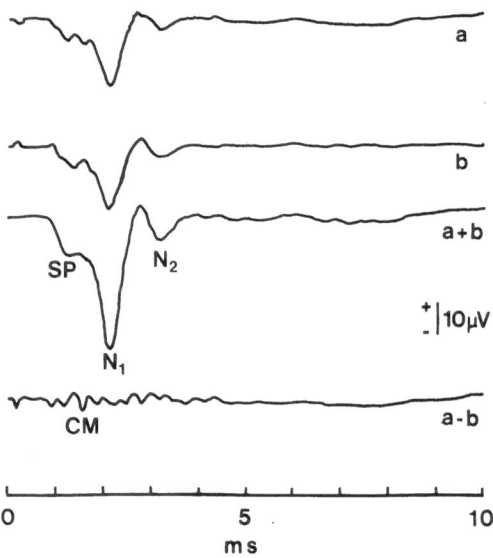

Fig. 1. The normal transtympanic electrocochleogram. Trace (a) is the evoked response to 80 dB intensity unipolarity stimuli, trace (b) to opposite polarity stimuli. Trace (a + b) enhances the compound action potential (CAP) comprising the negative waves N_1 and N_2 and also shows the summating potential (SP) generated by basilar membrane displacement. Trace (a − b) enhances the cochlear microphonics (CM), which are hair cell generated and mimic the frequency and polarity of the stimulus

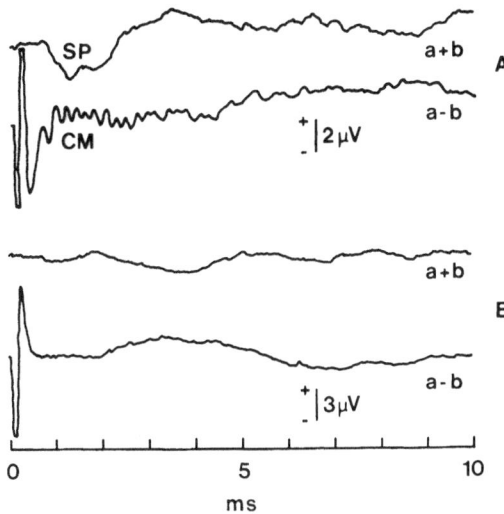

Fig. 3. Examples of postoperative fine needle transtympanic ECochG in two deaf patients, A and B – 15 months and 7 months after operation respectively (see Fig. 1 for explanation of derivation). Patient A had had a normal ECochG at the end of operation but had seen deaf on retesting immediately after operation. It can be seen that cochlear activity persists (*CM* and *SP*) but there is no true CAP (N_1 or N_2) which disappeared as the cochlear nerve degenerated after injury. Patient B has no remaining inner ear electrical activity, presumably secondary to cochlear ischaemia or direct labyrinthine injury

ECochG Lost Temporarily

In only one case was the ECochG lost temporarily with subsequent recovery. In this case the neuroma was large (> 3.0 cm) and cystic, and dissection from the brainstem caused loss of the potential, which reappeared after the cyst was punctured and the tumour decompressed. The speed of loss and recovery (within minutes) suggests cochlear ischaemia as the mechanism responsible, perhaps through traction on the labyrinthine artery.

ECochG Lost Permanently

In all our cases of ECochG loss the potentials disappeared over a maximum of 5–10 minutes, and generally occurred during dissection of the lateral cuff of tumour extening into the IAM. In four cases, obvious bleeding was noted at this time, suggesting labyrinthine artery damage, and in one case diathermy of a bleeding vessel after the tumour had been removed resulted in electrical silence. This type of rapid loss of potentials again suggests ischaemia[4] to us, although it is possible that intramodiolar haemorrhage may also have occurred[8]. An example of a "dead ear" is shown in Fig. 3, neither hair cell potentials, nor cochlear nerve action potentials are

seen. This type of vascular injury is obviously not reversible and all monitoring achieves is the ability to watch the potentials fade away and to suggest the likely cause of hearing loss. It is, however, conceivable that arterial spasm is present, and if prolonged this may lead to permanent cochlear damage. Reserach into this possibility and its pharmacological reversal would be interesting.

Facial Nerve Preservation

The facial nerve was preserved in 22 of the 24 patients (Table 2).

Table 2. *Facial Nerve Preservation Related to Tumour Size*

Tumour diameter	VII intact	VII divided
< 1.5 cm	7	0
1.5–3.0 cm	8	1
> 3.0 cm	7	1

Complications

The small tympanic perforations caused by this technique generally healed within 10 days, however, one patient had a torn tympanic membrane following a difficult electrode insertion and this was still present after three months, although eventually healed spontaneously.

Conclusions

Although intraoperative monitoring of ECochG has proved to be interesting, it probably does not help to save the hearing of the patient being monitored. It has, however, suggested some causes of deafness with an intact cochlear nerve and has made us wary of damage to the labyrinthine artery, especially with the use of diathermy at the end of operation. The results of hearing preservation seem more dependent on the tumour size, although this in itself is probably misleading as very small intracanalicular tumours may invade the modiolus[10] and thus inevitably lead to hearing loss on removal. In addition, although the numbers are small, the risks to the facial nerve seem little greater using this approach than using a translabyrinthine route, and seven patients now have hearing on the operated side who would not if that particular approach had been used.

References

1. Fisch UP, Ruben RJ (1962) Electrical acoustical response to click stimulation after section of the eighth nerve. Acta Otolaryngol 54: 532–542

2. Hardy RW, Kinney SE, Lueders H *et al* (1982) Preservation of cochlear nerve function with the aid of brainstem auditory evoked potentials. Neurosurgery 11: 16–19

3. Ojemann RG, Levine RA, Montgomery WR, *et al* (1984) Use of intraoperative auditory evoked potentials to preserve hearing in unilateral acoustic neuroma removal. J Neurosurg 61: 938–948

4. Perlman HB, Kimura R, Fernandez C (1959) Experiments on temporary obstruction of the internal auditory artery. Laryngoscope 69: 591–613

5. Ruben RJ, Hudson W, Chiong A (1963) Anatomical and physiological effects of chronic section of the eighth nerve in cat. Acta Otolaryngol 55: 473–484

6. Sabin HI, Bentivoglio P, Symon L, *et al* (1987) Intraoperative electrocochleography to monitor cochlear potentials during acoustic neuroma excision. Acta Neurochir (Wien) 85: 110–116

7. Sabin HI, Prasher D, Bentivoglio P, *et al* (1987) Preservation of cochlear potentials in a deaf patient fifteen month after excision of an acoustic neuroma. Scand Audiol (in press)

8. Sekiya T, Moller AR, Jannetta PJ (1986) Pathophysiological mechanisms of intraoperative and postoperative hearing deficits in cerebellopontine angle surgery: an experimental study. Acta Neurochir (Wien) 81: 142–151

9. Spoendlin H (1975) Retrograde degeneration of the cochlear nerve. Acta Otolaryngol 79: 266–275

10. Suga F, Lindsay JR (1976) Inner ear degeneration in acoustic neuroma. Ann Otol 85: 343–358

11. Tos M, Thomsen J (1982) The price of preservation of hearing in acoustic neuroma surgery. Ann Otol Rhinol Laryngol 91: 240–245

Address for correspondence: Lindsay Symon TD, FRCS (Ed), FRCS, Professor of Neurosurgery, The Gough-Cooper Department of Neurological Surgery, Institute of Neurology, The National Hospital, Queen Square, London WC1N 3BG, U.K.

Acta Neurochirurgica, Suppl. 42, 31–34 (1988)

The Use of EEG Spectral Analysis After Thiopental Bolus in the Prognostic Evaluation of Comatose Patients with Brain Injuries

H. J. Klein, S. A. Rath, and **F. Göppel**

Department of Neurosurgery, University of Ulm, Günzburg, Federal Republic of Germany

Summary

The value of EEG data and evoked potentials (EP), combined with the patient's clinical status, are important parameters used to document decerebration, but their reliability is at its best when the decerebration is nearly complete. Based on spectral analysis of the EEG we compiled criteria for the cerebral function with application of agents known to alter the normal EEG. We found very distinct and different pattern responses in spectral analysis of the EEG after the bolus injection of 100–500 mg Thiopental, correlating well with the patient's prognosis. In a study with 40 patients, those who showed a significant increase of power in all frequencies, especially a short lasting increase in beta (3–7 min) survived in 84%. On the other hand, all the patients who had a decrease of absolute power in all frequency band died, even when brainstem reflexes and various pain reactions were still present. Spectral analysis after Thiopental bolus injection also permits a very immediate assessment of slight improvements or deteriorations in the clinical course.

Keywords: EEG; evoked potentials; brain injury; prognosis; thiopental.

Introduction

The literature cites numerous procedures for evaluating the prognosis in patients with severe craniocerebral trauma and spontaneous intracerebral haemorrhages. These procedures are based on diverse clinical[1], morphological and electrophysiological parameters like the Glasgow coma score, computed tomography findings[9, 13], intracranial pressure recording[5, 14], EEG-analysis[3, 4, 12, 14] and evoked potential registration[6, 8] and are used either transiently or throughout the observational course.

The reliability of the prognostic evaluation of clinical, neuroradiological and electrophysiological parameters is measured by their correspondence to the actual later course as well as by the time these evaluations become reliable in a brain injury patient. This means that a high degree of reliability is of no clinical value if it is achieved too late. Thus, although the lack of intracerebral response to acoustically evoked potentials is identical with definitive decerebration of the patient, the phenomenon is of only diagnostic value at that late stage.

The question we were considering was whether an electrophysiological test of function could permit an early and reliable prognostic statement while allowing for a similar characterization of responses as other tests of function.

Material and Methods

The Fast-Fourier-transformation of the EEG in compressed spectral array presentation was used as a method for prognostic monitoring[2, 12]. The continuous presentation of the EEG spectra is monitored on the oscilloscope as an alternative to the EEG signal. 8-sec-spectra are related for a time period of 30 sec with an immediate plot of mean spectra in pseudo-three-dimensional display using variable line feed. A second printer displays the quantitative analysis of the individual spectral bands over time. The spontaneous frequency analysis of leads C_3 to F_3 and of leads C_4 to F_4 was compared with the changes occurring after a bolus injection of Thiopental in a dose of about 2–4 mg/kg body weight.

We based our prognostic EEG analysis on the changes known to occur in neurologically healthy patients (e.g., introduction of anaesthesia) immediately after administration of Thiopental; these changes are general stimulation of all EEG frequency bands as well as excessive beta-stimulation, termed primary stimulation in the following[11].

Results and Discussion

The EEG reaction was compared with the response pattern elicited by Thiopental in healthy individuals. The EEG response after a bolus injection of Thiopental was evaluated in 40 patients who had a Glasgow coma score of less than 7 points. 26 of these patients were males aged 8 to 72 with an average age of 39.5 years;

Table 1. *EEG—Response to Thiopental Bolus Injection (40 Patients)*

	General reduction, particularly alpha and beta	No change in activity	Beta-stimulation	General stimulation including beta-frequencies
Survivors 23	0	1	22	12
Fatal cases 17	13	0	4*	4

* 4 patients, fatal course despite beta-stimulation: 1 patient died of massive intracerebral pressure in decerebration, 3 patients died of other causes (pneumonia).

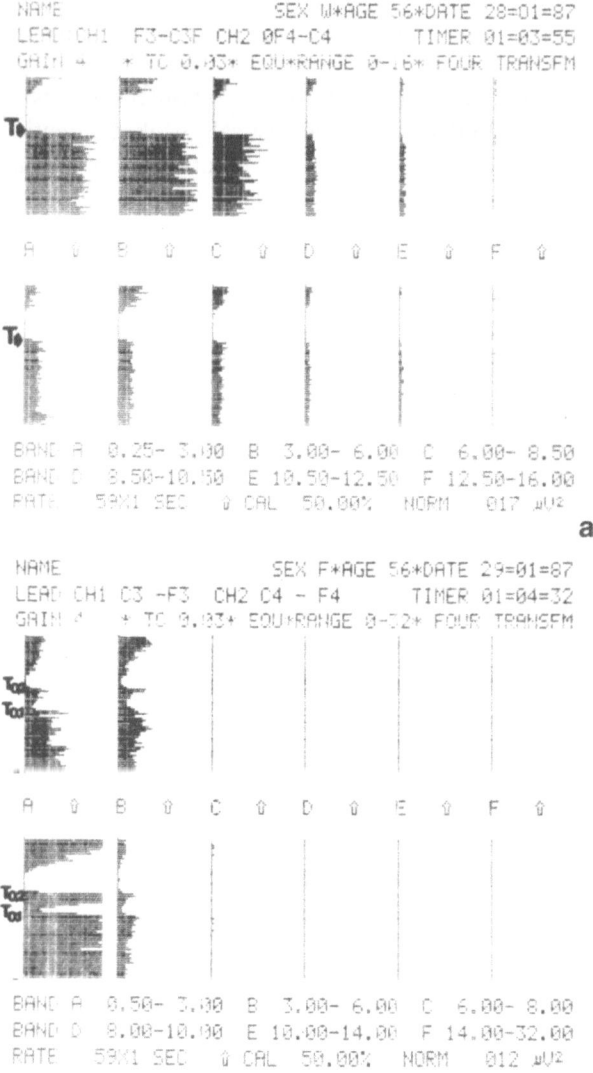

Fig. 1 a, b. Sequence of leads taken daily from a 56-year-old patient with severe craniocerebral lesion (acute right-side intracerebral haemorrhage): successive reduction of the higher frequencies (alpha and beta) after Thiopental administration (*T*), beginning with a general reduction and terminating in complete reduction of alpha and beta frequencies in the following days

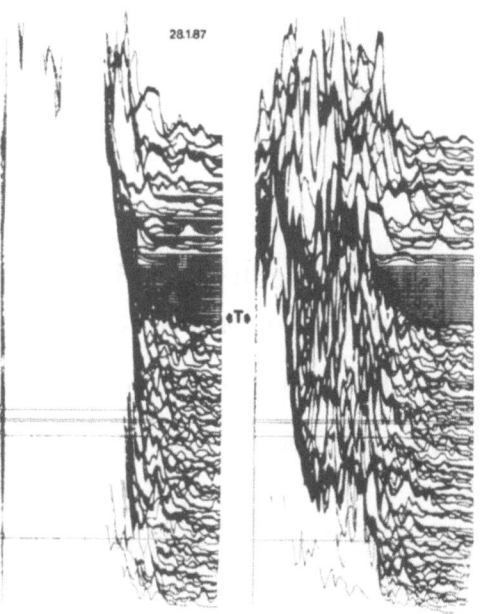

Fig. 1 c. General reduction of all frequency bands (with burst suppression) in compressed spectral array representation (same patient)

7 had a spontaneous intracerebral haemorrhage, the remainder had cerebrocranial trauma. The 14 female patients were 2 to 63 years old, averaging 44.5 years, of which 6 had a spontaneous intracerebral bleed. The evaluation shows 21 of 25 patients with primary stimulation to have a benign course. 2 of the remaining 4 patients died of protracted pneumonia after 2 and 6 weeks, respectively. Stimulation could not be tested in one small child who died in coma vigile after three months, because it had been transferred to another clinic early in its course. Only one female patient died of acutely increased cerebral pressure despite beta-stimulation. In contrast to this, 17 patients showed a general reduction of all EEG spectral bands after Thiopental bolus injection, some of them with burst-suppression. All patients of this group died in decerebration after their neurological status had progressively deteriorated (Table 1).

The lack of EEG stimulation after Thiopental administration is followed by a progressive loss of the rapid spectral bands. The lack of stimulation itself is irreversible. The inability of the cortex to produce barbiturate-induced fast activity has been described as a sign of considerable cerebral impairment[7, 10].

An EEG response showing a lack of beta or general stimulation and with a successive loss of the rapid frequencies (over 6 Hz) is obtained in all those patients dying subsequently with complete decerebration. The lack of EEG stimulation in the beta range or in the lower spectral bands can precede the frequency loss by

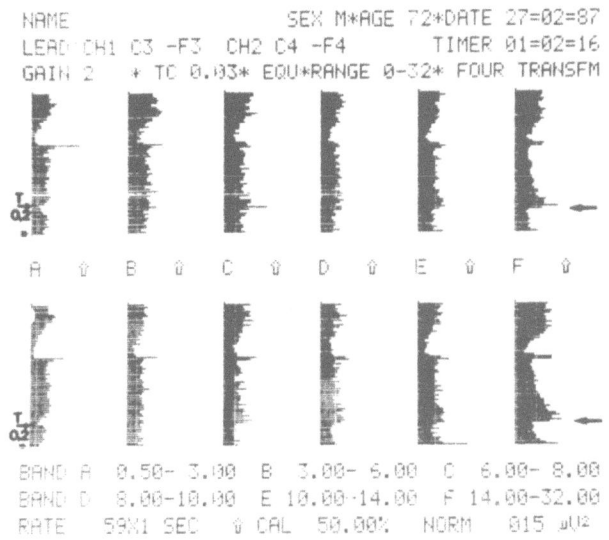

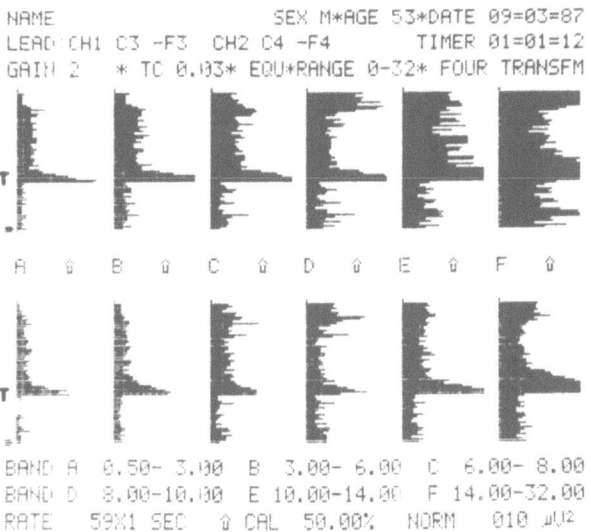

Fig. 2 a. Isolated beta-stimulation, with simultaneous reduction of the lower frequencies in parts

Fig. 3. Stimulation of all frequencies following Thiopental bolus injection

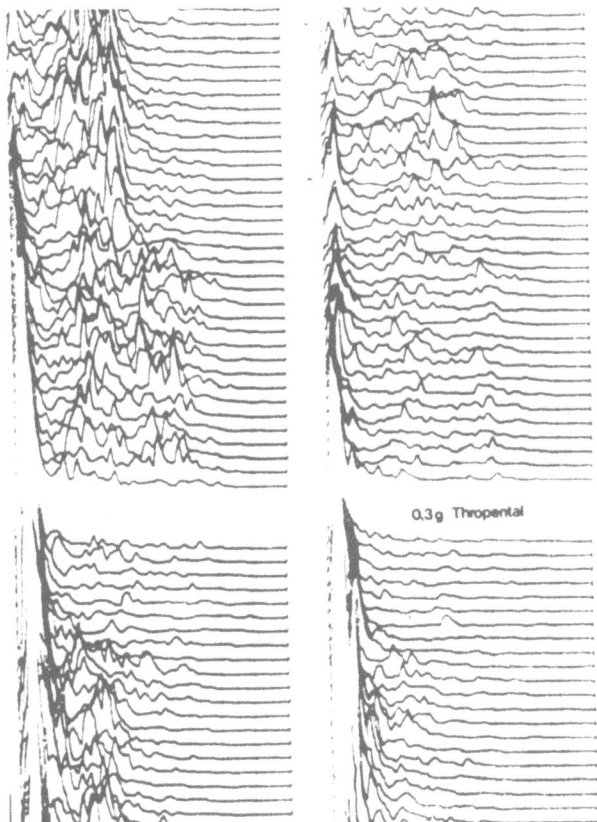

Fig. 2 b. Isolated beta-stimulation, with simultaneous reduction of the lower frequencies in compressed spectral array representation

several days without necessarily being reflected by a deterioration of the clinical condition of the patient at this time.

Patients with progressive decerebration show the same reaction in the EEG (reduction of the rapid frequencies) with progressively reduced doses of Thiopental. Occasionally, rapid alpha- and beta-stimulation was observed without concurrent stimulation of the other spectral bands. The additional activation of the lower frequency bands is a further sign for a good prognosis, and frequently occurs some time after isolated beta-stimulation. In patients with compensated hemisphere syndrome, the stimulation of the healthy side coupled with a loss of response in the affected side means that, over time, there is a reintegration of the physiological response on the affected side as well, but this does not need to be complete (with regard to bilateral response).

A comparison with the SEP-leads recorded simultaneously with the EEG shows that isolated cases of EEG stimulation occurred even when the cortical wave was missing or distinctly reduced over both hemispheres in medianus stimulation and, conversely, some leads showed a lack of EEG-stimulation when there was still a demonstrable central response.

Bearing in mind that 40 patients constitute only a small number of cases, a 100% correlation between lack of EEG-stimulation and clinically progressive decerebration with ultimate brain death was found in this study. A response to stimulation showed a 84% agreement with regard to a good prognosis and decerebration in only 4%.

References

1. Bauer G, Gerstenbrand F, Rumpl E (1979) Varieties of the locked-in syndrome. J Neurol 221: 77–91
2. Bickford R, Billinger TW, Fleming NI, Stewart L (1972) The

compressed spectral array (CSA-A pictorial EEG). Proc San Diego Biomed Symp 11: 365–370

3. Bricolo A, Turazzi S, Faccioli F, Odorizzi F, Sciaretta G, Erculiani P (1978) Clinical application of compressed spectral array in long-term EEG monitoring of comatose patients. Electroenceph Clin Neurophysiol 45: 211–225

4. Bricolo A, Turazzi S, Faccioli F (1979) Combined clinical and EEG examinations for assessment of severity of acute head injures. Acta Neurochir (Wien) [Suppl] 28: 35–39

5. Gaab MR (1981) Die Registrierung des intrakraniellen Druckes. Grundlagen, Ergebnisse und Möglichkeiten. Habil Schrift, Med Fakultät, Würzburg

6. Greenberg RP, Becker DP, Mayer DJ, Miller JD (1977) Evaluation of brain function in severe human head trauma with multimodality evoked potentials. Part 2: Localization of brain dysfunction and correlation with posttraumatic neurological conditions. J Neurosurg 47: 163–177

7. Hansen RB, Niedermeyer E (1987) Electroencephalography and Intensive Care Medicine. In: Landolt AM (ed) Intensive care and monitoring of the neurosurgical patient, progress in neurological surgery, vol 12. Karger, Basel, pp 105–145

8. Klug N (1982) Brainstem auditory evoked potentials in syndroms of decerebration, bulbar syndrome and in central death. J Neurol 227: 219–225

9. Lobato RD, Cordobes F, Rivas JJ, de la Fuente M, Montero A, Barcena A, Perez C, Cabrera A, Lamas E (1983) Outcome from severe head injury related to the type of intracranial lesion. A computerized tomography study. J Neurosurg 59: 762–774

10. Niedermeyer E, Yarworth S, Zobniw AM (1977) Absence of drug-induced beta-activity in the electroencephalogram. Eur Neurol 15: 77–84

11. Pichlmayr I, Lips U, Künkel H (1983) Das Elektroenzephalogramm in der Anästhesie. Springer, Berlin Heidelberg New York Tokyo, pp 213–224

12. Steudel WI, Krüger U (1979) Using the spectral analysis of the EEG for prognosis of severe brain injuries in the first posttraumatic week. Acta Neurochir (Wien) [Suppl] 28: 40–42

13. Turazzi S, Bricolo A, Pasut ML, Formenton A (1987) Changes produced by CT scanning in the outlook of severe head injury. Acta Neurochir (Wien) 85: 87–95

14. Wilson WP, Tindal GT, Greenfield JC (1965) Effects of an acute increase of intracranial pressure upon the electroencephalogram. Electroenceph Clin Neurophysiol 19: 184–186

Address for correspondence: Heinz J. Klein, M.D., Neurochirurgische Abteilung der Universität Ulm, Bezirkskrankenhaus, D-8870 Günzburg, Federal Republic of Germany.

Acta Neurochirurgica, Suppl. 42, 35–39 (1988)
© by Springer-Verlag 1988

Continuous versus Serial Global Cerebral Hemometabolic Monitoring: Applications in Acute Brain Trauma*

J. Cruz

The Division of Neurosurgery, Escola Paulista de Medicina, Hospital São Paulo, São Paulo, Brazil

Summary

Global cerebral hemodynamic and metabolic variables were assessed in twenty-eight adults with acute brain trauma. Systemic hemodynamics was also evaluated in twenty-one of the patients at that time. Up to six variables were simultaneously assessed, including systemic arterial pressure (SAP), intracranial pressure (ICP), expired or arterial carbon dioxide tension (P_ECO_2, $PaCO_2$), arterial, jugular bulb, and pulmonary artery oxyhemoglobin saturation percents (SaO_2, SjO_2, $S\bar{v}O_2$). All subjects having developed severe intracranial hypertension (SICH), those having also sustained systemic hypoxia presented with signs of neurological deterioration, significantly different from the hypoxic-free cases.

Keywords: Brain trauma; monitoring; hemometabolism; coma.

Introduction

Adequate tissue oxygenation requires two-way optimization. The first relies on the sufficiency of the arterial oxygen content (CaO_2), whereas the second depends on blood flow adequacy. Both being harmonically integrated, tissue oxygen delivery will ultimately reach ideal levels.

Nearly half a century has passed since global cerebral extraction of oxygen and glucose were first quantified in normal humans[8]. Shortly thereafter, another pioneering work disclosed a technique for quantification of cerebral blood flow (CBF) and oxygen consumption ($CMRO_2$), also in normal subjects[10]. Later, clinical continuous monitoring of ICP would flourish as a landmark contribution for patient care[9, 12].

In terms of continually monitorable hemodynamic variables, simultaneous SAP and ICP recordings have long been adopted in many centers as a routine tool for the assessment of cerebral perfusion pressure (CPP), the difference between mean SAP and ICP values. CPP being directly proportional to CBF, its monitoring

would appear to provide an index of tissue perfusion. However, the ultimate goal of tissue perfusion, namely tissue oxygenation, closely depends on CaO_2, as well as on the cerebral vascular resistance (CVR). Since neither CaO_2 nor CVR are inferrable from CPP, the need for alternative means of actually monitoring tissue oxygenation becomes evident. An extension of our previous experience on clinical cerebral hemometabolic monitoring[3, 4, 5], the present work was carried out in an attempt to compare continuous versus serial multivariate monitoring, and to correlate SICH and cerebral oxygenation with coma depth, in acute brain trauma.

Methods

Patient selection – As part of a broader investigation, twenty-eight prospectively observed adults entered this study fulfilling the criteria: a) acute closed brain trauma; b) coma depth reflected by low scores (4 to 7 points) on the Glasgow Coma Scale[18], after six post-accident hours; c) computed tomography (CT) scan of the head revealing predominantly diffuse brain insults (23 cases), or mass lesions (5 cases with acute subdural hematomas) which, after emergency evacuation, were replaced by visible re-expansion of the underlying brain; d) watertight dural closure and anatomical repositioning of the bone flaps in the operated cases; e) ICP monitoring carried out for no less than five days; f) SICH (as described below) having developed, and successfully managed.

Monitoring – Once in the surgical intensive care unit, all patients underwent routine monitoring of ECG, SAP, and ICP. SAP was recorded from axillary or radial 20 G cannulas in 16 cases, and from femoral 4 Fr fiberoptic saturometry catheters (Oximetrix, Inc.) in the other 12 patients. ICP was recorded from a subarachnoid bolt in 13 cases, lateral ventricular catheter in 9, and frontal subarachnoid catheter in the remainder. Recordings were 3 to 4 times daily checked for accuracy and recalibrations, against a centimetered column of physiologic saline solution. In addition, jugular bulb catheterization was performed percutaneously (from the level of the cricoid cartilage) with 19 or 16 G intracaths in 16 cases, and with 4 Fr fiberoptic lines in the remainder. The twelve subjects undergoing continuous saturometry also had continuous recordings of P_ECO_2 (Hewlett Packard Medical Products Division), thus constituting a multivariate continually monitored group (hereon called group CM). The other 16 cases (hereon called group IM) had intermittent measurements

---

* Work done, in parts, at the Divisions of Neurosurgery, The University of Texas Medical School at Houston, Hermann Hospital, Houston, Texas, U.S.A.; and Escola Paulista de Medicina, Hospital São Paulo, São Paulo, Brazil.

of $PaCO_2$, SaO_2, and SjO_2, at varying intervals from routinely 4 hours to more frequent off-schedule blood sampling whenever indicated. Supplementary monitoring of $S\bar{v}O_2$, along with pulmonary artery and pulmonary capillary – wedge – pressures, and cardiac output (CO) measurements, were also undertaken with conventional 7.5 Fr pulmonary artery Swan-Ganz catheters (American Edwards Laboratories) in 15 cases from both the CM and IM groups, and with fiberoptic 7.5 Fr lines[1] (Oximetrix, Inc.) in 6 cases of the CM group. The analog outputs from the recorded variables in group CM were converged to a multi-speed, six-channel strip chart recorder (Gould, Inc.).

Physiologic background – Seeking multivariate optimization, attempts were exaustively made to maintain most interrelated variables within their normal ranges, by means of physiologic and/or therapeutic manipulations. Hence, sedation (morphine sulfate), muscle paralysis (metubine iodide or pancuronium bromide), and hyperventilation were systemically employed for ICP control below 20 mmHg, as the three initial steps. Hypocapnia was titrated in order to avoid excessive increases in CVR. This would eventually avoid abnormally augmented cerebral oxygen extraction, consistent with oligemic cerebral hypoxia. Cerebral extraction of oxygen, the ratio between $CMRO_2$ and CBF, was directly monitored from the oxyhemoglobin saturation difference ($CEO_2 = SaO_2 - SjO_2$), whose normal (or normalization) range was adopted as 20 to 40%, from previously cited norms[8]. Figure 1 illustrates the above described manœuvres.

Should patients require more than the first three steps for ICP and CEO_2 optimization, and for over 24 hours, SICH was then recognized. In the present series, SICH therapy extended to mannitol boluses of 25 grams (or over) up to a serum osmolality of 315 mOsm, and/or cerebrospinal fluid (CSF) drainage should an intracranial catheter have been placed. These measures were adopted in a cumulative treatment protocol[4, 5] which, under circumstances of more aggressive ICP requirements involved pentobarbital coma, hypothermia, and even internal decompressive craniotomy (the latter in extreme instances). Regarding optimization of the relationship between CO and systemic consumption of oxygen ($\dot{V}O_2$), attempts were routinely made to avoid excessive increases in systemic extraction of oxygen ($SEO_2 = SaO_2 - S\bar{v}O_2$) over 30%, which approximates a lower range of 65–70% for $S\bar{v}O_2$. Titrating SEO_2 was usually accomplished by manipulations of cardiac output, via volume expansion.

Multivariate optimization was finally focused on the detection and the permissible reversal of arterial desaturation, primarily attributed to ventilatory and/or diffusional abnormalities in gas exchange. From previously cited norms[8], hypoxemic hypoxia was graded I, II, or III respectively, should SaO_2 fall into the ranges of 90–86, 85–81, and below 81%. While the on-line system frequently allowed sophisticated identification and management of hypoxemia in the CM group (see Figs. 2 and 3), standard measures were adopted for managing hypoxemia in group IM, from intermittent assessment of changes in clinical, radiographic, and laboratory findings. Measures were generally adopted when gross evidence of ventilatory disturbances became apparent, the first protective step being an increase in the fraction of inspired oxygen (FiO_2). Off-schedule blood gas and oximetry measurements would follow, along with additional measures for improving pulmonary care, suited for the individual needs. While in group CM the duration of prolonged hypoxemia could precisely be documented, a minimum 10-minute duration was estimated for hypoxemic events in the IM group. This was largely supported by the time constraints of laboratory oximetry and blood

gas analyses, which ultimately confirmed previously ongoing arterial desaturation. Once the occurrence of arterial hypoxemia was identified, patients were subdivided into the hypoxemic (H) and non hypoxemic (NH) categories. This led to a four subgroup distribution as follows: HCM, NHCM, HIM, NHIM, for hypoxemic or non hypoxemic, continually or intermittently monitored combinations.

At two weeks, SICH having subsided, changes in neurological status were assessed in relation to the patients' mean highest ICP values (while aggressively managed), as well as to the occurrence of prolonged hypoxemia (over 10 minutes). From the admission GCS scores, H and NH groups were compared in relation to the best GCS (BGCS) scores prior to the highest ICP and/or hypoxemic events, as well as to their "final" GCS (FGCS) scores at two weeks and the GCS differences.

Data analysis – Two-tailed t-test was used in assessing group differences for: a) highest ICP and corresponding CPP; b) lowest SaO_2 and corresponding SjO_2; c) BGCS and FGCS scores. Two-tailed Fisher's exact test was used in comparing the distribution of H versus NH cases in the CM and IM groups.

Results

While monitoring was required (average of 9 days), prolonged systemic hypoxemia was found at least once in 6 and 7 cases of groups CM and IM, respectively, superimposed to successfully managed SICH. Means and standard errors ($\pm$ S.E.M.) for the interrelated variables are as follows: a) highest ICP of 30.9 ± 1.46 mmHg and 29.1 ± 1.16 mmHg for H and NH, respectively, in association with CPP of 56.2 ± 2.52 mmHg and 58.9 ± 2.69 mmHg ($p = 0.6$ and $p = 0.8 - NS -$); b) lowest SaO_2 of $65.3 \pm 5.61\%$ and $80.3 \pm 2.35\%$, respectively, for HCM and HIM ($p < 0.05$), in association with SjO_2 of $33.2 \pm 1.9\%$ and $46 \pm 5.1\%$ ($p = 0.05$); c) BGCS score of 6.7 ± 0.36 and 5.3 ± 0.27 for H and NH, respectively ($p < 0.01$); d) FGCS score of 3.8 ± 0.17 and 8.7 ± 0.91 for H and NH, respectively ($p < 0.0005$); e) GCS differences between BGCS and FGCS scores were of -2.9 ± 0.43 and 3.5 ± 0.75 for H and NH, respectively ($p < 0.0005$).

While no negative GCS changes were found among NH cases, all patients in group H presented with worsening neurological status quantifiable on the coma scale, the 13 hypoxemic subjects being almost evenly distributed into groups CM and IM ($p = 1.0 - NS -$). In addition, abnormal neurological signs not quantifiable by GCS were found to develop in 1 and 4 cases among NHCM and NHIM, respectively. Those included changes in brainstem reflexes (slowing of oculocephalic and corneal responses). Adding up neurologically worsened cases with unchanged GCS scores to those with negative changes, 18 out of the 28 cases presented with – temporary – evidence of neurological impairment in the early recovery phase.

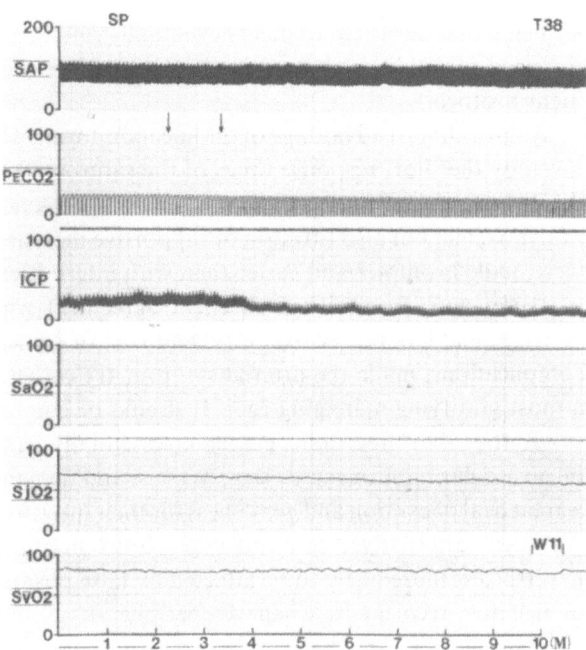

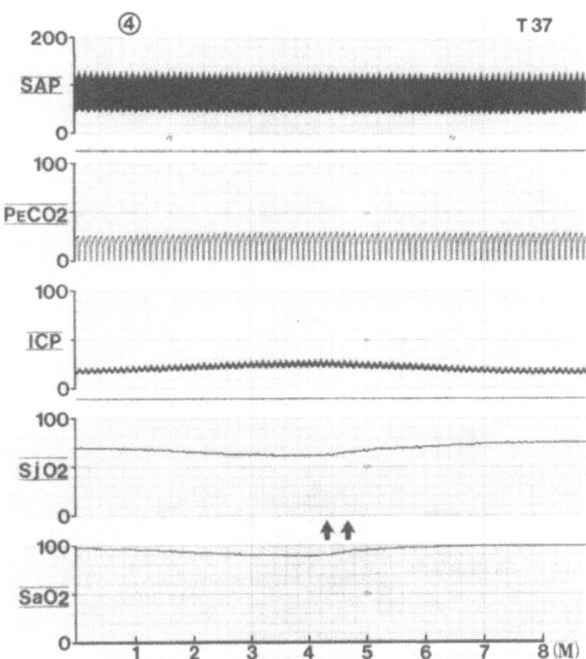

Fig. 1. SAP, P_ECO_2, ICP, SaO_2, SjO_2, and $S\bar{v}O_2$, with respective scales, recorded over 10 minutes during the early monitoring stage. The initial GCS score was 5, and the opening ICP of 35 mmHg had demanded standard hyperventilation for its temporary control. During the above period the patient was under hourly sedation and muscle paralysis (SP), rectal temperature being 38 °C (T 38). Despite optimized hypocapnia ($CEO_2 = 27\%$), a recurrent ICP rise to 24 mmHg induced a CPP fall to 59 mmHg and slight CEO_2 rise to 29%. A two-step increase in the patient's minute-volume (arrows) yielded a PCO_2 fall from 26 to 23 mmHg. This prompted an ICP drop to 11 mmHg, a concomitant CPP rise to 69 mmHg, but a further CEO_2 increase to 37%, still in the normal range. Simultaneously, SAP and SEO_2 remained in their normal ranges, in association with normal pulmonary capillary-wedge-pressure of 11 mmHg (W 11), as well as the pulmonary artery pressure (not illustrated), suggesting unchanged systemic hemometabolism

Fig. 2. SAP, P_ECO_2, ICP, SjO_2, and SaO_2, with respective scales, recorded over 8 minutes during the early monitoring stage. The initial GCS score was 4, and the opening ICP was 16 mmHg while the patient was moderately hyperventilated. Shortly prior to the above period, GCS score was still 4 (circled number at top). Rectal temperature was 37 °C (T 37). A period of optimized hypocapnia (CEO_2 of 27%) associated with ICP of 17 mmHg and CPP in the range of 58 mmHg. Rapid changes spontaneously occurred, involving simultaneous PCO_2 and ICP increases, as well as SaO_2 and SjO_2 decreases. As the SaO_2 fell into the alarm range of 90%, a two-step increase in FiO_2 from 50 to 80 then 100% (arrows), resulted in rapid arterial resaturation, followed by SjO_2 to slightly above baseline levels. At this point, PCO_2 being virtually constant, the ICP fall was given by arterial hyperoxia (SaO_2 up to 100%). On re-assessment, auscultation of the lungs was unremarkable, as was a chest X-ray done shortly thereafter. Stepwise FiO_2 decreases were subsequently started, restoring previous levels of O_2 delivery. The above recorded manœuvres demonstrate prompt resolution of arterial desaturation, which was routinely attempted. Notice that the lowest SjO_2, in the range of 62%, was still within normal limits

Discussion

Monitoring — Real-time assessment of changes in the interrelated variables was most useful in patient management. First, along with CPP optimization, indirect monitoring of PCO_2 related changes in CVR was also made possible. As illustrated in Fig. 1, it allowed fine titration of the $CMRO_2/CBF$ ratio, by physiologic modifications in CVR to optimize CEO_2. Quite clearly, neither CPP nor CBF measurements are capable of providing information on how CVR can be modified to allow optimal oxygen delivery. Although CVR and CBF values are unknown from CPP monitoring, CEO_2 reliably determines how adequately the changes in CVR (and consequently in CBF) will allow an optimal $CMRO_2/CBF$ coupling. Should continuous CBF measurement become available for as long as required by this group of patients, it would still lack the most cru-

cially needed information: how adequate is any CBF value to satisfy O_2 consumption at any level, at any particular point in time? While the answer to this question may be given by accompanying clinical changes to CBF modifications in awake subjects[6, 11] in deeply comatose ones the answer may virtually be given only by CEO_2[4].

A second practical feature of the system resided on unequivocal establishment of cause-effect relationships between rapidly occurring PCO_2 rises and subsequent ICP elevations. These events were more frequently observed than what could have been documented from routine blood sampling. Thus, minute-to-minute titration of the ICP to PCO_2 response was very effective,

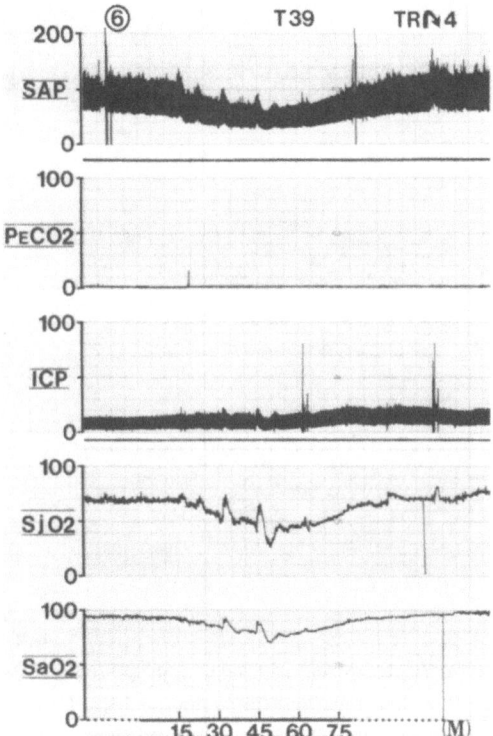

Fig. 3. SAP, ICP, SjO₂, and SaO₂, with respective scales, recorded over 80 minutes during the late monitoring stage. Long vertical lines correspond to artifacts generated by: blood sampling and line flushing-for SAP; head motion-for ICP; periodical self-checking of monitors – for SjO₂ and SaO₂. On a previous re-evaluation, GCS score was 6 (circled number at top), rectal temperature was 39 °C (T 39). After a period of over 24 hours of normal ICP in the presence of normocapnia, SICH having resolved, monitoring had partially been discontinued (notice P$_E$CO₂ flat line). The patient had developed extensive bilateral pneumonia, which progressively required increasing FiO₂ up to 90%, as well as positive end-expired pressure (PEEP) up to 12 cmH₂O. A chest X-ray done shortly prior to the above period, revealed extensive bilateral infiltrates and a pneumothorax. At 15 minutes, SAP, SaO₂, and SjO₂ gradually fell to levels of combined oligemic and hypoxemic cerebral hypoxia, induced by a tension pneumothorax. Further increasing the FiO₂ to 100% yielded no practical results. Between 45 and 50 minutes, following the placement of a chest tube, the three variables gradually returned to previous levels. The hypoxic episode lasted approximately 60 minutes, 30 of which at levels of cerebral hypoxia (SjO₂ below the range of 50%). Re-assessment of coma depth after the episode revealed a fall in GCS score to 4 points, by supraorbital trigeminal stimulation (TR 4). The curved arrow indicates a transient simultaneous rise in SAP, SjO₂, and ICP (the head motion artifact excluded) which, in conjunction with the spontaneously occurring changes during the hypoxic episode, resembled a defect in pressure autoregulation. The ICP rise from pre to post hypoxic was still within the normal range. Moreover, no changes in pupillary diameter were noticed, which further supported a cause-effect relationship between cerebral hypoxia and neurological deterioration. Notice a post hypoxic "narrowing" in CEO₂

while modulating CEO₂ as well. Indirectly, it prevented unnecessary use of mannitol, which would be employed for recurrent ICP rises not as well clarified. Moreover,

reducing mannitol requirements eventually minimized the need for additional steps in the cumulative treatment protocol.

A third clinical advantage of on-line monitoring was given by the short response times of the saturometers. As illustrated in Figs. 2 and 3, not only the detection of hypoxemia but also titration of supportive measures were made prompter and easier than with intermittent measurements. In addition, excellent agreement with conventional laboratory oximetry, and lack of clinical complications, made the arterial saturometry catheter a most gratifying managing tool. It should be emphasized, that 4 Fr fiberoptic catheters essentially limit intravascular monitoring to large vessels, and demand a great deal of caution and gentleness for their insertion. An ideal alternative, non invasive saturometry, is nevertheless more likely to be affected by false positive or negative recordings, whenever peripheral vascular resistance is altered. Therefore, should a less cumbersome technique become available for intraarterial saturometry, its routine use would be likely to flourish.

GCS changes – From a fine multicentric study, evidence has strongly been presented in favor of the type of traumatic brain lesion on the patients' outcome[7]. Perhaps most intriguing in correlating CT findings with clinical course is the fact that diffuse axonal injury may, more frequently than suspected on CT, superimpose on contusions, acute subdural hematomas, or even functional changes in brain swelling. Thus, given the current lack of well established diagnostic and therapeutic tools directed toward diffuse axonal brain damage, the treatment protocol previously described has essentially aimed at the – aggressive – prevention of secondary brain damage (intracranial herniation, oligemic and/or hypoxemic cerebral hypoxia). Unfortunately, adequate control of SICH and oligemic hypoxia, even though standardized, did not seen to suffice, for prolonged hypoxemia did strongly correlate with neurological impairment. The lowest SjO₂ values reported in both groups, mostly in HCM, ultimately supported the magnitude of arterial desaturation as being sufficient to induce additional brain damage.

The finding of lowest SaO₂ levels in group CM may imply a potential failure of the system in optimizing patient care. In fact, it most likely reflects increased accuracy of the system in demonstrating minute-to-minute changes. Those were clearly not given by intermittent monitoring, and most of the off-schedule blood samples were obtained after presumed supportive measures had been initiated. In addition to that, after preliminary evidence was provided from HCM cases, on the clinical relevance of prolonged hypoxemia, renewed directions were added to an already aggressive protocol for pulmonary care, delivered by the multidisciplinary intensive care team. As illustrated in Fig. 3, irrespective of how accurately and promptly the system would run, routinely available tools for the resolution of ventilatory problems may be time demanding, depending on the required intervention. This possibly limited, to a large extent, the potential for a better outcome in group H, at two weeks. Peculiar, however, was the observation on patients whose motor responses would deteriorate to just flickering or even flaccid after the occurrence of prolonged hypoxemia, and whom would slowly regain consciousness and other signs of recovery, over largely varying periods of time (a few weeks to months). Those were in sharp contrast to the ones who expired, invariably from progressive or recurrent systemic complications. In this respect, a long term follow-up study is currently underway.

The relevance of secondary insults to an injured brain has been clearly recognized as affecting the overall outcome of head-injured patients. On one hand, vigorous management of intracranial hypertension has widely been advocated[2, 4, 13, 14, 15]. A promising feature in the acute management of severe brain trauma would then focus, perhaps more aggressively, on the already known influence of secondary brain damage attributed to factors other than intracranial hypertension, such as cerebral hypoxia[3, 4, 16, 17]. Indeed, having available quite a sophisticated on-line system, one might envision the power for preventing or resolving most of the known problems related to the current clinical management of these patients. However, the preliminary results of this work have clearly shown that the prompt reversal of early "benign" oligemic and/or hypoxemic insults would frequently be offset later[4]. Hence, improved means of rapidly detecting and eventually resolving potentially aggravating pulmonary and hemodynamic instability factors would, a priori, appear as the future challenge. Admittedly, it would still contribute to no more than allowing for spontaneous neurological recovery.

Acknowledgement

Incentive and support from Michael E. Miner, M.D., Ph.D., and Fernando de Menezes Braga, M.D. were greatly appreciated. Highly meaningful was the contribution for patient care, delivered by the multidisciplinary intensive care teams of Hermann Hospital and Hospital São Paulo.

Dr. Cruz's Post-Graduate experience was supported, in its early phase, by a 1980–1981 Post-Graduate Fellowship Award from the Rotary Foundation of Rotary International.

References

1. Baele PL, McMichan JC, Marsh HM, Sill C, Southorn PA (1982) Continuous monitoring of mixed venous oxygen saturation in critically ill patients. Anesth Analg 61: 513–517
2. Becker DP, Miller JD, Ward JD, Greenberg RP, Young HF, Sakalas R (1977) The outcome from severe head injury with early diagnosis and intensive management. J Neurosurg 47: 491–502
3. Cruz J, Allen SJ, Miner ME (1985) Hypoxic insults in acute brain injury. Crit Care Med 13: 284
4. Cruz J, Miner ME (1986) Modulating cerebral oxygen delivery and extraction in acute traumatic coma. In: Miner ME, Wagner KA (eds) Neurotrauma treatment rehabilitation and related issues. Butterworths, Boston London Durban Singapore Sydney Toronto Wellington, pp 55–72
5. Cruz J (1987) Continuous monitoring of cerebral hemometabolic variables: preliminary therapeutic observations in acute brain trauma. J Cereb Blood Flow Metab 7 [Suppl] 1: 626
6. Finnerty AF, Witkin L, Fazekas JF (1954) Cerebral hemodynamics during cerebral ischemia induced by acute hypotension. J Clin Invest 33: 1227–1232
7. Gennarelli TA, Spielman GM, Langfitt TW, Gildenberg PL, Harrington T, Jane JA, Marshall LE, Miller JD, Pitts LJ (1982) Influence of the type of intracranial lesion on outcome from severe head injury. A multicenter study using a new classification system. J Neurosurg 56: 26–32
8. Gibbs EL, Lennox WG, Nims LF, Gibbs FA (1942) Arterial and cerebral venous blood. Arterial-venous differences in man. J Biol Chem 144: 325–332
9. Guillaume J, Janny P (1951) Manometrie intracrânienne continue: intérêt de la méthode et premiers résultats. Rev Neurol 84: 131–142
10. Kety SS, Schmidt CF (1948) The nitrous oxide method for the quantitative determination of cerebral blood flow in man: theory, procedure, and normal values. J Clin Invest 27: 476–483
11. Kety SS, King BD, Horvarth SM, Jeffers WA, Hafkenschiel JH (1950) The effects of an acute reduction in blood pressure by means of differential spinal sympathetic blockage on the cerebral circulation of hypertensive patients. J Clin Invest 29: 402–407
12. Lundberg N (1960) Continuous recording and control of ventricular fluid pressure in neurosurgical practice. Acta Psychiatr Neurol Scand 36 [Suppl] 149: 1–193
13. Marshall LF, Smith RW, Shapiro HM (1979) The outcome with aggressive treatment in severe head injuries. Part I: The significance of intracranial pressure monitoring. J Neurosurg 50: 20–25
14. Marshall LF, Smith RW, Shapiro HM (1979) The outcome with aggressive treatment in severe head injuries. Part II: Acute and chronic barbiturate administration in the management of head injury. J Neurosurg 50: 26–30
15. Miller JD, Becker DP, Ward JD, Sullivan HG, Adams WE, Rosner MJ (1977) Significance of intracranial hypertension in severe head injury. J Neurosurg 47: 503–516
16. Miller JD, Sweet RC, Narayan R, Becker DP (1978) Early insults to the injured brain. JAMA 240: 439–442
17. Miller JD (1985) Head injury and brain ischaemia – implications for therapy. Br J Anesth 57: 120–130
18. Teasdale G, Jennett B (1974) Assessment of coma and impaired consciousness. A practical scale. Lancet 2: 81–84

Address for correspondence: Julio Cruz, M.D., Serviço de Neurocirurgia, Escola Paulista de Medicina, Hospital São Paulo. Cx. Postal 57011, São Paulo, SP 04093 – Brasil.

Acta Neurochirurgica, Suppl. 42, 40–46 (1988)

Cerebral Hemisphere Swelling in Severe Head Injury Patients

R. Sarabia, R. D. Lobato, J. J. Rivas, F. Cordobés, J. Rubio, A. Cabrera, P. Gomez, M. J. Muñoz, and **A. Madera**

Service of Neurosurgery and Intensive Care Unit, Hospital 1° de Octubre and Faculty of Medicine, Madrid, Spain

Summary

The clinical course and the intracranial pressure (ICP) changes in 66 severe head injury patients presenting bulk enlargement of one cerebral hemisphere within a few hours of trauma have been analyzed. These patients represent 11% of a series of 589 severe head injury cases studied with computerized tomography (CT).

Cerebral hemisphere swelling, which was associated with an ipsilateral subdural haematoma of variable extent in 58 patients (88%), or a large epidural haematoma in 5 patients (7%), and occurred as an isolated lesion in 3 patients (4%), carried the highest incidence of uncontrollable intracranial hypertension, the highest mortality rate and the shortest survival period after trauma in the authors' severe head injury series.

The high incidence of arterial hypotension and/or hypoxaemia at admission (48% of cases), and the severity of clinical presentation (82% of patients scored 5 patients or less in the Glasgow Coma Scale, 77% had uni- or bilateral mydriasis and 82% initial ICP above normal limits) correlated with the very poor final outcome (85% mortality). Only one of the 12 patients with normal initial ICP continued to have low pressure throughout the course. High dose thiopental failed to control severe intracranial hypertension in 29 patients (44%) who had a fulminant, malignant course. A transient decrease in ICP elevation was achieved in 17 patients (26%) and a definitive control in 12 patients (18%), among them the 10 survivors in this series. In the authors experience once ICP is controlled, and unless haemodynamic instability compells action to the contrary, barbiturate should not be discontinued until a control CT scan shows complete disappearance of the mass effect.

Keywords: Barbiturate; brain oedema; brain swelling; coma; computerized tomography; intracranial pressure; severe head injury.

Introduction

The new insights provided by computerized tomography (CT) allow us to establish clinicopathological correlations in severe head injury patients and to delineate the prognostic significance of the major intracranial lesions seen during the immediate posttraumatic period[1, 3, 5, 14, 16–18, 23, 26, 29, 31]. One of the most threatening posttraumatic lesions is massive swelling (primarily defined in terms of volume) of a single cerebral hemisphere, which is most commonly associated with acute subdural haematoma and has a rapidly devastating course[9, 10, 14, 17, 18, 23]. Bulk enlargement of a cerebral hemisphere occurring within a few hours of injury may be due to increased intravascular blood volume, to increased brain water content or both. Until new developments in neuroimaging provide a means of determining the intrinsic mechanism of brain swelling in the clinical routine[14], we must rely on the findings of clinical-CT correlation in patients with this type of lesion and comparison of CT scans shortly before death with brain slices at autopsy.

In the present study we analyze the clinical, CT and intracranial pressure (ICP) findings in a series of 66 consecutive severe head injury patients who developed bulk enlargement of a cerebral hemisphere within a few hours of trauma.

Material and Methods

This study includes 66 patients selected from a total series of 589 severe head injury cases. The mechanism of trauma was a traffic accident in 52 patients, a fall in 9 and assault in 5. Nineteen patients had relevant associated extracranial injuries. Nearly two thirds of the patients arrived to our Unit directly from the scene of the accident, the others being referred from local hospitals.

All patients were comatose as defined by the Glasgow Coma Scale (GCS) and all underwent conventional X-ray studies and a CT immediately after admission. Control CT scans were performed in 42 cases, the first generally within 24 hours of admission and then every 2–3 days until midline normalization or death. Twenty four patients died too soon for the CT control. The radiological criterion for inclusion in this study was evidence of expansion of one cerebral hemisphere causing marked midline shift (septal and pineal displacement), either alone (3 cases) or in association with acute extraaxial haematoma (63 cases). Some patients had subarachnoid haemorrhage and 3 had intraventricular haemorrhage. We excluded patients who evidenced associated focal mass lesions, such as brain contusion or intraparenchymal haematoma, in the initial CT scan or subsequent CT controls. Cerebral angiography was not performed so the rare possibility of posttraumatic carotid artery occlusion was not ruled out.

All patients with acute extracerebral haematoma, except 16 showing very thin subdural collections, underwent haematoma evacuation immediately after admission. Four had a second operation several days later consisting of temporal decompressive craniectomy and temporal lobe tip resection. Removal of the extraaxial haematoma was performed within 4 hours of trauma in 20% of cases and 4 to 12 hours after trauma in the remainder. An intraventricular catheter or epidural fiber optic sensor was implanted for continuous measurement of ICP immediately after admission or operation in all cases. Intracranial hypertension was managed according to a standardized protocol reported elsewhere[18] which includes hyperventilation, osmotic agents and CSF drainage. When these conventional measures failed to keep ICP below 20 mmHg, highdose thiopental (4–10 g/day; average = 6 g/day) was administered in continuous infusion, doses being titrated according to ICP response. Patients so treated had pulmonary artery pressure monitored. If systemic arterial hypotension occurred, it was managed with dopamine and volume replacement. Thiopental plasma levels ranged from 2 to 18 mg% (average = 8 mg%).

The final result was graded in the categories of good recovery (GR), moderate disability (MD), severe disability (SD), vegetative status (VS) and death (D).

Results

Mortality rate in the 66 patients with cerebral hemispheric swelling (58 of whom had associated subdural haematoma) was 85%, while it was 41% for the entire severe head injury series (589 cases) and 63% in the total group of patients with acute subdural haematoma (120 cases). The 58 patients with hemispheric swelling and associated subdural haematoma had a younger average age, a higher incidence of peritraumatic hypotension-hypoxemia and were more frequently involved in traffic accidents than the other patients with acute subdural haematoma (cases of pure haematoma or haematoma associated with brain contusion).

The age distribution of the 66 patients in this study is shown in Table 1. Age ranged from 3 months to 69 years and 84% of the patients were under 40 years.

Skull X-ray evidenced a linear fracture in 31 patients. The fracture was ipsilateral to the affected cerebral hemisphere in 20 patients, among them the 5 patients with associated epidural haematoma, contralateral in 8 cases and bilateral in 3 cases.

Cerebral hemispheric swelling appeared as a single lesion in the admission CT scan in only 3 patients. In the other 63 patients it was associated with a thin subdural haematoma (36 cases), or a large subdural (22 cases), or epidural (5 cases) haematoma (Fig. 1). The involved cerebral hemisphere appeared either isodense (29 cases) or hypodense (37 cases) in comparison to the contralateral hemisphere. The amount of midline displacement seen in the admission CT scan and in subsequent CT controls is reflected in Table 2. It can be appreciated that 83% of the patients had marked (greater than 8 mm) midline displacement, which diminished in only 24% of the patients after extracerebral haematoma evacuation. Rapid intraoperative brain

Table 1. *Age Distribution in 66 Severe Head Injury Patients Showing Posttraumatic Cerebral Hemispheric Swelling*

Age (years)	Cases no. (%)
0–10	7 (12)
11–20	14 (21)
21–30	23 (35)
31–40	11 (17)
41–50	4 (6)
51–60	4 (6)
>60	3 (4)

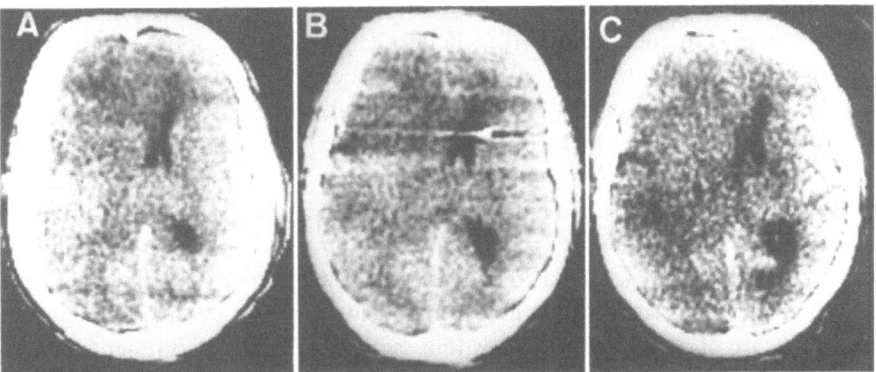

Fig. 1. (A) is the admission CT scan of a 22-year-old man showing a left acute subdural haematoma with marked midline shift. Although the haematoma was immediately removed, midline shift persisted unmodified because of bulk enlargement of the ipsilateral cerebral hemisphere. (B) and (C) are control CT scans performed 3 and 6 days after injury respectively. The GCS score at admission was 4 and the patient presented left mydriasis which did not disappear following haematoma evacuation. Initial ICP ranged from 30 to 50 mmHg, and could be controlled with thiopental until 6 days after surgery when it rose to uncontrollable levels

Table 2. *Midline Displacement Seen in the Initial CT Scan and in Subsequent CT Controls in 66 Patients with Posttraumatic Cerebral Hemispheric Swelling*

Initial CT scan	Cases no.	Midline displacement (mm)			
		Control CT scan			No CT control
		Diminished	Unchanged	Increased	
4–8	11 (17%)	3 (3)	2	6	—
9–12	34 (51%)	9 (7)	2	13	10
>12	21 (32%)	4	3	—	14
Total	66 (100%)	16 (10)	7	19	24

Figures in parentheses represent the number of survivors in each group.

Table 3. *Correlation Between the Admission Glasgow Coma Scale Score, the Occurrence of Lucid Interval, Hypotension-hypoxaemia at Admission, Pupillary Changes, Midline Displacement, ICP Level, and the Final Outcome in 66 Patients with Posttraumatic Cerebral Hemispheric Swelling*

GCS score	Cases no. (%)	Lucid interval	Hypotension -hypoxaemia	Pupillary changes			Midline shift (mm)			ICP (mmHg)			Outcome			
				None	Unilat mydr	Bilat mydr	4–8	9–12	>12	< 20	20–40	> 40	GR	MD	SD	D
3–5	54 (82)	7 (1)	28 (1)	5 (1)	29 (3)	20 (1)	7 (2)	26 (3)	21	10 (2)	14 (3)	30	3	1	1	49
6–9	12 (18)	2 (2)	4 (1)	10 (4)	2 (1)	—	4 (1)	8 (4)	—	2 (2)	6 (1)	4 (2)	2	2	1	7
Total	66 (100)	9 (3)	32 (2)	15 (5)	31 (4)	20 (1)	11 (3)	34 (7)	21	12 (4)	20 (4)	34 (2)	5	3	2	56

Figures in parentheses represent the number of survivors in each subgroup.

reexpansion with herniation through the craniotomy window or burr holes occurred in the majority of the patients operated for associated extraaxial clot, whose neurological status scarcely changed with operation.

Table 3 shows correlation between neurological status at admission, the occurrence of lucid interval, peritraumatic hypotension-hypoxemia, pupillary changes, midline shift, ICP level and final outcome. Fifty four (82%) patients scored 5 points or less in the GCS. Only 8 patients (12%) (including 3 of the 5 with associated epidural haematoma) had a short lucid interval, the rest remaining unconscious throughout the course. Arterial hypotension and/or hypoxemia occurred in 32 patients (48%) and 77% had uni or bilateral mydriasis at admission. The incidence of peritraumatic brain ischaemia, pupillary changes, marked midline displacement and high ICP, was significantly higher in patients with a GCS score of 3–5 than in those with a score of 6–8 (p < 0.001). The mortality rate of patients with 3–5 and 6–8 GCS scores respectively were 91 and 58% ($\chi^2 = 8.0$; p < 0.005). Only 8 patients (12%) made a functional recovery.

Initial ICP was above normal limits in 54 patients (82%) and there were 34 patients (51%) who had ICP over 40 mmHg. Only one of the 12 patients with normal initial ICP continued to have low pressure throughout the course.

Fifty-eight patients (88%) received high-dose thiopental to control raised ICP. This type of therapy was omitted in 8 patients because of impending arterial hypotension (7 cases) or normal ICP (1 case). ICP response to thiopental and the ICP profiles throughout the course are reflected in Table 4 and Fig. 2, respectively. ICP did not respond to thiopental in 29 patients (44%). Most of the non-responders were treated in 1977–1982, when a ceiling dosage of 4–6 g/day was established. Fulminant, uncontrollable intracranial hypertension killed 20 of these patients within 1–3 days of admission. Another 9 patients showed a more progressive ICP elevation, dying 5–8 days after initiation of treatment. Temporal lobe tip resection and decompressive craniectomy resulted in transient ICP normalization in 3 of these patients, but pressure rose to uncontrollable levels within 1–3 days of surgery.

Another 17 patients (26%) died several days after having had raised ICP that could initially be controlled

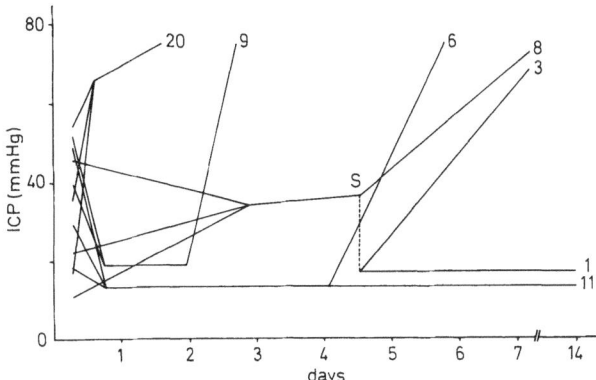

Fig. 2. Intracranial pressure (ICP) profiles in the 58 patients treated with thiopental. Four patients underwent temporal decompressive craniectomy between the third and fifth day after admission, but ICP decrease was very transient in 3 of them. The other patient survived as did 9 more patients in whom ICP could be controlled

with thiopental. Thiopental was tapered in 6 of these patients following 2–4 days of normal ICP, while there was still marked midline shift, and fatal pressure rebound occurred which could not be controlled in spite of resuming barbiturate infusion. Thiopental was reduced or discontinued in 7 more patients because of collateral side effects, also resulting in fatal pressure rebound. Finally, in another 4 patients, ICP progressively rose in spite of increased thiopental.

Raised ICP was definitively controlled in 12 patients (18%), among whom were the 10 survivors in this series. Thiopental withdrawal in these patients was not initiated until complete disappearance of mass effect with midline normalization were apparent in the control CT scan.

Table 4 shows the correlation between ICP response to thiopental administration, different clinical variables and final outcome. Most non-responders had a very low GCS score (90% scored 3–5), marked pupillary changes (83% had uni or bilateral mydriasis), marked midline shift (93% had more than 8 mm displacement), and very high initial ICP (over 40 mmHg in 52%). Patients showing transient response to thiopental were similar to non-responders in their bad clinical condition (82% scored 3–5 in the GCS, 70% had uni or bilateral mydriasis, 82% had midline shift greater than 8 mm, and 65% had ICP over 40 mmHg). All these patients died. Conversely, the patients who showed a definitive ICP response had a relatively better clinical status (58% with 3–5 score), less midline displacement (67% with shift greater than 8 mm), and lower initial ICP (25% with pressure over 40 mmHg).

Discussion

The mortality of severe head injury patients is still 20–50% in most series[1, 3, 5, 18, 26, 29] and the highest mortality is observed among patients with acute subdural haematoma so that in an given series, as the proportion of subdural haematoma cases increases, overall mortality also rises. The lethality of acute subdural haematoma is largely explained by its frequent association with primary (impact damage) and/or secondary brain damage consisting of contusion, laceration, swelling or oedema (the so-called complicated subdural haematoma)[3, 5, 6, 9, 18, 28]. The most severe lesion associated with subdural haematoma seems to be acute swelling of the ipsilateral cerebral hemisphere, which almost always runs with severe intracranial hypertension and leads to death within hours or a few days of trauma[6, 8–9, 14, 17, 18]. Since the radiological definition of this lesion is not yet well established, its true incidence in severe head injury patients is difficult to estimate. According to the conventional CT definition used in this study, 11% of our severe head injury patients had cerebral hemispheric swelling, which was associated to a

Table 4. *Intracranial Pressure (ICP) Response to Thiopental as Related to Different Clinical Variables and Survival in 66 Patients with Posttraumatic Cerebral Hemispheric Swelling*

Response to thiopental	Cases no.	GCS score		Mydriasis		Midline shift (mm)			Initial ICP (mmHg)		
		3–5	6–8	Unilat	Bilat	4–8	9–12	> 12	< 20	20–40	> 40
None	29	26	3	14	10	2	14	13	4	10	15
Transient	17	14	3	7	5	3	11	3	3	3	11
Sustained	12 (10)	7 (5)	5 (5)	8 (4)	3 (1)	4 (3)	7 (7)	1	4 (4)	5 (4)	3 (2)
Untreated	8	7	1	2	2	2	2	4	1	2	5
Total	66	54 (5)	12 (5)	31 (4)	20 (1)	11 (3)	34 (7)	21	12 (4)	20 (4)	34 (2)

Figures in parentheses represent the number of survivors in each subgroup.

more or less voluminous ipsilateral subdural haematoma in the majority of cases. It occurred as an isolated pathology in only 3 patients. Cerebral hemispheric swelling not only carried the highest mortality rate in our severe head injury series, but also the highest incidence of uncontrollable intracranial hypertension and the shortest survival period after injury. After examining the records of our patients, it becomes clear that independently of the role played by the extraaxial haematoma in the production of cerebral hemispheric swelling, it was the latter lesion that conditioned the clinical course. The limited or null effect of extraaxial haematoma evacuation on clinical status and midline displacement strongly suggests that it was bulk enlargement of the cerebral hemisphere which determined the outcome. This was particularly evident in nearly half of the cases in which the amount of extracerebral blood collection was rather insignificant.

Forty percent of the 553 traumatic subdural haematomas in the series of Jamieson and Yelland[9], were classified as "complicated haematoma". This type of lesion was most often observed after severe acceleration-deceleration injuries and carried the highest mortality in the series (2.5 times greater than simple subdural haematoma and 1.3 times more than haematoma plus brain contusion). According to these authors, "the swelling of the damaged brain added to the sometimes relatively minor accumulation of blood in the subdural space may explain the rapid course of complicated subdural haematoma". McKissock *et al.*[22] found hemisphere pulping and massive oedema in 11 fatal cases of their series of 125 patients with traumatic epidural haematoma. Heiskanen[7] observed widespread diffuse cerebral contusion of the ispilateral cerebral hemisphere in 7 out of the 10 patients who died in his series of epidural haematoma. However, neither these reports nor other studies on extracerebral haematoma performed during the CT era[5, 6, 17, 28] make a distinction between multifocal brain contusion-haematoma and diffuse swelling as a cause of cerebral hemisphere enlargement. Using the CT scan to differentiate these conditions, we excluded from this study patients showing brain contusive foci, haematoma or focal infarct. Unfortunately, the CT scan does not accurately distinguish whether hemispheric enlargement is due to increased water content or increased cerebral blood volume and cannot differentiate between brain oedema and the early stages of infarction.

The physiopathology of posttraumatic brain swelling is a matter of controversy[2, 10, 14, 17, 25, 31]. Taking into account its rapid development after injury, a sudden increase in cerebral blood volume resulting from cerebrovascular dilatation seems a more likely mechanism than acute oedema formation[1, 15, 24]. Diffuse swelling of a cerebral hemisphere occurring as early as 20–30 min after injury has been documented in two patients in whom the EMI numbers of the involved hemisphere were higher than in the contralateral unaffected hemisphere[13, 30]. The rapid resolution of the swelling effect in these two cases supports the interpretation that hemispheric enlargement was due to vascular engorgement of the type described by Langfitt and his colleagues in the early 1960s. However, laboratory work[12] and direct measurement of brain water content in head injury patients[2], have demonstrated that posttraumatic brain oedema may occur much more rapidly than is usually thought. Densitometric[8, 16, 17] and dynamic CT studies[31] seem to indicate that acute oedema formation is a common cause of bulk brain enlargement in fatal head injury cases. Although we did not perform planimetric analysis in our patients, it should be noted that in 56% of them the affected cerebral hemisphere appeared hypodense in comparison with the contralateral hemisphere. The prolonged duration of the mass effect and the null effect of hyperventilation on high ICP, suggest oedema formation as the mechanism of cerebral hemisphere enlargement in our patients. This type of brain swelling is probably different from the generalized swelling most commonly observed in children and young adults as a self-limiting phenomenon occuring with increased CT numbers and moderately raised ICP, apparently due to transient hyperaemia and easily controlled by hyperventilation[1, 14, 18, 24, 32]. These two basic mechanisms of brain expansion, i.e., hyperaemia and oedema, are not mutually exclusive and it may be that a vascular engorgement that is severe and persistent enough results in widespread brain hypoxia and oedema formation[10, 14, 23–25]. It has been postulated that severe brain trauma may damage the hypothalamic and brainstem motor centers triggering vasomotor paralysis with a sudden increase in cerebral blood volume which in turn would lead to a rise in ICP, compression of cerebral veins, increased cerebrovascular resistance, decrease of cerebral blood flow to ischaemic levels and oedema formation[10, 14, 24, 25]. Vasodilation might also be a direct response of cerebral vessels to mechanical injury[14]. In patients with associated large extracerebral haematoma, a critical reduction in cerebral blood flow may result from the local pressure exerted by the clot plus the reduction in cerebral perfusion pressure caused by ICP elevation to almost systemic blood pressure. A delay of more than

4 hours between trauma and haematoma evacuation occurred in 5 of the patients with epidural haematoma and in 82% of those with large subdural haematoma in this series. The fact that we never observed cerebral hemisphere swelling among patients operated for acute extracerebral haematoma within 2 hours of injury, corroborates the decisive influence of rapid haematoma evacuation[1, 18, 28]. Finally, peritraumatic hypotension-hypoxaemia, which was recorded in 48% of our patients, and posttraumatic cerebrovascular spasm[19], may also contribute to oedema formation[11].

High ICP, usually accompanying cerebral hemispheric swelling, is extremely difficult to control[18]. Posttraumatic brain swelling is not amenable to surgical treatment, the results of large decompressive craniectomy are useless or even detrimental[4]. Temporal lobe tip resection with temporal craniectomy decreased ICP for only a short period of time in our patients. High dose barbiturate given immediately after diagnosis produces a rapid ICP decrease followed by overall inhibition of the metabolic activity with the consequent reduction in cerebral oxygen substrate requirements and seems the most effective available therapy[11, 20]. High dose thiopental prevents elevation of blood pressure, which may increase intraluminal pressure in brain areas with altered blood brain barrier permeability aggravating vasogenic oedema[21, 27], but it requires meticulous control of extracerebral parameters since arterial hypotension may intensify interference with cerebral blood flow in ischaemic regions[11, 20]. The 10 survivors in this study were among the 12 responders to thiopental. The majority of patients (70%) showed only a transient response or none, but a retrospective analysis indicates that some non-responders received either small doses of thiopental or adequate doses for only a short time. According to our experience, once ICP is controlled, thiopental should not be discontinued until complete disappearance of the mass effect is seen in the control CT scan, an event which may occur as late as 15 days after onset of therapy.

References

1. Bruce DA, Alavi A, Bilaniuk L, Dolinskas C, Obrist W, Uzzell B (1981) Diffuse cerebral swelling following head injuries in children: the syndrome of "malignant brain edema". J Neurosurg 54: 170–178
2. Bullock R, Smith R, Farier J, Trevou M, Blake G (1985) Brain specific gravity and CT scan density measurement after human head injury. J Neurosurg 63: 64–68
3. Clifton GL, Grossman RG, Makela ME, Miner ME, Handel S, Sadhu V (1980) Neurological course and correlated tomography findings after severe closed head injury. J Neurosurg 52: 611–624
4. Cooper PR, Rovit RL, Ransohoff J (1976) Hemicraniectomy in the treatment of acute subdural haematoma. Surg Neurol 5: 25–28
5. Gennarelli TA, Spielman GM, Langfitt TW, Gildenberg PL, Harrington T, Jane JA, Marshall LF, Miller JD (1982) Influence of the type of intracranial lesion on outcome from severe head injury. A multicenter study using a new classification system. J Neurosurg 56: 26–32
6. Gennarelli TA, Thibault LE (1982) Biomechanics of acute subdural haematoma. J Trauma 22: 680–686
7. Heiskanen O (1975) Epidural haematoma. Surg Neurol 4: 23–26
8. Ito U, Tomita H, Yamazaki S, Takada Y, Inaba Y (1986) Brain swelling and brain oedema in acute head injury. Acta Neurochir (Wien) 79: 120–124
9. Jamieson KG, Yelland JDN (1972) Surgically treated traumatic subdural haemtaomas. J Neurosurg 37: 137–149
10. Jennett B (1981) Clinical brain swelling: oedema or engorgement. In: Vlieger M, Lange SA, Beks JW (eds) Brain oedema. John Wiley and Sons, New York, pp 61–65
11. Klatzo I (1985) Brain oedema following brain ischemia and the influence of therapy. Br J Anesth 57: 18–22
12. Kobrine AI, Kempe LG (1973) Studies in head injury. Part I: An experimental model of closed head injury. Surg Neurol 1: 34–37
13. Kobrine AI, Timmins E, Rajjoub RK, Rizzoli HV, Davis DO (1977) Demonstration of massive traumatic brain swelling within 20 minutes after injury. J Neurosurg 46: 256–258
14. Langfitt TW, Gennarelli TA, Obrist WD, Bruce D, Zimmerman RA (1982) Prospects for the future in the diagnosis and management of head injury: Pathophysiology, brain imaging and population-based studies. Clin Neurosurg 29: 353–376
15. Langfitt TW, Tannanbaum HM, Kassell NF (1966) The etiology of acute brain swelling following experimental head injury. J Neurosurg 24: 47–56
16. Lanksch W, Grumme T, Kazner E (1978) Correlations between clinical symptoms and computerized tomography findings in closed head injury. In: Frowein RA, Wilcke O, Karimi-Nejad A (eds) Advances in neurosurgery, vol 5. Springer, Berlin Heidelberg New York, pp 27–30
17. Lanksch W, Baethmann A, Kazner E (1981) Computed tomography of brain oedema. In: Vlieger M, Lange SA, Beks JW (eds) Brain oedema. John Wiley and Sons, New York, pp 67–98
18. Lobato RD, Cordobés F, Rivas JJ, de la Fuente M, Montero A, Bárcena A, Pérez C, Cabrera A, Lamas E (1983) Outcome from severe head injury related to the type of intracranial lesion. A computerized tomography study. J Neurosurg 59: 762–774
19. MacPherson P, Graham DI (1978) Correlation between angiographic findings and the ischemia of head injury. J Neurol Neurosurg Psychiatry 41: 122–127
20. Marshall LF (1980) Treatment of brain swelling and brain oedema in man. In: Cervos-Navarro J, Ferszt R (eds) Advances neurology, vol 28. Raven Press, New York, pp 459–469
21. Marshall WJS, Jackson JLF, Langfitt TW (1969) Brain swelling caused by trauma and arterial hypertension. Arch Neurol 21: 545–553
22. McKissock W, Taylor JC, Bloom WH, Till K (1960) Extradural haematoma. Observations on 125 cases. Lancet 2: 167–172

23. Miller JD, Corales RL (1981) Brain oedema as a result of head injury: Fact or fallacy? In: Vlieger M, Lange SA, Beks JW (eds) Brain oedema. John Wiley and Sons, New York, pp 99–115

24. Obrist WD, Langfitt TW, Jaggi JL, Cruz J, Gennarelli TA (1984) Cerebral blood flow and metabolism in comatose patients with severe head injury. J Neurosurg 61: 241–253

25. Overgaard J, Tweed WA (1976) Cerebral circulation after head injury. Part 2: The effects of traumatic brain oedema. J Neurosurg 45: 292–300

26. Saul TG, Ducker TB (1982) Effect of intracranial pressure monitoring and aggressive treatment on mortality in severe head injury. J Neurosurg 56: 498–503

27. Schutta HS, Kassell NF, Langfitt TW (1968) Brain swelling produced by injury and aggravated by arterial hypertension. A light and electron microscopic study. Brain 91: 281–294

28. Seelig JM, Becker DP, Miller JD, Greenberg RP, Ward JD, Choi SC (1981) Traumatic acute subdural haematoma: Major mortality reduction in comatose patients treated within four hours. N Eng J Med 304: 1511–1518

29. Tamas LB, Dacey RG, Winn HR (1985) Studies of severe head injury: an overview. In: Dacey RG, Winn HR, Rimel RW, Jane JA (eds) Trauma to the central nervous system. Raven Press, New York, pp 103–121

30. Waga S, Tochio H, Sakakura M (1979) Traumatic cerebral swelling developing within 30 minutes after injury. Surg Neurol 11: 191–193

31. Yoshino E, Yamaki T, Higuchi T, Horikawa Y, Hirakawa K (1985) Acute brain oedema in fatal head injury: Analysis by dynamic CT scanning. J Neurosurg 63: 830–839

32. Zimmerman RA, Bilaniuk LT, Bruce D, Dolinskas C, Obrist W, Kuhl D (1978) Computed tomography of pediatric head trauma. Acute general cerebral swelling. Radiol 126: 403–408

Address for correspondence: Dr. Maria Sarabia, Instituto Nacional de la Salud. Hospital 1° de Octubre, Carretera de Andalucia km 5,400, E-28041 Madrid, Spain.

II. Cerebro-vascular Lesions

Acta Neurochirurgica, Suppl. 42, 49–52 (1988)

Italian Cooperative Study on Giant Intracranial Aneurysms: 1. Study Design and Clinical Data

R. Battaglia, A. Pasqualin, and **R. Da Pian***

Department of Neurosurgery, City Hospital, Verona, Italy

Summary

10 Italian centres joined together for a retrospective Cooperative Study aimed at evaluating clinical and radiological data, various modalities of treatment and clinical outcome in patients with giant intracranial aneurysms, observed from 1976 to 1986. Various clinical data were collected through a questionnaire and evaluated through computer analysis. Two size categories were considered: 2–2.5 cm in diameter (A 1 group) and over 2.5 cm (A 2 group). A total of 240 cases were evaluated: 110 in A 1 group and 130 in A 2 group. As regards clinical history, intracranial (mainly subarachnoid) haemorrhage was observed in 70% of A 1 patients and in 45% of A 2 patients ($p < 0.001$) and was more severe (Hunt's grades III–V) in A 2 patients. Symptoms of an expanding mass lesion were observed in 15% of A 1 patients and in 39% of A 2 patients ($p < 0.001$). Sudden deficits of cranial nerves were observed in 11% of A 1 and 12% of A 2 patients. Ischaemic episodes were rare. On admission, Glasgow Coma Score (GCS) was 15 in 62% of cases. Regardless of treatment employed, patients in A 1 group presented a slightly better (N.S.) outcome than patients with larger aneurysms (A 2 group). The presence of intracranial haemorrhage in the clinical history increased significantly the mortality rate ($p < 0.001$); symptoms of an expanding mass lesion were associated with a significant increase in the disability rate ($p < 0.001$). A direct influence on outcome was played by GCS on admission, with a recovery rate of 51% for GCS 15, decreasing progressively for lower scores and reaching 4% for GCS 3–6.

Keywords: Giant aneurysm; outcome; cooperative study.

Introduction

Giant intracranial aneurysms are classically defined as larger than 2.5 cm in diameter (i.e., larger than 1 inch, according to anglo-saxon measures)[2,4]. We believe that

a clearcut separation between "giant" and "normal" aneurysms is not correct on a scientific basis, owing to the presence of large aneurysms with features more similar to giant than to normal aneurysms: over 2 cm, these lesions already pose serious problems in management.

Based on the previous points, 10 Italian centres have joined together in a retrospective Cooperative Study evaluating clinical and management data of patients with very large or giant aneurysms, admitted to these centres from 1976 to 1986.

Study Design

10 Italian centres participated in this Cooperative Study: Neurosurgical Departments (or Divisions) of Verona, Vicenza, Florence (Careggi Hospital), Rome (Regina Elena Institute), Bologna (Bellaria Hospital), Brescia, Teramo, Naples (II University), Udine, and Milan (Niguarda Hospital: Division of Neurosurgery and Service of Neuroradiology).

The 10 centres agreed to evaluate retrospectively patients with aneurysms measuring 2.0 cm in diameter or more – admitted in the years 1976–1986 – through a computer analysis, based on a questionnaire comprehending the following items: (1) size classification (and means of size evaluation); (2) anagraphical/anamnestical data; (3) clinical manifestations; (4) neurological examination on admission; (5) findings at CT scan; (6) findings at angiography; (7) other findings (skull X-rays, NMR, CBF, transcranial Doppler); (8) adopted treatment or treatments; (9) preoperative neurological examination; (10) intraoperative findings; (11) additional surgical procedures; (12) medical therapy (pre- and postoperative); (13) posttreatment radiological findings; (14) results of treatment; (15) postoperative complications; (16) causes of morbidity and mortality. The information required by the questionnaire was decided during group sessions held at the beginning of the Study. Computer analysis was made through the PFS File program, using the chi-square test for statistical evaluation. Throughout the whole Study, the patients were divided into 2 groups, according to aneurysmal size: A 1 group (2.0–2.5 cm in maximum diameter), and A 2 group (over 2.5 cm in maximum, diameter). A total of 240 cases were accepted in this Study. The largest contribution was from the centres of Milan (Niguarda

* Cooperative Group: R. Da Pian, A. Pasqualin, R. Scienza/A. Benedetti, D. Curri, L. Volpin/P. Mennonna, F. Ammannati, D. Serino/E. Morace, F. Cattani, R. Mastrostefano/V. A. D'Angelo, C. Corona, V. Monte/F. Filizzolo, M. Collice, P. Versari/G. Scialfa, G. Scotti, E. Boccardi/A. Andreoli, P. Limoni, C. Testa/M. Bortoluzzi, G. Marini/R. Galzio, D. Lucantoni/E. De Divitiis, R. Spaziante, F. D'Andrea/C. Ceccotto, P. P. Janes, A. Cramaro.

Table 1. *Hunt & Hess Grading in Patients with Haemorrhage, According to Aneurysm Size (A 1 or A 2)*

	I–II	III	IV	V
A 1 group (77 cases)	42 (54%)	14 (18%)	11 (14%)	8 (10%)
A 2 group (59 cases)	14 (24%)	14 (24%)	14 (24%)	12 (20%)
Total (136 cases)	56 (41%)	28 (21%)	25 (18%)	20 (15%)
Significance	p < 0.001		p < 0.01	

Hospital) and Verona, which accounted together for 50% of cases. Evaluation of aneurysmal size was obtained through angiography in 145 patients (60%), through CT scan in 58 patients (24%), and only through direct operative inspection in 37 patients (16%): 110 cases (45%) belonged to the A 1 group (2.0–2.5 cm), and 130 (55%) to the A 2 group (over 2.5 cm).

Clinical Data

Out of the 240 patients studied, 57% were female and 43% male. The decade most frequently involved was 51–60 years (35%), followed by 41–50 years (24%); below 41 years giant aneurysms were observed in 20% of cases, and below 21 years in 5% of cases (12 patients).

In their *previous medical history*, 22% of A 1 and 27% of A 2 patients had suffered from headache; 21% of A 1 and 22% of A 2 patients had been hypertensive; 4% of A 1 and 2% of A 2 patients were known diabetics. No other significant anamnestic data were observed.

As regards clinical manifestations, one or more intracranial haemorrhages (mainly subarachnoid) were observed in 70% of A 1 and in 45% of A 2 patients (p < 0.001). 28 patients (12% of the total number, and 21% of patients with haemorrhage) suffered from more than one haemorrhage before admission. The haemorrhage was clinically more severe (Hunt and Hess grades III to V) in A 2 than in A 1 patients (Table 1).

Symptoms of an expanding mass lesion (not caused by haemorrhage) were observed in 15% of A 1 patients (16 cases) and in 39% of A 2 patients (44 cases) (p < 0.001). Sudden deficits of cranial nerves (not caused by haemorrhage) were observed in 11% of A 1 patients (12 cases) and in 12% of A 2 patients (16 cases).

Less frequent clinical manifestations included ischaemic episodes (2 cases in A 1, and 6 cases in A 2, with more than 1 episode in 3 cases), epilepsy (3 cases in A 2), and other less frequent symptoms. The aneu-

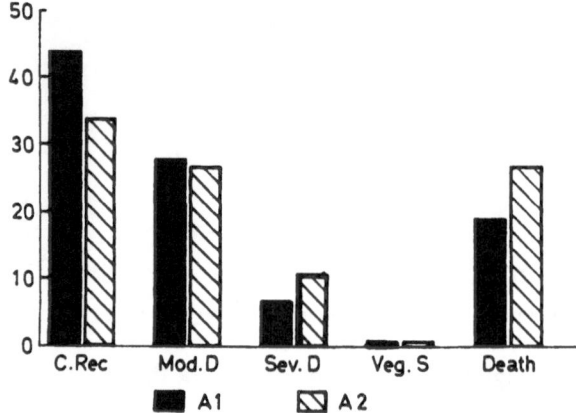

Fig. 1. Relation between aneurysmal size and outcome, according to the Glasgow Scale: no significance (numbers are given in percent)

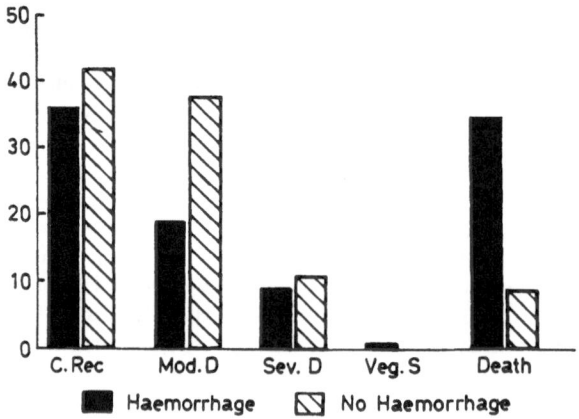

Fig. 2. Relation between presence or absence of haemorrhage and outcome, according to the Glasgow Scale: p < 0.01 for disability, p < 0.001 for death (numbers are given in percent)

rysm was discovered incidentally in 6 cases (2 cases in A 1, and 4 cases in A 2).

On admission to the neurosurgical centre, oculomotor deficits were observed in 22% of A 1 and in 28% of A 2 patients (N.S.); visual disturbances were observed in 9% of A 1 and in 26% of A 2 patients (p < 0.001); 14% of A 1 patients and 22% of A 2 patients (N.S.) presented with hemiparesis or hemiplegia; speech disturbances were noted in both groups with the same incidence (5%); the Glasgow Coma Score was 15 in 63% of A 1 and 61% of A 2 patients, 13–14 in 12% of A 1 and 12% in A 2 patients, between 12 and 7 in 17% of A 1 and 12% of A 2 patients, 3–6 in 8% of A 1 and 15% of A 2 patients.

Relationships with the outcome: Patients with aneurysms between 2.0 and 2.5 cm in diameter (A 1 group) presented a slightly better (N.S.) outcome than patients with larger aneurysms (A 2 group), as shown in Fig. 1.

Table 2. *Relation Between Glasgow Coma Scale (GCS) at Admission and Glasgow Outcome Scale*

	C. rec.	Mod. d.	Sev. d.	Veg. s.	Death
GCS: 15 (148 cases)	76 (51%)	46 (31%)	12 (8%)	1 (1%)	13 (9%)
GCS: 13–14 (29 cases)	10 (34%)	7 (24%)	4 (13%)	1 (4%)	7 (24%)
GCS: 7–12 (35 cases)	6 (17%)	11 (32%)	5 (14%)	—	13 (37%)
GCS: 3–6 (28 cases)	1 (4%)	2 (7%)	2 (7%)	—	23 (82%)
Significance*	$p < 0.001$		$p < 0.05$		$p < 0.001$

* GCS 15, vs. 7–14, vs. 3–6

In patients aged 21–60 years, the recovery rate was 41% and the mortality 25%; in patients aged 61 or more, the recovery rate was 26% ($p < 0.05$), and the mortality rate 22% (N.S.).

The presence of one or more haemorrhages in the clinical history influenced outcome, as shown in Fig. 2 ($p < 0.001$ for death). Symptoms of an expanding mass lesion were associated with a disability rate of 60%; in patients without these symptoms, the disability rate was 30% ($p < 0.001$).

A direct influence on outcome was obviously played by the Glasgow Coma Score on admission – as seen in Table 2 – with a recovery rate of 51% for GCS 15, decreasing progressively for lower scores and reaching 4% for GCS 3–6.

Independently of treatment employed, causes of morbidity were mainly constituted by the direct effect of haemorrhage (or rebleeding), by the effects of an expanding mass lesion, and by ischaemic disturbances without vasospasm, while causes of mortality were constituted mainly by the direct effect of haemorrhage (or rebleeding).

Discussion

In the current literature, interest for giant aneurysms has been limited to lesions with a maximum diameter over 2.5 cm[2, 4], in spite of the fact that a clearcut margin between less difficult and more difficult lesions – from a surgical view point – can hardly be defined. As recently reported, aneurysms between 1.5 and 2.5 cm in diameter may already pose therapeutic difficulties[3]. In the present study the critical size has been fixed at 2.0 cm of maximum diameter, and any clinical and therapeutic differences or similarities between patients with aneurysms of 2–2.5 cm and patients with "giant" aneurysms have been investigated.

The rarity of giant aneurysms – around 3–5% of all aneurysms[2, 4, 6] – prevents a valid comparison between various modalities of treatment. To circumvent this problem, 3 years ago a Cooperative Study held by French centres collected a large number of patients[9]. With the same aim, we have collected cases in which the definition of aneurysmal size – through CT scan, angiography, or direct inspection during surgery – was very rigorous in order to avoid any possible bias, sometimes present in other reported series on giant aneurysms.

The incidence of intracranial haemorrhage in our study is high, confirming the impressions of other authors[2, 10]; however, there is a significant difference in the bleeding rate between the A 1 group – more prone to bleeding – and the A 2 group (the classic "giant" aneurysms). The bleeding rate of our A 2 group (45%) is almost the same as that reported by the French study[9]. Even if larger aneurysms bleed less frequently, haemorrhages are more severe in this group, possibly due to the negative influence of the aneurysmal mass effect on the clinical evolution of haemorrhage.

Symptoms of an expanding mass lesion depended strictly upon aneurysmal size, being rare in our A 1 group (15%) and quite common in larger aneurysms (39% in our series, and 47% in the French study). In our study – contrary to other large studies[9] – sudden deficits of cranial nerves have been considered separately from pseudo-tumoral symptoms. The sudden onset of the deficit may be explained by rapid thrombosis in the aneurysmal lumen[7, 12], causing sudden enlargement of the sac and ischaemia of the adjacent nerve[8];

also small localized haemorrhages may be responsible for this symptomatology[8].

Ischaemic episodes not associated with subarachnoid haemorrhage have been reported only for giant, but also for smaller aneurysms[1, 5, 11] and are thought to be due to thromboembolism form the aneurysmal lumen to the distal arteries. In the French and in the present study, this symptomatology has been rarely observed (5 and 3% respectively).

As regard outcome – independently from adopted treatment – there does not seem to be a significant difference in prognosis for patients with very large (2–2.5 cm), and patients with giant aneurysms, thus justifying our inclusion of the former group (A 1 patients) in this study on giant aneurysms. The presence of subarachnoid haemorrhage in patients with very large or giant aneurysms seems to play a determining role on outcome, and particularly on mortality. Symptoms of an expanding mass lesion have more influence on the disability rate – as could be expected – than on the recovery or mortality rate. From our data, morbidity due to mass effect appears directly related to the aneurysmal size.

References

1. Antunes JL, Correll JW (1976) Cerebral emboli from intracranial aneurysms. Surg Neurol 6: 7–10
2. Drake CG (1979) Giant intracranial aneurysms: experience with surgical treatment in 174 patients. Clin Neurosurg 26: 12–95
3. Heros RC, Nelson PB, Ojeman RG, Crowell RM, Debrun G (1983) Large and giant paraclinoid aneurysms: surgical techniques, complications and results. Neurosurgery 12: 153–163
4. Locksley HB (1966) Report on the cooperative study of intracranial aneurysms and subarachnoid haemorrhage. Section V, part II. Natural history of subarachnoid haemorrhage, intracranial aneurysms, and arteriovenous malformations. J Neurosurg 25: 321–368
5. Mehdorn HM, Chater NL, Townsend JJ, Darroch JD, Perkins RK, Lagger R (1980) Giant aneurysm and cerebral ischaemia. Surg Neurol 13: 49–57
6. Pia HW, Zierski J, Peerless SJ, Drake CG, Kodama N, Suzuki J, Sundt TM, Handa H, Hashimoto N, Yonekawa Y, Sengupta RP (1982) Direct management for large and giant aneurysms. Neurosurg Rev 5, no 4: 115–178
7. Pozzati E, Fagioli L, Servadel F, Gaist G (1981) Effect of common carotid ligation of giant aneurysms of the internal carotid artery: computerized tomography study. J Neurosurg 55: 527–531
8. Sarwar M, Batnitzky S, Schechter M (1976) Tumorous aneurysms. Neuroradiology 12: 79–97
9. Sindou M, Keravel Y *et al* (1984) Les anévrysmes géants intracrâniens. Approches thérapeutiques. Intracranial giant aneurysms. Therapeutic approaches. Neurochirurgie 30 [Suppl] 1
10. Sonntag VKH, Yuan RH, Stein BM (1977) Giant intracranial aneurysms: a review of 13 cases. Surg Neurol 8: 81–83
11. Whittle IR, Dorsch NW, Besser M (1984) Giant intracranial aneurysms: diagnosis, management and outcome. Surg Neurol 21: 218–230
12. Wilmink JT, Vencken LM (1979) Complementary values of static and dynamic scintigraphy, computerized tomography and angiography in the diagnosis of a partially thrombosed giant intracranial aneurysm. Clin Neurol Neurosurg 81: 87–96

Address for correspondence: Dr. R. Battaglia, Department of Neurosurgery, City Hospital, Verona, Italy.

Acta Neurochirurgica, Suppl. 42, 53–59 (1988)

Italian Cooperative Study on Giant Intracranial Aneurysms: 2. Radiological Data

L. Rosta, R. Battaglia[1], A. Pasqualin[1], and **A. Beltramello***

Service of Neuroradiology and [1] Department of Neurosurgery, City Hospital, Verona, Italy

Summary

240 patients with giant aneurysms admitted to 10 Italian centres were evaluated regarding radiological features and relationships to the outcome.

Visualization of the aneurysm without contrast was obtained on CT scan in 49% of patients with aneurysms between 2 and 2.5 cm in diameter (A 1 group) and in 80% of patients with larger aneurysms (A 2 group). Contrast enhancement was homogeneous in 49% of patients, not homogeneous – with central or peripheral hypodensity – in 47% of patients, absent in 4% of patients. Ventricular shift was present in 17% of A 1 patients and in 36% of A 2 patients. Hypodense areas were observed in 12% of cases, and calcifications in 19% of cases; bone erosions were rare.

On angiography, the most common aneurysmal locations were the intracavernous carotid (21% of cases) and the middle cerebral artery (23% of cases). A neck could be identified on angiography in only 14% of patients, and stenosis of afferent vessel in 16% of patients. Vasospasm was rarely observed on angiography (17% of A 1 and 10% of A 2 patients). Aneurysmal thrombosis (partial, subtotal or total) was present in 48% of A 1 and 76% of A 2 patients ($p < 0.001$); partial peripheral thrombosis was most commonly observed (one half of cases). Round or oval shapes were most commonly observed, while fusiform, irregular, serpentine or doughnut shapes were rare.

A few clinical/radiological relations were considered, such as the relation between partial thrombosis and haemorrhage, symptoms of expanding mass lesion, ischaemic episodes, aneurysmal location.

In regard to outcome – and independently form treatment employed – the highest recovery rate was observed for intracavernous and carotid-ophthalmic aneurysms, and the lowest for carotid bifurcation aneurysms.

Keywords: Giant aneurysm; cooperative study; radiological data; outcome.

* Cooperative Group: R. Da Pian, A. Pasqualin, R. Scienza/A. Benedetti, D. Curri, L. Volpin/P. Mennonna, F. Ammannati, D. Serino/E. Morace, F. Cattani, R. Mastrostefano/V. A. D'Angelo, M. Ferrara, E. Fiumara/F. Filizzolo, M. Collice, G. D'Aliberti/G. Scialfa, G. Scotti, E. Boccardi/A. Andreoli, P. Limoni, C. Testa/M. Bortoluzzi, G. Marini/R. Galzio, D. Lucantoni/E. De Divitiis, R. Spaziante, F. D'Andrea/C. Ceccotto, P. P. Janes, A. Cramaro.

Introduction

Radiological findings in giant aneurysms present a wide variation, mainly due to the fact that very large aneurysms have a tendency to thrombosis of various extent, as a result of flow abnormalities within the sac. Moreover, the presence of an associated intracranial haemorrhage can alter the visualization of the aneurysm.

In this section of the cooperative study, radiological features of very large or giant aneurysms have been carefully evaluated, and related to clinical data and outcome.

Radiological Data

Out of the total 240 patients, 204 (85%) were submitted to CT scan (94 in A 1 and 110 in A 2 group). Contrast enhancement was carried out in 126 patients (53%). Angiography was performed in 238 patients (99%).

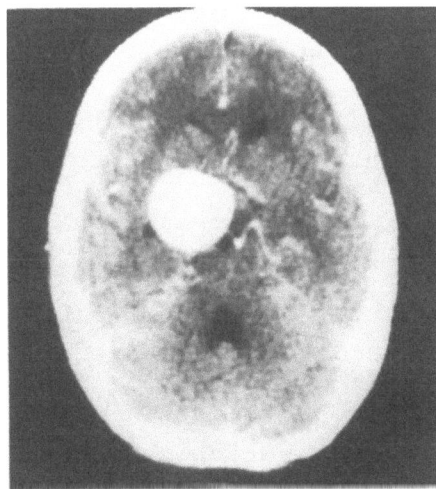

Fig. 1. Giant carotid supraclinoid aneurysm with homogeneous enhancement on CT scan

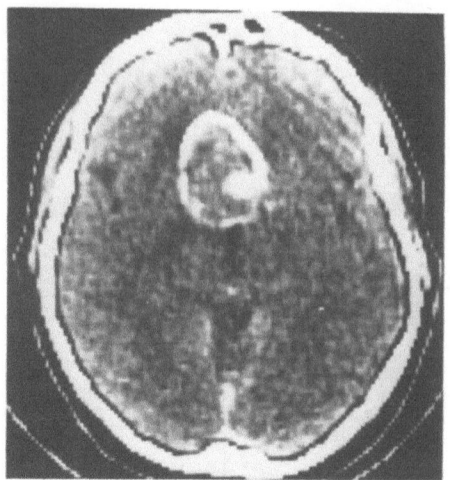

Fig. 2. Giant anterior communicating aneurysm with non-homogeneous enhancement and central hypodensity on CT scan

As regard *CT scan*, the following findings were observed:

1. The aneurysm was visualized on non-enhanced CT scan in 66% of cases: in 49% of A1 and in 80% of A2 patients (p < 0.001). After contrast injection, the aneurysm was visualized in 96% of patients: in 93% of A1 and in 97% of A2 patients. Contrast enhancement was homogeneous in 49% of patients (Fig. 1); not homogeneous with central hypodensity in 16% of patients (Fig. 2); not homogeneous with peripheral hypodensity in 31% of patients (Fig. 3a).

2. 79 patients were evaluated with CT scan within 7 days of haemorrhage: 23 (29%) presented a thin or absent cisternal deposition, and 56 (71%) a consistent or thick deposition, according to a practical scale published by our group[5].

3. Ventricular shift was present on CT scan in 27% of cases: in 17% of A1 patients and in 36% of A2 patients (p < 0.01).

4. Hydrocephalus was observed in 11% of cases: it was obstructive in 2% and post-haemorrhagic in 9% of cases. No patient with obstructive hydrocephalus was found in A1 group.

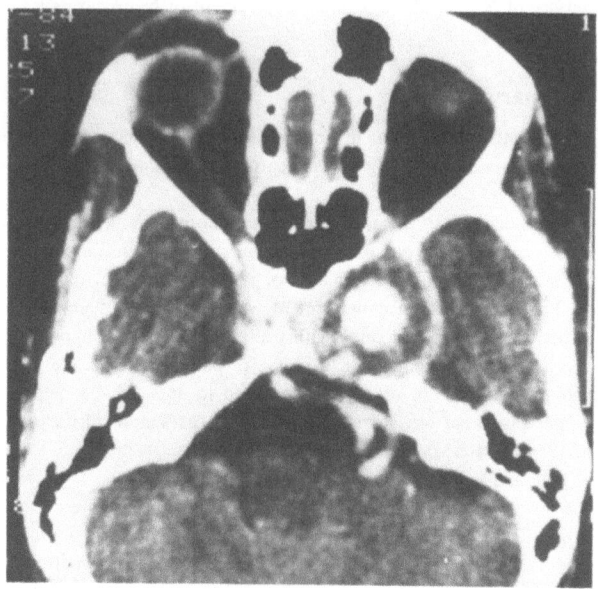

a

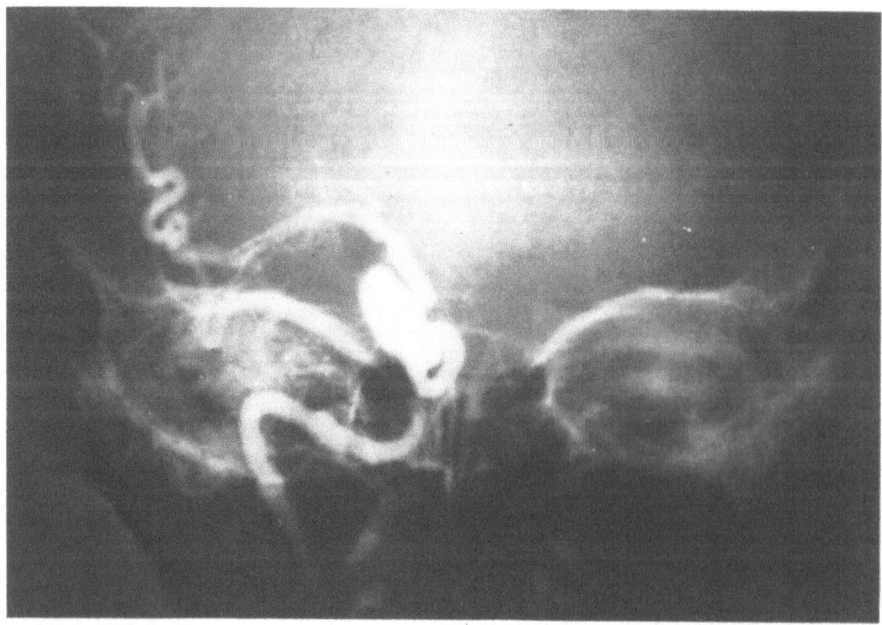

b

Fig. 3. (a) Giant intracavernous aneurysm with non-homogeneous enhancement on CT scan and peripheral hypodensity; (b) only the central part of the aneurysm is filling on angiography, suggesting partial peripheral thrombosis

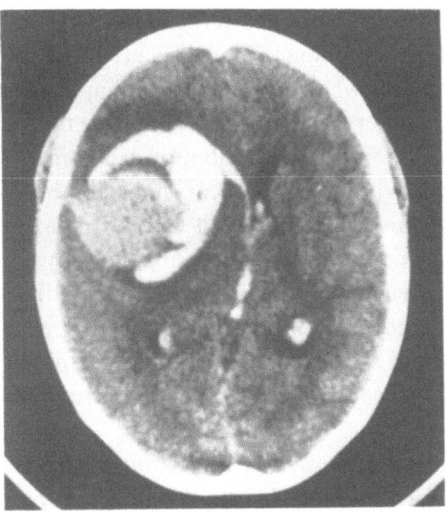

Fig. 4. Giant middle cerebral aneurysm, surrounded by intraparenchymal and intraventricular haemorrhage (CT scan without enhancement)

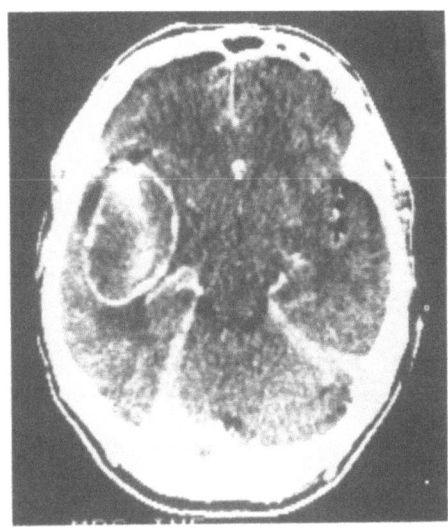

Fig. 5. Completely thrombosed giant middle cerebral aneurysm, not filling on angiography, with wall enhancement on CT scan (central hyperdensity seen also without contrast)

Table 1. *Relationship Between Aneurysmal Location and Aneurysmal Size*

	A 1 size	A 2 size	A 1 + A 2
Intracavernous	24 (22%)	27 (21%)	51 (21%)
Carotid ophth.	13 (12%)	22 (17%)	35 (15%)
Carotid supracl.	12 (11%)	13 (10%)	25 (11%)
Carotid bifurc.	3 (3%)	9 (7%)	12 (5%)
Middle cerebral	22 (20%)	32 (25%)	54 (23%)
Ant. cerebral complex	28 (25%)	16 (13%)	44 (18%)
Post. circulation	8 (7%)	9 (7%)	17 (7%)

5. 13% of patients presented a ventricular haematoma (with the same incidence in A 1 and A 2 groups), 9% a parenchymal haematoma (13% in A 1 and 6% in A 2 group) (Fig. 4), and 1% a subdural haematoma (only patients in A 2 group).

6. Hypodense areas were observed in 12% of patients: in 8% following subarachnoid haemorrhage (SAH), and in 4% unrelated to SAH. Hypodense areas following SAH were prevalent in A 1 patients (13% against 4%); if unrelated to SAH, they had a similar incidence in the two groups (2% for A 1, and 5% for A 2).

7. Calcifications were observed on CT scan in 19% of cases: in 9% of A 1 and in 29% of A 2 patients ($p < 0.01$). Bone erosions were rare in both groups (4% in A 1 and 9% in A 2 group).

As regards *angiography*:

1. The most frequent aneurysmal locations were the intracavernous carotid (21% of cases) and the middle cerebral artery (23% of cases), as shown in Table 1. 45% of the aneurysms were on the right, 34% on the left, and 21% were in the midline.

Table 2. *Degree of Aneurysmal Thrombosis, According to Aneurysmal Size*

	Partial		Subtot.	Compl.	Absent
	Central	Periph.			
A 1 size (108 cases)	4 (4%)	43 (40%)	5 (4%)	—	56 (52%)
A 2 size (119 cases)	5 (4%)	70 (59%)	13 (11%)	3 (3%)	28 (23%)
Total (227 cases)	9 (4%)	113 (50%)	18 (8%)	3 (1%)	84 (37%)
Significance	N.S.	$p < 0.001$	N.S.	N.S.	$p < 0.001$

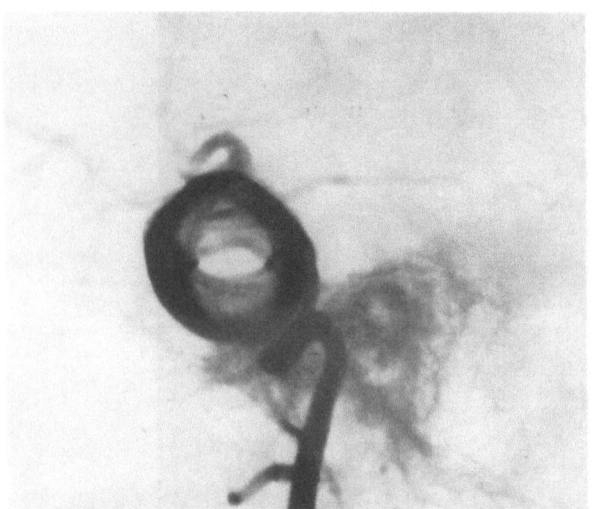

Fig. 6. Giant intracavernous aneurysm with intraluminar filling defect ("doughnut shape") on angiography

2. A neck on angiography was evident in only 14% of patients, with the same incidence in A 1 and A 2 groups.

3. Stenosis of the afferent vessel was observed on angiography in 16% of patients: in 20% of A 1 and in 12% of A 2 patients (N.S.).

4. Angiographical spasm was observed only in 13% of cases: in 17% of A 1 and 10% of A 2 patients; it should be remembered that patients with SAH constituted 57% of the study population, and that only 53% of patients with SAH were evaluated with angiography between 4 and 15 days from haemorrhage.

5. We classified the degree of thrombosis in 4 subgroups: (a) partial central, (b) partial peripheral, (c) subtotal, (d) total (or complete) (Fig. 5). As a whole, aneurysmal thrombosis (partial, subtotal or total) (was observed in 63% of cases; partial peripheral thrombosis (50% of cases) was the most common finding (Fig. 3 h) (Table 2). Thrombosis (partial, subtotal or total) was present in 48% of A 1 patients and in 76% of A 2 patients (p < 0.001).

6. 5 different morphological shapes were considered:

(a) round or oval, (b) fusiform, (c) irregular, (d) serpentine, (e) with intraluminar filling defect ("doughnut shape") (Fig. 6). As a whole, round or oval shapes were most commonly observed (Table 3). There were no significant differences in shape distribution between A 1 and A 2 patients. Variations in aneurysmal shape and degree of thrombosis were observed before treatment in 7 out of 61 patients who repeated angiography.

7. Multiple aneurysms were observed in 29 patients (12% of cases): they were most commonly located on the internal carotid artery (13 cases) and on the middle cerebral artery (10 cases). Only one patient presented an associated AVM.

As regards *magnetic resonance tomography*, our data are too scant for any consideration (only 5 patients evaluated). A wider use of this new examination in giant aneurysms is probably warranted.

Clinical/radiological relations:

Subarachnoid haemorrhage was relatively more frequent in patients with aneurysms of the anterior communicating artery (85% of cases) and of the internal carotid bifurcation (75% of cases) (Table 4). As shown by Table 5, partial angiographic thrombosis was more frequently seen in patients without, than in patients with, a previous haemorrhage. Symptoms of an expanding mass lesion were most frequent for patients

Table 4. *Relationship Between Aneurysmal Location and Incidence of Haemorrhage* (in Brackets the Total Number of Patients for each Location)

	A 1 size	A 2 size	A 1 + A 2
Intracavernous	5 (24)	2 (27)	7 (51)
Carotid ophth.	9 (13)	8 (22)	17 (35)
Carotid supraol.	9 (12)	8 (13)	17 (25)
Carotid bifurc.	3 (3)	6 (9)	9 (12)
Middle cerebral	18 (22)	19 (32)	37 (54)
Ant. cerebral complex	26 (28)	11 (16)	37 (44)
Post. circulation	7 (8)	3 (9)	19 (17)

Table 3. *Distribution of Morphological Shapes, According to Aneurysmal Size*

	Round	Fusiform	Irregular	Serpent.	Doughnut
A 1 size (107 cases)	71 (66%)	2 (2%)	32 (30%)	2 (2%)	—
A 2 size (115 cases)	74 (64%)	7 (6%)	30 (26%)	1 (1%)	3 (3%)
Total (222 cases)	145 (65%)	9 (4%)	62 (28%)	3 (1%)	3 (1%)

Table 5. *Incidence of Haemorrhage, According to Degree of Aneurysmal Thrombosis*

	Partial	Subtotal/ complete	Absent
Haemorrage (134 cases)	64 (48%)	13 (10%)	57 (42%)
No haemorrhage (104 cases)	58 (56%)	8 (8%)	27 (26%)
Significance	N.S.	N.S.	p < 0.01

Table 6. *Relationship Between Aneurysmal Location and Failure of Aneurysmal Visualization on Plain CT Scan* (in Brackets the Total Number of Patients for Each Location)

	A 1 size	A 2 size	A 1 + A 2
Intracavernous	8 (24)	2 (27)	10 (51)
Carotid ophth.	1 (13)	3 (22)	4 (35)
Carotid supracl.	6 (12)	3 (13)	9 (25)
Carotid bifurc.	1 (3)	— (9)	1 (12)
Middle cerebral	13 (22)	6 (32)	19 (54)
Ant. cerebral complex	17 (28)	1 (16)	18 (44)
Post. circulation	2 (8)	1 (9)	3 (17)

with intracavernous aneurysms (45% of cases), followed by patients with carotid/ophthalmic aneurysms (22%). These symptoms were present in 26% of patients with aneurysmal thrombosis and in 17% of patients without thrombosis (N.S.). Ischaemic episodes were not influenced by intra-aneurysmal thrombosis (6% in the absence of thrombosis, and 8% in the presence of thrombosis).

On non-enhanced CT scan, failure of visualization was rare for intracavernous aneurysms (20% of cases) and especially for carotid-ophthalmic aneurysms (11% of cases) (Table 6).

Stenosis of afferent vessel was observed with almost the same incidence in patients with angiographic thrombosis (16%) and in patients without thrombosis (14%). When angiographic thrombosis was absent, aneurysmal shape was round-oval in 90% of cases. Intracavernous aneurysms had the highest incidence of partial thrombosis (69%), while aneurysms in the other locations exhibited this phenomenon with a similar incidence: from 38 to 51% in the different locations, with an average of 46% (p < 0.01).

Relationships to the outcome:

Excluding obvious relations with outcome (such as ventricular shift, hydrocephalus, intracranial haema-

toma), and independently of treatment, the following features are stressed:

(a) extensive or thick cisternal depositions of blood were associated with a significantly higher mortality rate (p < 0.001) than thin or absent depositions;

(b) the highest recovery rate was observed for patients with intracavernous (47%) and carotid-ophthalmic aneurysms (43%), and the lowest for patients with internal carotid bifurcation aneurysms (8%);

c) stenosis of the afferent vessel was associated with a significantly worse outcome than absence of stenosis (p < 0.05).

Discussion

On plain *CT scan*, visualization of the aneurysm seems to depend upon its size: in our study, A 1 aneurysms were visualized in 49% of cases, while A 2 in 80% of cases. The aneurysm – when not thrombosed – showed a slightly increased homogeneous density, as compared to surrounding brain; in cases with partial thrombosis, a mixed density could be observed[8, 13], often with ring-shaped irregular hyperdensity and wall calcifications[7]. After contrast injection, half of our cases have shown homogeneous enhancement, against 60% of cases in the French study[10]. Non-homogeneous enhancement is typical of partially thrombosed aneurysms; another morphological characteristic of partial thrombosis is wall enhancement (the so-called "target sign")[2, 7, 10]; sometimes this feature can be observed also in completely thrombosed aneurysms (see Fig. 5)[13], although more frequently absence of enhancement is a distinctive sign of complete thrombosis[7].

Calcifications have been observed on CT scan in 29% of our patients with aneurysms over 2.5 cm in diameter, confirming the impression of other authors[3, 7, 13]; they are predominantly observed on the aneurysmal wall, and almost never inside the lesion[7].

Bone erosions are rare (9% of our "giant" aneurysms); they occurred chiefly in the sellar region, and were mainly associated with intracavernous aneurysms[2, 3].

As regards other CT features, ventricular shift was a relatively common finding (36% in our "giant" aneurysms). Hydrocephalus was rare, and was generally post-haemorrhagic. Hypodense areas were also rare and were most frequently due to post-haemorrhagic vasospasm.

As regards *angiography*, the most common location for giant aneurysms in the paraclinoid portion of the internal carotid artery[1, 6, 11]; although intracavernous aneurysms seem to be the commonest variety of giant aneurysms[6, 10], in our study middle cerebral aneurysms have been even more common (23% of cases, against 21%). Anterior communicating aneurysms have been relatively frequent in our series (18%) as well as in Suzuki's experience[4], in spite of their reported rarity[1]. Posterior circulation aneurysms constitute 7% of all giant aneurysms in our study and 6% in the French study[10], although a much higher incidence has been reported by Drake, mainly due to a particular referral pattern[1].

It is generally agreed that the neck of a giant aneurysm is often undetectable on angiography, although a small neck can be found at surgery; in our series, a clear neck could be demonstrated on angiography in only 14% of cases. Another rare finding was stenosis of the afferent vessel; it can be due to vasospasm, or more frequently to the proximal spread of intra-aneurysmal thrombus[9, 13].

A point to be stressed is that vasospasm is not observed in giant aneurysms as frequently as in smaller aneurysms; the rarity of this phenomenon is dependent upon the lower incidence of subarachnoid haemorrhage in this group of patients.

As regards degree of thrombosis, the most common finding in our study has been arterial peripheral thrombosis (half the cases). Thrombosis is more common with larger aneurysms: up to 76% in our "giant" group, from 52% to 83% in other series[2, 7, 13], and from 28 to 36% in the French study[10]. Complete thrombosis has been observed in only 8% of cases in the French study[10] and in 1% of our cases.

The shap of the aneurysm depends mainly on the degree and the distribution of thrombosis within the lumen, and it is generally round/oval in non-thrombosed aneurysms. A rare morphological appearance is constituted by serpentine aneurysms (1% in our series and 3% in the French study)[10].

Variations in aneurysmal shape and degree of thrombosis can be sometimes observed [9, 11, 12, 13] — 7 cases in our study — and constitute proof of the dynamic state of thrombosis.

A few interesting facts regarding *clinical/radiological* relationships have been brought out by our study. In relation to clinical manifestations:

(a) intracavernous aneurysms can bleed, even if rarely[10] (14% in our series);

(b) non-thrombosed aneurysms are more prone to bleed than partially thrombosed aneurysms;

(c) symptoms of an expanding mass lesion are chiefly observed in patients with intracavernous or carotid-ophthalmic aneurysms;

(d) pseudo-tumoral symptoms are not strictly related to presence and degree of thrombosis, at least in our study.

In relation to radiological features:

(a) intracavernous and carotid/ophthalmic aneurysms have been visualized on plain CT scan in 80% of our cases, thus constituting the most easily detected types of giant aneurysms;

(b) partial thrombosis is significantly more common with intracavernous aneurysms (69% of cases).

In relation to outcome:

(a) a severe haemorrhage on CT scan (extensive or thick deposition) is linked with a poor prognosis;

(b) the prognosis is generally benign in patients with intracavernous and carotid-ophthalmic aneurysms, and very poor in patients with carotid bifurcation aneurysms[10];

(c) no significant influence on outcome seems to be exerted by intra-aneurysmal thrombosis, while stenosis of the afferent vessel seems to be a bad prognostic factor.

References

1. Drake CG (1979) Giant intracranial aneurysms: experience with surgical treatment in 174 patients. Clin Neurosurg 26: 12–95
2. Lavyne MH, Kleefield J, Davis KR, Ojemann RG, Crowell RM (1978) Giant intracranial aneurysms of the anterior circulation: clinical characteristics and diagnosis by computed tomography. Neurosurgery 3, no 3: 356–363
3. Macpherson P, Anderson DE (1981) Radiological differentiation of intrasellar aneurysms from pituitary tumours. Neuroradiology 21: 177–183
4. Onuma T, Suzuki J (1979) Surgical treatment of giant intracranial aneurysms. J Neurosurg 51: 33 36
5. Pasqualin A, Rosta L, Da Pian R, Cavazzani P, Scienza R (1984) The role of computed tomography in the management of vasospasm following subarachnoid haemorrhage. Neurosurgery 15: 344–353
6. Pia HW, Zierski J, Peerless SJ, Drake CG, Kodama N, Suzuki J, Sundt TM, Handa H, Hashimoto N, Yonekawa Y, Sengupta RP (1982) Direct management for large and giant aneurysms. Neurosurg Rev 5, no 4: 115–178
7. Pinto RS, Kricheff II, Buttler AR, Murali R (1979) Correlation of computed tomographic, angiographic and neuropathological changes in giant cerebral aneurysms. Radiology 132: 85–92
8. Pozzati E, Fagioli L, Servadel F, Gaist G (1981) Effect of common carotid ligation on giant aneurysms of the internal carotid artery: computerized tomography study. J Neurosurg 55: 527–531
9. Robbins J, Fein JM, Lantos J, Hooshangi N (1984) Reflow into a thrombosed giant middle cerebral artery aneurysm after extracranial – intracranial bypass. Neurosurgery 15, no 1: 120–124

10. Sindou M, Keravel Y *et al* (1984) Les aneurysmes geants intra-craniens. Approches therapeutiques. Intracranial giant aneurysms. Therapeutic approaches. Neurochirurgie 30 [Suppl] 1

11. Sonntag VKH, Yuan RH, Stein BM (1977) Giant intracranial aneurysms: A review of 13 cases. Surg Neurol 8: 81–83

12. Tognetti F, Limoni P, Testa C (1983) Aneurysm growth and hemodynamic stress. Surg Neurol 20: 74–78

13. Whittle IR, Dorschn W, Besser M (1982) Spontaneous thrombosis in giant intracranial aneurysms. J Neurol Neurosurg Psychiatry 45: 1040–1047

Address for correspondence: Dr. L. Rosta, Service of Neuroradiology, City Hospital, Verona, Italy.

Acta Neurochirurgica, Suppl. 42, 60–64 (1988)

Italian Cooperative Study on Giant Intracranial Aneurysms: 3. Modalities of Treatment

A. Pasqualin, R. Battaglia, R. Scienza, and **R. Da Pian***

Department of Neurosurgery, City Hospital, Verona, Italy

Summary

240 patients with giant aneurysms were treated in 10 Italian centres with various therapeutic modalities: out of them, 50 patients were conservatively treated (19 with a severe intracranial haemorrhage). Direct surgery was performed in 140 patients: 67% of patients with aneurysms between 2 and 2.5 cm (A 1 group) and 50% of patients with larger aneurysms (A 2 group). The aneurysm could be secured by clip in 102 cases (56% of A 1 and 31% of A 2 patients). In patients with subarachnoid haemorrhage, surgery was done within 3 days in 24 cases, between 4 and 14 days in 21 cases, and later in 52 cases. In patients operated on directly, brain swelling was observed in 39% of cases; controlled hypotension was employed in 56 cases, and temporary vessel occlusion (mainly of M 1 tract) in 33 cases; removal of intra-aneurysmal thrombi was done in 18 cases, and intraoperative aneurysmal rupture occurred in 39 cases. Carotid ligation was performed in 31 patients, and was associated with a by-pass in 17 cases. Balloon occlusion was performed in 23 cases, and was associated with a by-pass in 10 cases. As regards aneurysmal location, intra-cavernous aneurysms were treated mainly by balloon occlusion or carotid ligation, while carotid/ophthalmic, middle cerebral and anterior communicating aneurysms were treated prevalently by direct surgery. 60% of treated patients were submitted to postoperative angiography, and 54% to postoperative CT scan; total obliteration of the aneurysm was documented in 83% of patients submitted to postoperative angiography.

Keywords: Giant aneurysm; treatment; outcome; cooperative study; clipping; carotid ligation; by-pass; balloon occlusion.

Introduction

Modalities of treatment for giant aneurysms can be very different according to aneurysmal location and features, condition of the patient, surgical skills, pos-sibility of interventional radiological management, and other factors. Also conservative treatment constitutes a reasonable choice in selected patients, for whom active management entails a considerable risk.

In this section of the Cooperative study, the various forms of treatment are evaluated, operative findings are presented, and a few clinical and radiological relationships are considered.

Modalitites of Treatment and Findings

Modalities of treatment are summarized in Table 1. 50 patients were not treated: 26 had suffered from an intracranial haemorrhage, severe in 19 cases (GCS 3–6 on admission in 15 patients, and 7–2 in 4 patients) and 24 were treated conservatively for various reasons. No treatment was undertaken in 14% of A 1 and in 27% of A 2 patients.

Direct surgery was performed in 140 patients: in 67% of A 1 and in 50% of A 2 patients (p < 0.01). The aneurysm could be secured by clip in 102 patients (42% of cases): in 56% of A 1 and in 31% of A 2 patients (p < 0.001). Wrapping of the aneurysm was done in 4

* Cooperative Group: R. Da Pian, A. Pasqualin, R. Scienza/A. Benedetti, D. Curri, L. Volpin/P. Mennonna, F. Ammannati, D. Serino/E. Morace, F. Cattani, R. Mastrosetfano/V. A. D'Angelo, C. Corona, V. Monte/F. Filizzolo, M. Collice, P. Versari/G. Scialfa, G. Scotti, E. Boccardi/A. Andreoli, P. Limoni, C. Testa/M. Bortoluzzi, G. Marini/R. Galzio, D. Lucantoni/E. De Divitiis, R. Spaziante, F. D'Andrea/C. Ceccotto, P. P. Janes, A. Cramaro.

Table 1. *Modalities of Treatment According to Aneurysmal Size and Glasgow Coma Scale (GCS) on Admission*

	Total (A 1 + A 2)	GCS 3–6
Conservative treatment	50 (21%)	15 (30%)
Carotid ligation	31 (13%)	—
EC-IC bypass	32 (13%)	—
Balloon occlusion	23 (10%)	—
Intraluminar thrombosis	2 (1%)	—
Direct surgery	140 (58%)	13 (9%)
Total	240	28 (12%)

cases. Closure of the afferent vessel was effected in 7 cases (mainly A 2 patients). A trapping procedure was carried out in 6 patients. Aneurysmal resection was performed in 9 cases, with associated end-to-end anastomosis in 1 case. Exploration without exclusion was done in 5% of cases (4 A 1 and 8 A 2 patients).

Carotid ligation was performed in 31 patients: in 11% of A 1 and in 15% of A 2 patients. It was done as the only procedure in 12 cases, and was associated with a by-pass in 17 cases, and with balloon occlusion or with direct surgery in 2 cases. Carotid occlusion was performed gradually – through a Selverstone clamp – in 81% of cases, and abruptly in 19% of cases.

Balloon occlusion was performed in 23 patients: in 9% of A 1 and 10% of A 2 patients. It was done as the only procedure in 12 cases, in association with a by-pass in 10 cases and with carotid occlusion in 1 case. The balloon was directly inflated in the lumen of the aneurysm in 4 cases, and in the proximal internal carotid artery in 19 cases.

STA-MCA anastomosis was performed in 32 patients: in 11% of A 1 and in 15% of A 2 patients. The anastomosis was done as the only procedure in 2 cases; it was associated with carotid ligation or balloon occlusion in 26 cases, and with direct surgery in 4 cases. The anastomosis anticipated carotid ligation or balloon occlusion by 10 days or less in 55% of cases, and by 1 month or more in 17% of cases.

Apart from patients submitted to a by-pass in association with carotid ligation or balloon occlusion, 2 patients were submitted to 2 direct procedures and 9 patients were submitted both to direct and to indirect procedures.

Intraluminar thrombosis with fine wire (Gianturco coils) was effected in 2 cases (both intracavernous aneurysms).

Timing: Among the 97 patients operated on after SAH, surgery was done within 3 days in 24 cases, between 4 and 14 days in 21 cases, and later in 52 cases.

Several patients were submitted to *further operative procedures*, independently from the one aimed at the exclusion of the giant aneurysm: a shunt was inserted in 13 patients, removal of infected bone flap was performed in 13 patients, exclusion of multiple aneurysms in 9 patients, evacuation of a postoperative haematoma in 5 patients, and other procedures in 6 patients.

Preoperative medical management consisted of antifibrinolytics in 47% of patients, steroids in 59% of patients, antiplatelet agents in 8% of patients; there were no substantial differences between A 1 and A 2 patients, except for a higher use of antifibrinolytics in

A 1 patients who were more prone to SAH. Postoperatively, steroids were used in 84% of patients, antiplatelet agents in 19% of patients, hypervolaemia and/or induced hypertension in 12% of patients; there were no differences between A 1 and A 2 patients.

Intraoperative Findings

The brain was swollen in 39% of patients submitted to direct surgical exposure of the aneurysm: 42% of A 1 and 38% of A 2 patients. Perioperative lumbar drainage was used in only 14 patients. During the procedure, cisternal drainage proved satisfactory in 72% of cases. Deep controlled hypotension was used in 56 patients. EEG monitoring was performed in only 6 cases. Temporary vessel occlusion was used in 33 cases. The artery most frequently closed was the middle cerebral (17 cases), followed by the anterior cerebral (9 cases) and the internal carotid (7 cases). The minimum time of temporary occlusion was 3 minutes, the maximum 60 minutes. Carotid compression was done in 8 cases: percutaneously in 6 cases, directly in 2 cases. Suction-decompression of the sac was performed in 53 patients. Removal of intra-aneurysmal thrombi was performed in 18 patients. Intraoperative aneurysmal rupture occurred in 39 cases: it was deliberate in 17 cases, accidental in 22 cases.

Clinical/Radiological Relations

After haemorrhage, direct surgery was performed in 71% of cases; in patients with symptoms of an expanding mass lesion, it was performed in only 37% of cases (p < 0.001); there were significant differences in regard to clinical symptomatology in those patients

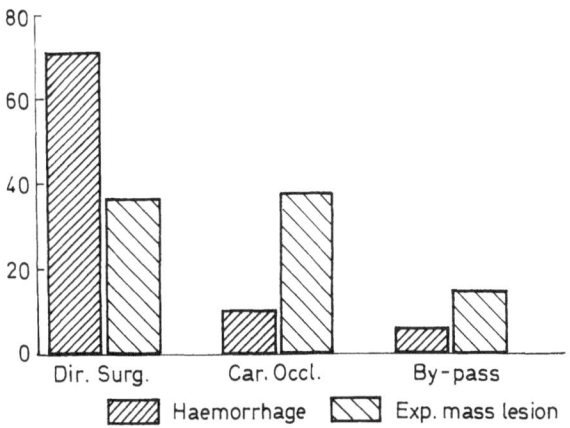

Fig. 1. Relation between clinical symptomatology and modality of treatment: p < 0.001 for direct surgery, p < 0.001 for carotid occlusion, p < 0.05 for by-pass (numbers are given in percent)

Table 2. *Relation Between Aneurysmal Location and Type of Active Treatment*

	Dir. surg.	Carotid lig. Balloon occl.	By-pass
Intracavernous (51 cases)	1 (2%)	36 (70%)	19 (37%)
Carotid ophth. (35 cases)	27 (77%)	10 (28%)	5 (14%)
Carotid supracl. (35 cases)	13 (52%)	7 (28%)	6 (24%)
Carotid bifurc. (12 cases)	7 (58%)	1 (8%)	—
Middle cerebral (54 cases)	47 (85%)	1 (2%)	2 (4%)
Ant. cerebral complex (44 cases)	36 (83%)	—	—
Post. circulation (17 cases)	9 (53%)	1 (6%)	—
Total (249 cases)	140 (58%)	56 (23%)	32 (13%)

Table 3. *Relation Between Angiographic Features and Type of Active Treatment*

	Dir. surg.	Carotid lig. Balloon occl.	By-pass
Neck present (33 cases)	24 (73%)	3 (9%)	2 (6%)
Neck absent (202 cases)	114 (56%)	53 (26%)	30 (15%)
Significance	N.S.	p < 0.05	N.S.
Thrombosis present (122 cases)	67 (55%)	32 (26%)	23 (19%)
Thrombosis absent (84 cases)	55 (65%)	21 (25%)	8 (9%)
Significance	N.S.	N.S.	N.S.

submitted to carotid occlusion or to by-pass surgery (Fig. 1).

As regards aneurysmal location (Table 2), intracavernous aneurysms were treated mainly by carotid ligation or balloon occlusion, while carotid ophthalmic, internal carotid bifurcation, middle cerebral, anterior communicating and posterior circulation aneurysms were treated mainly by direct surgery.

Table 3 shows the lack of any significant relation between such angiographic findings as presence of a clear aneurysmal neck or presence of intra-aneurysmal thrombosis, and modality of treatment. Direct surgery was performed in 69% of patients with stenosis and 58% of patients without stenosis of afferent vessel (N.S.).

Postoperative Radiological Data

A total of 115 patients were submitted to postoperative angiography out of 190 treated patients (60%). Post-

operative CT scan was done in 102 patients (54%). Total obliteration of the aneurysm was documented in 83% of patients — 85% of A 1 and 81% of A 2 patients — and partial obliteration in 10% of patients — 8% of A 1 and 13% of A 2 patients —; in 7% of patients, the examination was not conclusive. Direct surgery was linked with 87% of total, and 13% of partial, obliteration; indirect methods of treatment were associated with total obliteration in 75% of cases, partial obliteration in 6% of cases, and inconclusive examinations in 19% of cases.

Discussion

According to some authors, *conservative treatment* of giant aneurysms should be reserved for patients with asymptomatic intracavernous aneurysms[1, 6] and to patients with unclippable aneurysms, for whom carotid ligation or balloon occlusion are contraindicated according to the criteria of the Glasgow group[5, 9]; in these cases, symptomatic medical management constitutes a reasonable alternative choice[13].

Exclusion of the aneurysm by clip is still considered the treatment of choice by most authors[3, 6, 15]. Direct clipping is certainly more difficult in larger aneurysms: in our study, only 31% of A 2 aneurysms have been excluded by clip, versus 56% of A 1 aneurysms. In Drake's experience, neck occlusion could be achieved in only 38% of cases (44% of anterior circulation and 35% of posterior circulation aneurysms)[3]. When clipping of the neck is unfeasible, occlusion of the parent artery — whether or not associated with aneurysmal resection — is a reasonable alternative[3]; this method was used in 16% of our operated patients. In these cases, the use of tourniquet in the awake patient has been strongly recommended by Drake[3], in order to test tolerance to the occlusion postoperatively.

Carotid ligation has been for many years the treatment of choice for unclippable giant aneurysms[3]. Safe criteria for carotid ligation have been established by the Glasgow group[2, 5, 9] and will not be reviewed here. Carotid ligation does not always prevent rebleeding[3, 6, 14], although it often improves the clinical symptoms[3, 8].

Balloon occlusion constitutes a new method of treatment for giant aneurysms[1, 7, 12]. Direct occlusion of the aneurysm – only 4 cases in our series – is a high-risk procedure, especially considering the danger of aneurysm rupture during balloon inflation; moreover, the aneurysm can present delayed filling, owing to progressive deflation of the balloon[1].

STA-MCA anastomosis – associated with carotid ligation or with balloon occlusion – has been recommended as a safety measure, in order to avoid ischaemic disturbances due to decreased regional perfusion[5, 14, 15]. However, there is currently insufficient support to recommend prophylactic by-pass in patients with sufficient collateral circulation, assessed through the Glasgow method and/or angiographical documentation[6]. The timing of direct or balloon carotid occlusion is generally established within the first week after by-pass surgery[1, 2, 6], although it was more delayed in our study. The by-pass can be also associated with parent artery occlusion (or trapping) in giant middle cerebral aneurysms, as a prophylactic measure against distal postoperative ischaemia[3, 15].

As regards *medical management*, antiplatelet agents have been used in our study in a minority of patients submitted to carotid ligation or balloon occlusion; anticoagulants have never been administered postoperatively to our patients, but only during balloon occlusion. According to recent reports the use of anticoagulants shold be encouraged[1, 2, 3, 6] considering the high risk of thromboembolic complications linked to indirect methods of treatment. The role of antiplatelet drugs in carotid occlusion remains to be determined[2, 6].

As regards *intraoperative data*, lumbar drainage has been rarely adopted in our study (14% of cases). Excluding patients with ventricular shift – in whom the insertion of a drain can be dangerous – quite a large number of patients did not receive perioperative drainage; this factor may account for the high percent of brain swelling at operation with negative consequences on final outcome[6]. Temporary vessel occlusion has been adopted in 24% of our cases: while some authors advise temporary occlusion in every giant aneurysm[10], others reserve this technique mainly to giant MCA aneu-

rysms[15]. Suction/decompression of the aneurysmal sac[4] is a very convenient procedure, especially if associated with deep hypotension. In our study, it was done in 38% of operated cases.

The importance of *postoperative radiological evaluation* in giant aneurysms is related to the fact that obliteration is not always complete (89% of our cases), and partial obliteration can be linked to a substantial risk of future haemorrhage[8]. Follow-up CT scan can be considered a valuable tool for detection of aneurysmal thrombosis following treatment[11], although this examination cannot be accepted as a definitive proof of complete aneurysmal obliteration.

References

1. Debrun G, Fox AJ, Drake CG, Peerless S, Girvin J, Ferguson G (1981) Giant unclippable aneurysm: treatment with detachable balloons. AJNR 2: 167–173
2. Diaz FG, Ausman JI, Pearce JE (1982) Ischemic complications after combined internal carotid artery occlusion and extracranial-intracranial anastomosis. Neurosurgery 10: 563–570
3. Drake CG (1979) Giant intracranial aneurysms: experience with surgical treatment in 174 patients. Clin Neurosurg 26: 12–95
4. Flamm ES (1981) Suction decompression of aneurysms. Technical note. J Neurosurg 54: 275–276
5. Gelber BJ, Sundt TM (1980) Treatment of intracavernous and giant carotid aneurysms by combined internal carotid ligation and extra- to intracranial bypass. J Neurosurg 52: 1–10
6. Heros RC (1984) Thromboembolic complications after combined internal carotid ligation and extra- to intracranial bypass. Surg Neurol 21: 75–79
7. Hieshima GB, Grinnell VS, Mehringer CM (1981) A detachable balloon for therapeutic transcatheter occlusions. Radiology 138: 227–228
8. Matsuda M, Shiino A, Handa J (1985) Rupture of previously unruptured giant carotid aneurysm after superficial temporal – middle cerebral artery bypass and internal carotid occlusion. Neurosurgery 16, no 2: 177–184
9. Miller JD, Jawad K, Jennett B (1977) Safety of carotid ligation and its role in the management of intracranial aneurysms. J Neurol Neurosurg Psychiatry 40: 64–72
10. Onuma T, Suzuki J (1979) Surgical treatment of giant intracranial aneurysms. J Neurosurg 51: 33–36
11. Pozzati E, Fagioli L, Servadel F, Gaist G (1981) Effect of common carotid ligation on giant aneurysms of the internal carotid artery: computerized tomography study. J Neurosurg 55: 527–531
12. Scialfa G, Vaghi A, Valsecchi F *et al* (1982) Neuroradiological treatment of carotid and vertebral fistulas and intracavernous aneurysms. Technical problems and results. Neuroradiology 24: 13–25
13. Slosberg PS (1982) Symptomatic unruptured giant aneurysms: medical treatment. Acta Neurochir (Wien) 62: 207–218

14. Spetzler RF, Schuster H, Roski RA (1980) Elective extracranial-intracranial arterial bypass in the treatment of inoperable giant aneurysms of the internal carotid artery. J Neurosurg 53: 22–77

15. Sundt TM, Piepgras DG (1979) Surgical approach to giant in-tracranial aneurysms: Operative experience with 80 cases. J Neurosurg 51: 731–742

Address for correspondence: Dr. A. Pasqualin, Department of Neurosurgery, City Hospital, Verona, Italy.

Acta Neurochirurgica, Suppl. 42, 65–70 (1988)

Italian Cooperative Study on Giant Intracranial Aneurysms: 4. Results of Treatment

A. Pasqualin, R. Battaglia, R. Scienza, and **R. Da Pian***

Department of Neurosurgery, City Hospital, Verona, Italy

Summary

240 patients with giant aneurysms admitted to 10 Italian centres were evaluated in regard to results of treatment and postoperative complications. As a whole, a complete recovery was observed in 39% of cases, various degrees of disability in 38% of cases, and death in 23% of cases, considering also patients admitted in Glasgow Coma Scale (GCS) 3–6. By excluding patients in GCS 3–6, active treatment was linked with a recovery rate of 47% and a mortality rate of 15%; the worst outcome was observed for carotid bifurcation aneurysms, the best for intracavernous aneurysms. In patients with giant carotid/ophthalmic or supraclinoid aneurysms, the outcome was similar after early surgery and after indirect methods of treatment. Factors playing a negative influence on outcome were mainly early exclusion after haemorrhage, operative brain swelling and cisternal tamponade; induced hypotension was associated with a significant decrease in mortality.

Postoperative complications were observed in 48% of cases, and were mainly caused by ischaemic disturbances not associated with vasospasm (17% of cases), followed by surgical trauma and cerebral oedema; in 32 patients the postoperative neurological deterioration was fully reversible. In patients submitted to carotid occlusion association with a by-pass did not decrease the rate of ischaemic complications.

In patients submitted to active treatments (open surgery or indirect methods of exclusion) the causes of morbidity were mainly: – deficits due to mass lesion, surgical complications, and ischaemic disturbances without vasospasm; the causes of mortality were mainly surgical trauma or medical complications.

Keywords: Giant aneurysm; cooperative study; treatment; outcome; complications.

---

* Cooperative Group: R. Da Pian, A. Pasqualin, R. Scienza/A. Benedetti, D. Curri, L. Volpin/P. Mennonna, F. Ammannati, D. Serino/E. Morace, F. Cattani, R. Mastrostefano/V. A. D'Angelo, M. Ferrara, E. Fiumara/F. Filizzolo, M. Collice, G. D'Aliberti/G. Scialfa, G. Scotti, E. Boccardi/A. Andreoli, P. Limoni, C. Testa/M. Bortoluzzi, G. Marini/R. Galzio, D. Lucantoni/E. De Divitiis, R. Spaziante, F. D'Andrea/C. Ceccotto, P. P. Janes, A. Cramaro.

Introduction

The treatment of giant intracranial aneurysms is still linked with a considerable incidence of postoperative complications. In patients submitted to direct surgery, considerable morbidity is due to injury of adjacent vessels; in the many cases in which direct exclusion of the aneurysm is not feasible, indirect methods of exclusion – such as clipping of parent vessel, carotid ligation, and balloon occlusion – must be balanced very carefully against the risk of ischaemic complications, which are often unpredictable.

In this section of the Cooperative study, the results of treatment are presented and postoperative complications, as well as causes of morbidity and mortality, are evaluated and discussed.

Results of Treatment

In the 240 patients evaluated – independently of treatment employed – a complete recovery was observed in 39% of cases, a moderate disability in 27%, a severe disability in 10%, a vegetative state in 1%, and death in 23%. It should be noted that 28 patients (12% of the whole number) were admitted in a moribund condition (Glasgow Coma Score 3–6) with very limited life expectancy; therefore, in the following evaluation these patients have been excluded from the comparison of the results obtained by various treatment modalities.

The results of active treatment (direct surgery, carotid ligation, balloon occlusion, by-pass surgery) are presented in Fig. 1; as a whole, 47% of patients made a good recovery, and 15% died. In comparison with active treatment, conservative management led to a significantly lower recovery rate (26%) (p < 0.05) and

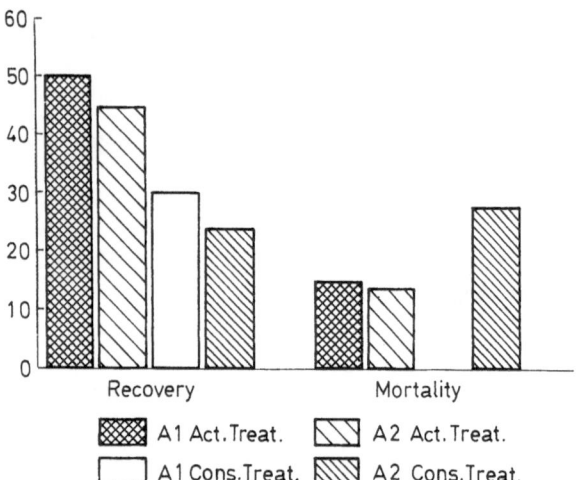

Fig. 1. Relation between modality of treatment (active or conservative) and outcome, according to aneurysmal size, and excluding patients in GCS 3–6 (numbers are given in percent)

a slightly higher mortality rate (20%). The conservative group amounted to only 35 patients, and the follow-up was limited to a maximum of 6 years: thus this group cannot be taken as representative of the natural history of giant aneurysms.

With regard to the various aneurysmal locations, active treatment (Table 1) was linked with the worst outcome in patients with a carotid bifurcation aneurysm and with the best outcome in patients with an intracavernous aneurysm. Comparison with conservative treatment was prevented due to the small numbers of conservatively treated patients, except for 14 conservatively treated intracavernous aneurysm exhibiting a 14% recovery rate and a 86% disability rate, with no mortality.

Comparing direct surgery to carotid ligation or balloon occlusion, the former (Fig. 2) was associated with a slightly lower recovery rate (N.S.) and a slightly higher mortality rate (p < 0.05). Owing to the lack of

comparable cases for the individual aneurysmal locations, a direct comparison between open surgery and indirect treatments was possible only for carotid-ophthalmic and carotid supraclinoid aneurysms (33 patients treated directly and 18 indirectly): the recovery rate was 42% for open surgery and 44% for indirect surgery, and the mortality rate 12% and 11% respectively.

As regards direct surgery, some factors exerted a negative influence on outcome:

(a) in patients with a previous SAH, surgery within 3 days of haemorrhage was associated with a significantly lower recovery rate (p < 0.05) and a significantly higher mortality rate (p < 0.05) than surgery after 14 days from haemorrhage (Fig. 3);

(b) operative brain swelling was associated with a recovery rate of 33% – against 53% (p < 0.05) for patients without brain swelling – and with a mortality rate of 39%, against 4% (p < 0.001);

(c) absence of intraoperative cisternal drainage was associated with a recovery rate of 23% – against 53% for patients with good cisternal drainage (p < 0.01) – and with a mortality rate of 53%, against 5% (p < 0.001);

(d) induced hypotension was associated with a significant decrease in mortality rate (14%, against 28% for patients not submitted to induced hypotension; p < 0.05), while the recovery rate was the same in patients submitted and not submitted to induced hypotension (39% and 40% respectively);

(e) accidental intraoperative rupture of the aneurysm was associated with a recovery rate of 28% and a mortality rate of 28%, against 46% and 16% respectively for patients without intraoperative rupture (N.S.).

Among intraoperative factors with no influence on outcome: temporary vessel occlusion was associated

Table 1. *Active Treatment: Relation Between Aneurysmal Location and Clinical Outcome (GCS 3–6 Excluded)*

	Recovery	Disability	Mortality
Intracavernous (37 cases)	22 (59%)	14 (38%)	1 (3%)
Carotid ophth. (32 cases)	14 (44%)	15 (47%)	3 (9%)
Carotid supracl. (19 cases)	8 (42%)	8 (42%)	3 (16%)
Carotid bifurc. (7 cases)	1 (14%)	3 (43%)	3 (43%)
Middle cerebral (37 cases)	17 (46%)	14 (38%)	6 (16%)
Ant. cerebral complex (34 cases)	15 (45%)	10 (27%)	9 (27%)
Post. circulation (11 cases)	6 (54%)	4 (37%)	1 (9%)
Total (177 cases)	83 (47%)	68 (38%)	26 (15%)

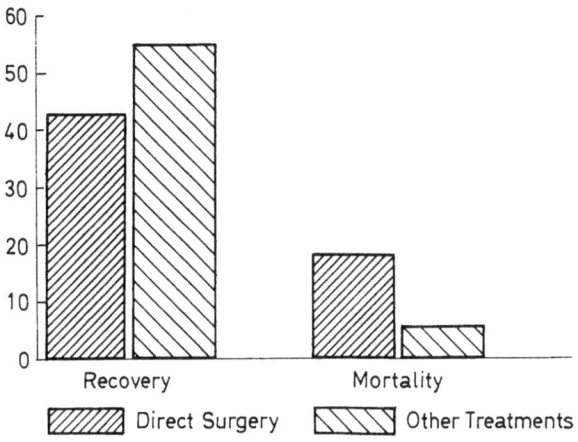

Fig. 2. Relation between type of active treatment and outcome, excluding patients in GCS 3–6 (numbers are given in percent)

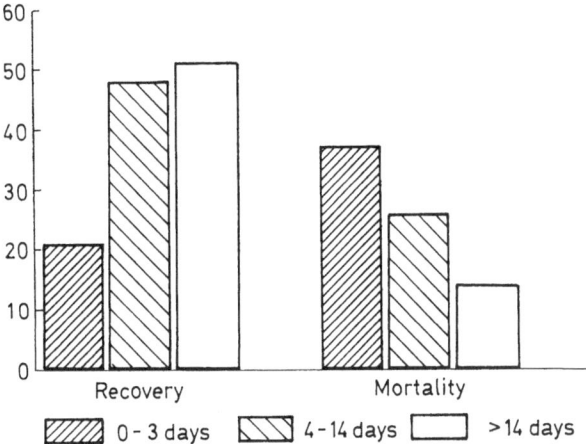

Fig. 3. Relation between timing of surgery after SAH and outcome, excluding patients in GCS 3–6: $p < 0.05$ for recovery, and for mortality between timing 0–3 and timing over 14 days (numbers are given in percent)

with a recovery rate of 41% and a mortality rate of 21%, against 44% and 17% for patients not submitted to this technique.

Postoperative complications of varying severity were observed in 48% of cases: 46% of A 1 patients, and 50% of A 2 patients (N.S.). The most frequent postoperative complication (Table 2) was ischaemic deterioration without vasospasm, followed by surgical trauma and cerebral oedema.

Considering indirect methods of treatment, gradual occlusion of the carotid artery was followed by ischaemic complications in 24% of cases, and abrupt occlusion in 19% of cases (N.S.), independently of presence or absence of a by-pass. Patients with intracavernous aneurysms treated with indirect methods presented ischaemic complications in 19% of cases; patients with carotid-ophthalmic and carotid supraclinoid aneurysms also treated with indirect methods presented a similar incidence of ischaemic complications (24%). The rate of ischaemic complications was the same in patients with carotid occlusion only (22%), and in patients with carotid occlusion associated to by-pass (21%).

Transient postoperative neurological deficits were observed in 32 patients: these deficits were most commonly due to surgical trauma (12 cases) or to ischaemic deterioration without vasospasm (12 cases), and were observed in 21% of A 1 and in 15% of A 2 patients.

In treated patients, *morbidity* was mainly due to deficits due to mass lesion, surgical complications, and ischaemic disturbances without vasospasm (Table 3). In untreated patients morbidity was usually attributable to deficits due to mass lesion.

As regards *mortality* (Table 4), in treated patients the chief causes were surgical complications, direct ef-

Table 2. *Postoperative Complications, According to Aneurysmal Size (GCS 3–6 Excluded)*

	A 1 (91 cases)	A 2 (86 cases)	Total (177 cases)
Ischemic deficit, no spasm	15 (16%)	15 (17%)	30 (17%)
Surgical trauma	11 (12%)	14 (16%)	25 (14%)
Cerebral oedema	7 (7%)	8 (9%)	15 (8%)
Medical complications	7 (7%)	3 (3%)	10 (5%)
Haematoma	6 (7%)	3 (3%)	9 (5%)
Ischemic deficit from spasm	5 (5%)	4 (5%)	9 (5%)
Scalp infection	2 (2%)	7 (8%)	9 (5%)
Hydrocephalus	5 (5%)	1 (1%)	6 (3%)
Cerebral infection	—	1 (1%)	1 —
Other complications	1 (1%)	2 (2%)	3 (2%)
Total	42 (46%)	43 (50%)	85 (48%)

Table 3. *Causes of Morbidity in Treated and Untreated Patients, According to Aneurysmal Size (GCS 3–6 Excluded)*

	Treated (177 cases)		Untreated (35 cases)	
	A 1 (91 cases)	A 2 (86 c)	A 1 (10 cases)	A 2 (25 cases)
Expand. mass lesion	7 (8%)	14 (16%)	5 (50%)	9 (36%)
Surgical trauma	7 (8%)	9 (10%)	—	—
Isch. deficit, no spasm	8 (9%)	6 (7%)	—	2 (8%)
Effect of initial bleed	4 (4%)	4 (5%)	1 (10%)	1 (4%)
Hydrocephalus	4 (4%)	1 (1%)	—	—
Clinical vasospasm	—	1 (1%)	1 (10%)	—
Rebleeding	—	1 (1%)	—	—
Postop. haematoma	1 (1%)	—	—	—
Other complications	1 (1%)	—	—	—
Total	32 (35%)	36 (42%)	7 (70%)	12 (48%)

Table 4. *Causes of Mortality in Treated and Untreated Patients, According to Aneurysmal Size (GCS 3–6 Excluded)*

	Treated (177 cases)		Untreated (35 cases)	
	A 1 (91 cases)	A 2 (86 c)	A 1 (10 cases)	A 2 (25 cases)
Surgical trauma	2 (2%)	6 (7%)	—	—
Effect of initial bleed	1 (1%)	1 (7%)	—	3 (12%)
Rebleeding	3 (3%)	1 (1%)	—	3 (12%)
Medical complications	4 (4%)	2 (2%)	—	—
Clinical vasospasm	3 (3%)	—	—	—
Hydrocephalus	1 (1%)	—	—	1 (4%)
Isch. deficit, no spasm	—	1 (1%)	—	—
Postop. haematoma	—	1 (1%)	—	—
Total	14 (15%)	12 (14%)	—	7 (28%)

Table 5. *Causes of Morbidity After Direct Surgery and After Indirect Methods of Exclusion, According to Aneurysmal Size (GCS 3–6 Excluded)*

	Direct surg. (127 cases)		Indirect meth. (56 cases)	
	A 1 (71 cases)	A 2 (56 cases)	A 1 (20 cases)	A 2 (36 cases)
Surgical trauma	7 (10%)	9 (16%)	—	1 (3%)
Expand. mass lesion	3 (4%)	7 (12%)	4 (20%)	8 (22%)
Isch. deficit, no spasm	6 (8%)	3 (5%)	2 (10%)	4 (11%)
Effect of initial bleed	3 (4%)	4 (7%)	1 (5%)	—
Hydrocephalus	4 (6%)	—	—	1 (3%)
Clinical vasospasm	—	1 (2%)	—	—
Rebleeding	—	1 (2%)	—	—
Postop. haematoma	1 (1%)	—	—	—
Other complications	—	—	1 (5%)	—
Total	24 (34%)	25 (45%)	8 (40%)	14 (39%)

Table 6. *Causes of Mortality After Direct Surgery and After Indirect Methods of Exclusion, According to Aneurysmal Size (GCS 3–6 Excluded)*

	Direct surg. (127 cases)		Indirect meth. (56 cases)	
	A 1 (71 cases)	A 2 (56 cases)	A 1 (20 cases)	A 2 (36 cases)
Surgical trauma	2 (3%)	5 (9%)	—	1 (3%)
Medical complications	4 (6%)	—	—	2 (6%)
Rebleeding	3 (4%)	1 (2%)	—	—
Clinical vasospasm	3 (4%)	—	—	—
Effect of initial bleed	1 (1%)	1 (2%)	—	—
Isch. deficit, no spasm	—	1 (2%)	—	—
Hydrocephalus	1 (1%)	—	—	—
Postop. haematoma	—	1 (2%)	—	—
Total	14 (20%)	9 (16%)	—	3 (8%)

fects of bleeding, medical complications, and clinical vasospasm. In untreated patients, mortality resulted chiefly from the direct effects of bleeding.

As shown by Table 5, the most common causes of morbidity after open surgery were surgical trauma, deficits due to mass lesion, ischaemic disturbances without vasospasm, and direct effects of bleeding. After indirect treatments, morbidity was mainly accounted for by deficits from mass lesion and by ischaemic disturbances without vasospasm. When evaluating only the patients treated with indirect methods:

(a) patients with carotid/ophthalmic and carotid supraclinoid aneurysms presented a slightly higher morbidity rate from ischaemia than patients with intracavernous aneurysms (18% against 11%, N.S.);

(b) patients treated with gradual occlusion of the carotid artery presented a morbidity rate from ischaemia similar to that of patients treated with abrupt occlusion (12% against 10%);

(c) patients submitted to a by-pass presented a slightly lower morbidity rate from ischaemia than patients without a by-pass (8% against 13%, N.S.).

Finally, the most common causes of death after open surgery were surgical trauma, medical complications and rebleeding; after indirect treatments, mortality was only due to medical complications and to surgical trauma (Table 6).

Discussion

Conservative treatment of giant aneurysms is debatable. Non-surgically treated patients are reported to have a poor prognosis[4, 7, 12]; also the morbidity rate is high[15]. Intracavernous aneurysms constitute an excep-

tion to the rule, owing to their relatively favourable natural prognosis. Although our conservative group is small, non-homogeneous, and with a short follow-up, the outcome has been less favourable than in patients with active treatment; in particular, intracavernous aneurysms have exhibited a higher disability rate if treated conservatively than if submitted to surgery.

Active treatment is reportedly linked with an unfavourable outcome in patients with internal carotid bifurcation aneurysms[4, 13]: the mortality rate has been 77% in the French study[13] and 43% in our series, and the recovery rate has been less than 20% in both studies. Also giant posterior circulation aneurysms are linked with a relatively unfavourable outcome according to Drake: more than 50% of patients die or become severely disabled after surgery[4]; our group is too small to confirm these data. Fortunately, for most giant aneurysms the results of active treatment are more favourable than for the previous locations; as a whole, the mortality rate of active treatment has ranged between 4% and 21% in the large reported series[4, 7, 10, 14], and has been 15% in our group.

A comparison between the results of direct surgery versus indirect methods has been possible in our study only for carotid/ophthalmic and carotid supraclinoid aneurysms, and has revealed no substantial difference between the two groups. The improvement in symptoms seen after carotid or balloon occlusion[2, 4, 8] is possibly due to elimination of the expansile pressure and/or shrinkage of the thrombosed aneurysm[4].

Among prognostic factors in patients submitted to direct surgery, timing of surgery seems to be very important, as do also operative brain swelling and the degree of cisternal drainage. The significantly more favourable results obtained in our patients after delayed surgery should be balanced against the possibility of aneurysmal rupture during the waiting period; nevertheless, it is likely that early surgery constitutes a risky procedure for giant aneurysms. The influence of brain swelling and impaired cisternal drainage on outcome has been already stressed for smaller aneurysms[1], and is crucial in the surgery of these large lesions[5]. A surprising finding in our study has been the significant decrease in mortality for patients submitted to intraoperative hypotension.

It is well known that the most frequent cause of morbidity and complications in the treatment of giant aneurysms is cerebral ischaemia[3, 4, 6, 14]. After direct surgery, immediate ischaemic complications are due to major vessel occlusion, retraction injury, or lesion of perforating vessels, while delayed ischaemic complications are rare and mostly due to vasospasm[14]. After indirect methods of treatment, ischaemia can result from decrease in cerebral blood flow, and more commonly from thromboembolism[2, 3, 6, 8] generally originating from the stump of the occluded vessel. While some authors believe that ischaemic deficits are more common after abrupt than after gradual carotid occlusion[3, 9], others are convinced of the opposite[6]; no significant difference in morbidity from ischaemia has been noted in our study between these two modalities. Peerless has recently suggested a lower incidence of ischaemic complications for patients with intracavernous aneurysms treated by carotid ligation[11]; we have noted no difference in the incidence of ischaemic complications as between intracavernous and paraclinoid aneurysms treated by indirect methods, although the permanent morbidity from ischaemia has been slightly lower for intracavernous aneurysms. The use of a bypass has not prevented ischaemic complications in our study, as well as in other series[3, 6, 7, 8]; however, after a by-pass permanent morbidity from ischaemia has been reduced in our series, and has almost disappeared in the series from Diaz[3]. The ischaemic disturbances seen in the presence of a functioning by-pass are reportedly due to embolism from the carotid stump[3, 6, 11].

Apart from ischaemia, another important cause of morbidity – observed only after direct surgery – is that of surgical trauma, often determined by the difficulties faced in the exposure and exclusion of these giant lesions.

References

1. Da Pian R, Pasqualin A, Scienza R, Cavazzani P (1985) Early surgery for intracranial aneurysm: influence of clinical and operative findings on final results. In: Auer LM (ed) Timing of aneurysm surgery. De Gruyter, Berlin, pp 115–124
2. Debrun G, Fox AJ, Drake CG, Peerless S, Girvin J, Ferguson G (1981) Giant unclippable aneurysm: treatment with detachable balloons. AJNR 2: 167–173
3. Diaz FG, Ausman JI, Pearce JE (1982) Ischaemic complications after combined internal carotid artery occlusion and extracranial-intracranial anastomosis. Neurosurgery 10: 563–570
4. Drake CG (1979) Giant intracranial aneurysms: experience with surgical treatment in 174 patients. Clin Neurosurg 26: 12–95
5. Heros RC, Kolluri S (1984) Giant intracranial aneurysms presenting with massive cerebral oedema. Neurosurgery 15, no 4: 572–577
6. Heros RC, Kolluri S (1984) Giant intracranial aneurysms presenting with massive cerebral oedema. Neurosurgery 15, no 4: 572–577
7. Hosobuchi Y (1979) Direct surgical treatment of giant intracranial aneurysms. J Neurosurg 51: 743–756
8. Matsuda M, Shiino A, Handa J (1985) Rupture of previously unruptured giant carotid aneurysm after superficial temporal – middle cerebral artery bypass and internal carotid occlusion. Neurosurgery 16, no 2: 177–184
9. Nishioka H (1966) Report on the cooperative study of intracranial aneurysms and subarachnoid haemorrhage: section VIII, part I. Results of the treatment of intracranial aneurysms by occlusion of the carotid artery in the neck. J Neurosurg 25: 660–682
10. Onuma T, Suzuki J (1979) Surgical treatment of giant intracranial aneurysms. J Neurosurg 51: 33–36
11. Peerless SJ (1982) Comment. Neurosurgery 10: 570
12. Pia HW, Zierksi J, Peerless SJ, Drake CG, Kodama N, Suzuki J, Sundt TM, Handa H, Hashimoto N, Yonekawa Y, Sengupta RP (1982) Direct management for large and giant aneurysms. Neurosurg Rev 5, no 4: 115–178
13. Sindou M, Keravel Y *et al* (1984) Les anévrysmes géants intracrâniens. Approches thérapeutiques. Intracranial giant aneurysms. Therapeutic approaches. Neurochirurgie 30 [Suppl] 1
14. Sundt TM, Piepgras DG (1979) Surgical approach to giant intracranial aneurysms: Operative experience with 80 cases. J Neurosurg 51: 731–742
15. Whittle IR, Dorsch NW, Besser M (1984) Giant intracranial aneurysms: diagnosis, management and outcome. Surg Neurol 21: 218–230

Address for correspondence: Dr. A. Pasqualin, Department of Neurosurgery, City Hospital, Verona, Italy.

Acta Neurochirurgica, Suppl. 42, 71–74 (1988)

Surgical Approach to Giant Aneurysms of the Anterior Circulation

S. Giombini, C. L. Solero, S. Ferraresi, A. Melcarne, G. Broggi, and **F. Pluchino**

Neurosurgical Department, Istituto Neurologico "C. Besta", Milan, Italy

Summary

The surgical approach to cerebral giant aneurysms is still a source of great concern. We describe our experience with giant aneurysms of the anterior circulation and discuss the different surgical techniques adopted.

During the period January 1972–December 1985, a total of 33 patients were operated upon at the Istituto Neurologico "C. Besta" of Milan for a giant aneurysm of the anterior circulation. Nineteen cases had suffered subarachnoid haemorrhage before admission; in 14 cases the hospitalization was due to evidence of mass effect on the surrounding neurovascular structures. All aneurysms were directly approached: in 24 cases the neck was occluded by a suitable clip, in 4 cases intramural thrombosis was attempted, in 3 cases the aneurysms were definitively trapped and in one case aneurysmorrhaphy was performed after resection of the sac.

Operative mortality was 12%. Long-term follow-up shows good results whenever exclusion of the aneurysm from cerebral circulation had been achieved, either after removal of the sac or not; on the contrary, only fair or poor results were evident when other surgical techniques were adopted, either electively or out of necessity.

The importance of intraoperative protection and monitoring of brain function is stressed.

Introduction

Giant intracranial aneurysms are commonly defined as aneurysms whose major diameter is over 2.5 cm. This unusual size is probably the result of the progressive enlargement of saccular aneurysms or of the atherosclerotic weakening of the wall in conducting vessels of the brain; the latter condition is often seen on the arteries of the posterior circulation (fusiform serpentine variety).

Giant aneurysms may rupture, or may become symptomatic because of direct compression of surrounding nervous and/or vascular structures. Furthermore, to compound the problem, small perforating vessels may arise from or tightly adhere to the sac; the parent vessel may be stretched around the surface of the aneurysm giving rise to great difficulties in the definition of the neck; the wall of the aneurysm may be thick and the sac partially thrombosed, a condition which often makes the occlusion by a normal clip impossible. These features imply particular surgical problems in the treatment of giant aneurysms, consequently several strategies have been proposed so far: simple closure of the feeding vessel, "trapping" or "wrapping" of the aneurysm, extra-intracranial arterial bypass (EIAB) followed by direct attack on the aneurysm, occlusion of the neck and/or of the aneurysm by balloon catheterization, intramural thrombosis, direct occlusion of the neck with or without removal of the sac. This broad range of surgical procedures emphasizes both the complexity of every single case and the improvements in neuroradiological and neurosurgical technology.

At least for giant aneurysms of the anterior circulation, the only effective policy is, in our opinion, the direct attack on the aneurysm and occlusion of its neck in order to exclude it from the cerebral circulation and to obtain the collapse of the sac.

The purpose of this paper is to review our surgical experience and confirm this statement.

Clinical Materials and Methods

From January 1972 until December 1985, 483 aneurysms of the anterior circulation were operated upon at the Istituto Neurologico of Milan. Giant aneurysms, which were found in 33 patients, represent 7% of this series. Four cases had multiple aneurysms, which were treated during the same surgical procedure.

The male/female ratio was 18:15. The patients were aged between 29 and 59 years, the median age being 46.5 years.

In 14 patients the giant aneurysm presented as an intracranial mass with visual field deficits (5 cases), seizures (4 cases), hemiparesis (2 cases), oculomotor nerves impairment (2 cases), trigeminal neuralgia (1 case). Subarachnoid haemorrhage (SAH) was the presenting symptom in 19 cases: 10 patients were admitted in grade I–II (Hunt and Hess grading), 6 patients in grade III, 3 more patients in grade IV. One patient rebled while in grade II and was operated on in grade IV. About one half of the SAH group also showed objective

Table 1. *Location of 33 Anterior Circulation Giant Aneurysms Related to Symptomatology at Admittance*

Aneurysm location	S.A.H.	No S.A.H.
Cavernous carotid	0	4
Carotid-ophthalmic	1	3
Carotid-post. comm.	5	0
Carotid bifurcation	3	0
Ant. cer./ant. comm. art.	3	1
Middle cer. art.	7	6
Totals	19	14

signs of compression of intracranial structures due to the bulk of the aneurysm (hemiparesis in 5 cases, visual impairment in 3 patients, psycho-organic syndrome in one case).

Table 1 shows the site of the aneurysms related to their recent story.

Until 1977, in our Institution, the diagnosis of giant aneurysm was made after angiography or directly at operation; after that date, when CT scan became available, a sure preoperative diagnosis was possible and adjunctive information were moreover available, like the presence and extent of SAH and satellite haematoma, the precise localization of the mass, the detection of any intramural thrombus.

All patients were operated upon directly on the aneurysm; surgery was performed adopting microsurgical technique under general anaesthesia, normothermia and slight hypocapnia; at the onset of surgery all patients received 150 mg/kg of Mannitol and 8 mg of Dexamethasone. Giant cavernous carotid aneurysms (aneurysms of the supraclinoid and subclinoid internal carotid artery) were treated by intramural thrombosis according to Mullan[8] and Hosobuchi[5] technique, also performing intraoperative serial angiography.

Giant aneurysms in other sites were approached with the intention of obliterating their neck. The exposure of cervical carotid artery and the resection of the anterior clinoid process were utilized in cases of carotid-ophthalmic and paraclinoid aneurysms.

The neck of the aneurysms was directly clipped in 24 cases, in 6 of them the sac was removed; wrapping of the sac in 1 case and aneurysmorraphy after temporary clipping of the parent vessel in another were also performed. The aneurysm was trapped in 3 patients; one of them was managed by combining malformation entrapment with extracranial to intracranial microvascular by-pass procedure.

Controlled systemic arterial hypotension was used only at the time dissection of the neck, while during or following the application of temporary clips or after permanent trapping the blood pressure was raised to levels normal for the patient, or even 10–20 mm Hg higher.

Systemic hypotension was induced by intravenously injected sodium-nitroprusside or by increasing anaesthetic regimen. In no case mean arterial pressure (MAP) was lowered below 65 mm Hg, and occlusion time never prolonged for more than 50′.

Results

Operative mortality was 12% (4 deaths).

Two patients, operated upon in poor conditions, underwent temporary trapping of the aneurysm be-

came of unexpected rupture during isolation manœuvres. Direct clipping of the neck was at last obtained, but interruption of circulation in the parent vessel (carotid siphon) lasted 30′ and 45′ respectively. Neurological status worsened rapidly until death a few days after the operation. The third patient harboured a carotid ophthalmic aneurysm, about 5 cm in diameter. In this case also the aneurysm ruptured during isolation and it was necessary to ligate the parent vessel: the patient died on the second postoperative day because of massive brain infarction. The 4th patient who died was a 58-year-old woman who had experienced rapid visual impairment caused by a huge aneurysm of the right carotid siphon. She underwent a by-pass procedure together with a staged ligation of cervical carotid artery by a Selverstone clamp, which was advanced to total occlusion over a period of 4 days. Angiograms performed during the last phase of this procedure demonstrated the patency of the by-pass. The aneurysm was attacked the day after. The broadness of the aneurysmal neck compelled us to place the intracranial clip distal to the origin of the posterior communicating artery. The patient recovered from anaesthesia but her clinical status rapidly worsened thereafter because of ischaemia of the right hemisphere, confirmed by CT scan. The lady died on the 6th postoperative day.

The survivors were periodically controlled as outpatients. Follow-up ranged from 1 to 9 years (mean: 3 years). 14 patients fully recovered (good results); 9 patients, with minor neurological impairment, were able to return to normal activities of daily living (fair results). One patient of this group survived 2 years after the operation, but than was killed by massive bleeding form an unrecognized berry aneurysm of the posterior circulation. Poor results were achieved in the last group of 6 patients. In 3 of them intramural thrombosis of the aneurysms was attempted: the procedure did not succeed in improving preoperative complaints, on the contrary, additional neurological deficits were evident both at discharge and at follow-up. At this point, we must acknowledge that our experience in intramural thrombosis, even if small, has not been satisfactory. As mentioned above, we treated 4 patients by this technique: from a clinical point of view, only one case had good result (improvement of declining visual acuity); angiographically, the aneurysm was completely thrombosed only in one patient in whom the internal carotid artery also became occluded. In the others, angiographic follow-up revealed a slight increase in the residual lumen of the aneurysm as compared with the early postoperative angiogram.

Discussion

The natural history of unoperated extracavernous giant aneurysms is poor[4, 7, 11], since 4/5 of patients die within a few years after discovery of the aneurysm, either from rupture or from progressive enlargement. SAH was considered a rare complication of giant aneurysms[6], since calcified walls and laminated thrombi within the sac appeared to behave like a " shield" against haemorrhage. In world series, SAH is found with variable incidence (from 81 to 31%)[1, 5, 9, 10, 12, 13], due to often misdiagnosed previous bleeding symptoms and also to the fact that the mass effect of the aneurysm presents opportunities for early diagnosis. In our series, 60% presented with SAH, and probably 2 more cases of the "no SAH" group had had previous leaks.

Despite some reports of spontaneous decrease in size or even disappearance of giant aneurysms[3, 7], we believe that most of them will continue to enlarge with time and do rupture. Therefore radical surgery is strongly recommended whenever possible. Even intracavernous giant aneurysms may bleed, as anecdotally reported by the literature[7], and as happend to one patient of our series while awaiting operation.

Drake[1] stated that about 1/3 of giant aneurysms may be clipped at their neck; this proportion can be increased, at least for the ones of the anterior circulation (e.g., 24/33 cases of our series), by resorting to technical expedients such as suitable clips, temporary major vessel occlusion under intraoperative monitoring of evoked potentials and brain protection, resection of anterior clinoid process, fragmentation of mural thrombus by ultrasonic aspiration, creating of a smaller neck with bipolar coagulation, closure of a tourniquet in an awake patient, etc, etc ... In our hands, some of these procedures were successful in isolating and then closing the aneurysmal neck in selected cases where the neck itself was not appreciable both in the angiograms and at initial operative exploration.

Temporary clipping of parent vessel is a rather risky procedure; it was adopted in 6 of our patients. Two of them died, as above reported, and another patient had a poor outcome because of infarction in the territory of the middle cerebral artery. In 3 more cases temporary clipping was well tolerated, in one patient removal of the sac and aneurismorraphy were performed meanwhile, the parent vessel being patent at postoperative angiography.

Intraoperative monitoring now available will enable quite prolonged occlusion to be safely carried out[13]. Whenever it is impossible to expose and occlude the neck of a giant aneurysm, alternative procedures such as permanent trapping and intramural thrombosis may be proposed. The former, like the classic cervical carotid ligation according to the Hunterian principle, harbours high risk of early or delayed cerebral ischaemia, even if some protective measures like EIAB, intraoperative monitoring of evoked potentials and cerebral blood flow, and pharmacological "cocktails" are adopted. In our series, total entrapment for the aneurysm was performed in 3 cases, two of them died soon after operation from brain infarction, even though one of them was previously treated with a bypass procedure. We found other cases[2, 5] in which a patent bypass failed to protect the brain from ischaemia after major vessel occlusion: therefore, in every patient it is mandatory to single out spontaneous collateral channels that should be saved during operation. The results after attempted intraluminal thrombosis are often neither clinically satisfactory nor angiographically complete, as happened in our cases and in those of other authors, who reported both rebleeding[5] and further enlargement of the sac[1]. We agree with Hosobuchi's statement that intramural thrombosis is best for giant aneurysms of the posterior circulation[5].

Once the giant aneurysm is excluded from blood stream, it is necessary to complete the surgical procedure by eliminating the mass effect due to its inherent bulk. We believe it is not always wise to remove the entire sac: it is often enough to empty its contents (blood, thrombi), removing its wall may prove to be very dangerous and not rewarding. Nevertheless, we had no problems in our cases (6 patients) of sac extirpation, on the contrary it happened that in 2 of 3 patients who were cured of preoperative seizures the aneurysmal sac had been radically removed.

References

1. Drake CG (1979) Giant intracranial aneurysms: experience with surgical treatment in 174 patients. Clin Neurosurg 26: 12–95
2. Gelber BR, Sundt TM Jr (1980) Treatment of intracavernous and giant carotid aneurysms by combined internal carotid ligation and extra- to intracranial bypass. J Neurosurg 52: 1–10
3. Carlson DH, Thomson D (1976) Spontaneous thrombosis of a giant cerebral aneurysm in five days. Report of a case. Neurology 26: 334–336
4. Heiskanen O, Nikki P (1962) Large intracranial aneurysms. Acta Neurol Scand 38: 195–208
5. Hosobuchi Y (1979) Direct surgical treatment of giant intracranial aneurysms. J Neurosurg 51: 743–756
6. Jain KK (1965) Surgery of intracranial berry aneurysms. A review. Can J Surg 8: 172–187
7. Morley TP, Barr HWK (1969) Giant intracranial aneurysms: diagnosis, course, and management. Clin Neurosurg 16: 73–94

8. Mullan S (1974) Experiences with surgical thrombosis of intracranial berry aneurysms and carotid cavernous fistulas. J Neurosurg 41: 657–670

9. Onuma T, Suzuki J (1979) Surgical treatment of giant intracranial aneurysms. J Neurosurg 51: 33–36

10. Sindou M, Keravel Y (1984) Les anévrysmes géants intracrâniens. Approches thérapeutiques. Neurochirurgie 30 [Suppl] 1

11. Sonntag VKH, Yuan RH, Stein BM (1977) Giant intracranial aneurysms: a review of 13 cases. Surg Neurol 8: 81–84

12. Sundt TM Jr, Piepgras DG (1979) Surgical approach to giant intracranial aneurysms. Operative experience with 80 cases. J Neurosurg 51: 731–742

13. Symon L, Vajda J (1984) Surgical experience with giant intracranial aneurysms. J Neurosurg 61: 1009–1028

Address for correspondence: S. Giombini, M.D., Div. di Neurochirurgia, Istituto Neurologico "C. Besta", via Celoria 11, I-20133 Milano, Italy.

Acta Neurochirurgica, Suppl. 42, 75–80 (1988)

The Microvascular Doppler – an Intraoperative Tool for the Treatment of Large and Giant Aneurysms

G. Laborde, J. Gilsbach, and A. Harders

Department of Neurosurgery, University of Freiburg, Federal Republic of Germany

Summary

One of the problems especially associated with large and giant aneurysms is the control of the patency of the parent artery and the exclusion of the aneurysm. While the exclusion can be tested by puncture (which may sometimes be problematic), the patency can only be controlled by intraoperative angiography or recently, by microvascular intraoperative Doppler. The device we use has a high resolution and is equipped with probes as small as 0.3 mm with which all visible vessels with diameters of more than 0.1 mm can be investigated. Local stenoses above a diameter reduction of 40% can be easily detected by localized accelerations and changes in the pulse curves. Our recent experience with 11 giant and 13 large aneurysms has revealed marked discrepancies between an apparently well placed clip and an obviously open vessel on the one hand, and the haemodynamic reality revealed by the Doppler on the other. We were able to decide whether the flow was undisturbed, or whether there was an haemodynamically non-effective lumen reduction due to a tight clip, or if there was severe stenosis or a total occlusion. In five cases with severely disturbed flow after clipping, we had to change our strategy: we resected the aneurysms and sutured the neck, performed an end-to-end anastomosis, coated the aneurysm, or repositioned the clips. In cases in which we did not have full view of the neck, the Doppler guided us to a proper clip position.

We conclude that especially for large and giant aneurysms, the microvascular interoperative Doppler is a valuable tool.

Keywords: Microvascular Doppler; intracranial aneurysm; giant aneurysm; intraoperative Doppler.

Introduction

One of the problems in the surgery of aneurysms, especially in large and giant ones, is proving the patency of the parent artery and the exclusion of the aneurysm. While the exclusion can be tested by puncture (which sometimes can present problems like bleeding), the patency can only be controlled intraoperatively by angiography[9, 11, 12, 15, 16] or, for a few years now, with the aid of a microvascular Doppler system[4, 5, 6] initially instigated by Nornes[13]. The Doppler method is simple (and inexpensive), and can be applied atraumatically, directly, and repeatedly on the vessels operated on. For these reasons in our institution we prefer it to angiography, which is expensive, invasive, and not easy to perform.

The usefulness of the microvascular Doppler has already been demonstrated in aneurysm surgery in general[4, 6]. In the following investigation our objective was to judge the significance of the method in the treatment of large and giant aneurysms.

Patients and Methods

Between 1982 and 1986, 22 consecutive patients with 24 aneurysms underwent intracranial operation on 11 giant ($\geqslant 2,5$ cm) and 13 large ($1-2,5$ cm) aneurysms (Table 1). Thirteen of the patients had a history of aneurysm rupture, eight of them had been operated on in the acute stage. Nine other patients had never experienced rupture and presented symptoms of a space-occupying effect of the lesion or of embolic complications. All of the patients underwent preoperative four vessel angiography and CT. In selected cases postoperative control angiography was also performed. In most cases for haemodynamic question we relied only on the intraoperative microvascular[4, 6], and postoperative transcranial[7], Doppler investigation.

The intraoperative recordings were performed with microvascular Doppler equipment consisting of a 20 MHz pulsed Doppler system including a built-in real time spectrum analyzer and of miniaturized probes with a diameter of 0.3 and 1 mm which could be sterilized by gas*. The recordings were performed after the parent arteries had been dissected for the purpose of obtaining basic values. After the various clipping steps had been performed, repeated controls were carried out. We looked for occlusions and local narrowings detectable by increased flow velocities, and for haemodynamically significant stenoses characterized by a reduced proximal and distal flow velocity and altered flow patterns.

---

* Distributed by EME, Ueberlingen, West Germany.

Table 1. *Synopsis of the Doppler and Clinical Findings in 22 Patients with Large and Giant Aneurysms*

Patient	Aneurysm	Diameter	Preop. (H & H)	Outcome (GOS)	Discrepancy outer aspect Doppler	Finally undisturbed patency	Doppler induced measures	Comments
B. K. 50 m*	MCA	2.5	IV	5	stenosis	—	frustrane clip repositioning	—
B. E. 62 m	ACoA	1.5	Ia	2	—	+	—	—
B. H. 44 m	MCA	3.0	0	1	—	+	—	—
B. W. 37 m*	MCA	4.0	III	3	MCA-branch occlusion	—	frustrane neck resection, frustrane clip repositioning	distal vessel occlusion despite open neck region (non reflow phenomenon?)
C. M. 39 m	Opth MCA	1.8 0.5	0	1	— —	+ +	—	—
E. G. 48 m*	MCA	2.0	IV	2	MCA-branch occlusion	+	successful clip repositioning	new deficit due to prolonged temporary clipping or embolism
F. G. 35 m	A2	2.0	0	1	—	—	—	stenosis visible, clip removal, coating
G. G. 42 m*	MCA A2	1.5 2.0	II	5	A2 occlusion	+ —	frustrane suture revision	neck suture due to unclippable aneurysm, distal vessel occlusion, despite open anastomosis (non reflow phenomenon?)
H. C. 43 m*	Opth ACoA	1.8 0.6	III	1	— —	+ +	—	—
H. H. 76 f*	ACoP	2.0	III	3	—	+	—	capsular infarction (A. ChorA), due to prolonged temporary clipping
K. P. 39 m	Opht	2.0	II	1	—	—	—	visible stenosis without haemodynamic effect, no clip repositioning
K. M. 61 f	BA	2.8	IV	3	—	—	—	confirmation of plenned occlusion and of collateral flow
R. A. 55 f	PICA	2.5	Ia	2	—	+	—	unclippable neck, aneurysm resection, end-to-end suture
S. K. 34 m	ACoA	1.5	0	3	—	+	—	—
S. W. 42 m	ACoP	3.0	0	1	ICA stenosis	+	no clip repositioning stenosis not effective	—
S. I. 41 f*	Opht	3.0	IV	5	—	+	—	incomplete aneurysm occlusion visible, corrected; despite patent vessels (TCD), hemiparesis due to prolonged temp. clipping
S. E. 20 m*	MCA	2.6	IV	1	MCA-branch occlusion	+	aneurysm resection, end-to-end suture, successful suture revision	unclippable neck, suture at first occluded despite normal outer appearance
S. R. 63 m	ACoA	2.6	0	1	—	+	—	—
V. H. 46 m	MCA A2	4.0 1.5	Ia	3	A 2 stenosis	+	sucessful clip repositioning	unclippable aneurysm, STA-MCA anastomosis
S. G. 69 f	ACoA	1.5	0	3	incomplete occlusion	+	successful clip repositioning	—
W. W. 35 m	ACoA	3.0	Ia	3	—	—	frustrane anastomosis, frustrane revision	unclippable neck, aneurysm resection, parent artery occlusion
V. M. 64 f	ICA-Bif	1.8	0	1	—	+	—	—

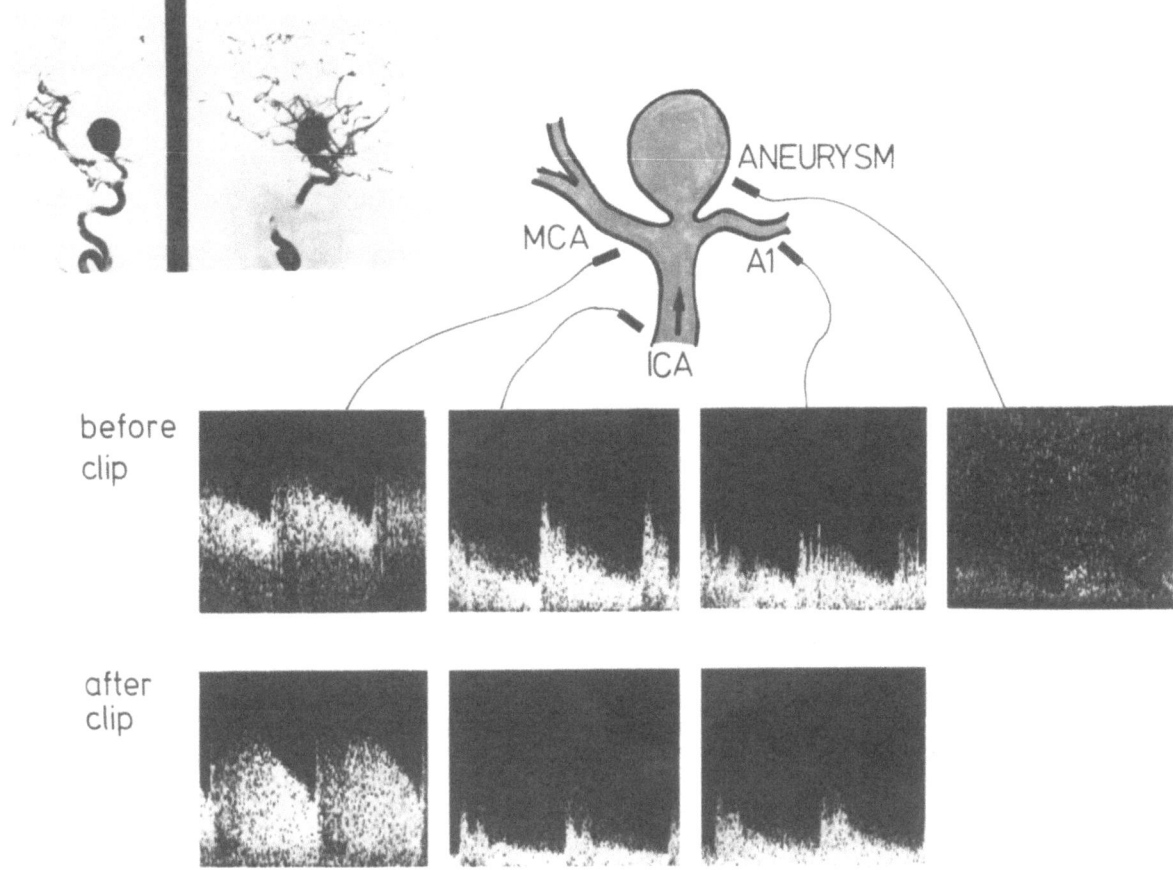

Fig. 1. 44-year-old man with visual disturbances and chronic headache caused by an aneurysm of the internal carotid artery (ICA) bifurcation. After clipping, velocities were not significantly changed in the middle cerebral artery (MCA), in the ICA, and in the anterior cerebral artery (ACA) as a sign of a satisfactory clip position with an undisturbed flow

Results

All the vessels which were associated with the aneurysms could be studied. In 12 aneurysms the clip appeared to be positioned correctly. The Doppler confirmed undisturbed flow in the parent arteries and that the aneurysm was occluded (Table 1, Fig. 1).

In four aneurysms the outer aspect of the clip position was not satisfactory and accordingly, the Doppler findings indicated a disturbed patency. In three of these four cases, however, we were unable to improve the haemodynamic situation (severe stenosis or occlusion during probatory clipping) because the neck could not be clipped any better. In one middle cerebral artery aneurysm, we performed a STA-MCA anastomosis. One distal anterior aneurysm was coated and in one anterior communicating aneurysm the parent arteries were occluded. In the last of the four cases the Doppler showed that the narrowing was only moderate with no haemodynamic effect and needed no clip correction. This patient showed only localized acceleration as a sign of stenosis; the distal flow velocities were not significantly reduced (Fig. 2).

In one patient, the aneurysm seemed to be occluded, but the Doppler revealed a residual flow within the sac (Fig. 3). In this patient the clip had to be repositioned.

In eight aneurysms the Doppler revealed discrepancies between the surgeon's view of the situation and the haemodynamic reality. The flow disturbances could not be detected from outside and could only be confirmed by the Doppler. In four out of these eight aneurysms, the information provided by the Doppler signal enabled us to improve the local haemodynamics by repositioning the clip (three times) or suturing the aneurysm neck (one case). In three further cases, we were technically not able to restore normal flow conditions. In one last case, the Doppler revealed flow disturbances which were haemodynamically unimportant and needed no correction.

In the four cases in which anastomosis of the parent artery or a suture of the resected aneurysm neck had to be performed, the Doppler revealed an initial oc-

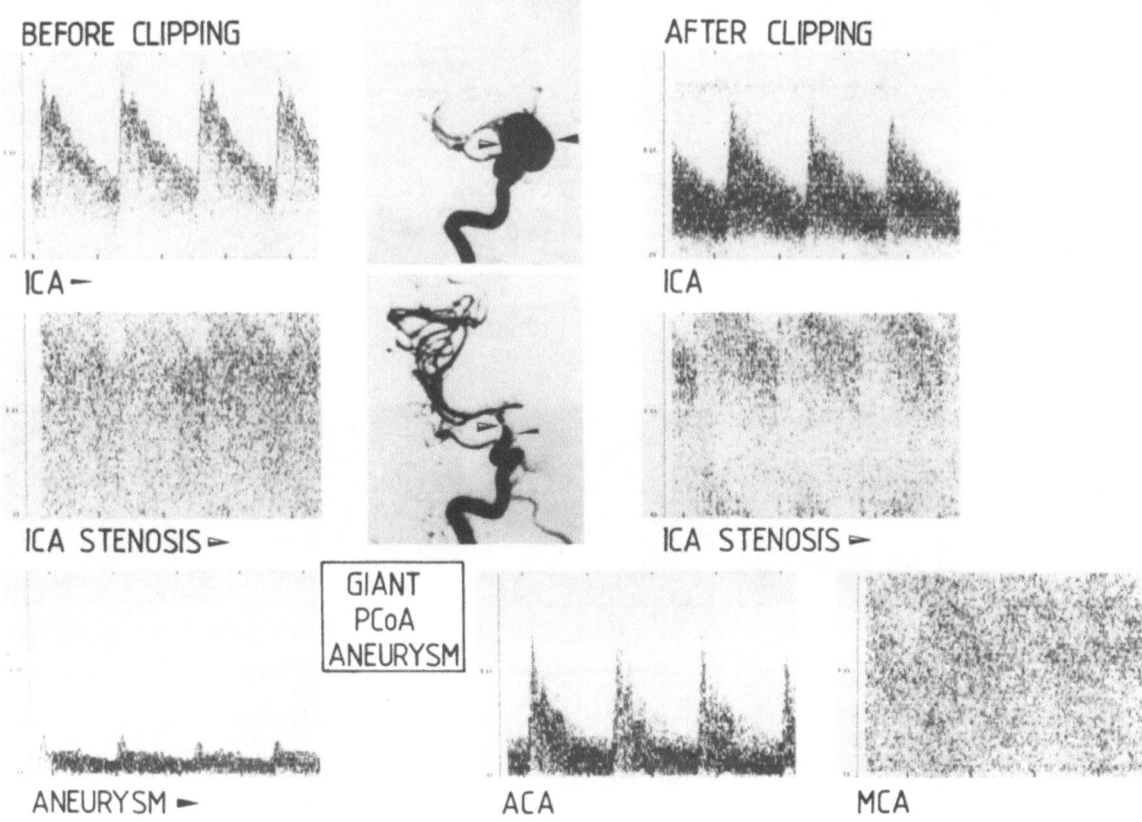

Fig. 2. 42-year-old man with a visual deficit, chronic headache caused by a posterior communicating artery (PCoA) aneurysm. The Doppler revealed a pre-existing stenosis with localized acceleration of the ICA. After clipping, the velocities had increased as a sign of further lumen reduction. The anterior cerebral artery flow velocities remained in the normal range as were the carotid flow pattern. Therefore, the clip could remain in place with a moderate stenosis

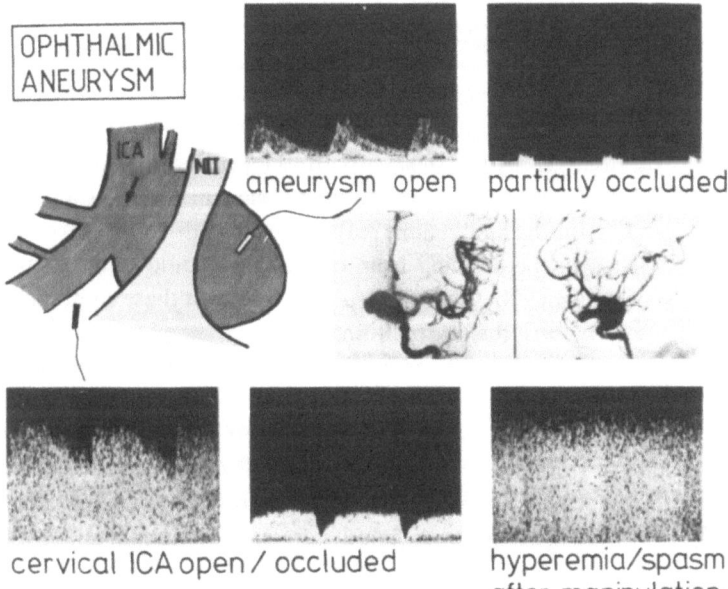

Fig. 3. 41-year-old woman with an ophthalmic aneurysm, operated on in grade III, 48 hours after the rupture. Typical slow flow velocities within the aneurysm with superimposed pulse curves[6]. Aneurysm neck dissection during occlusion of the cervical part of the ICA. After positioning of the first clip the Doppler confirmed an incomplete occlusion of the aneurysm neck. Hyperaemic reaction after removal of the temporary ligature of the ICA

clusion of the vessel three times, which could only be corrected once. After continuous milking of the end to end anastomosis, we were able to restore the flow under sonographic control. In one case, after resection of the aneurysm and repeated clipping of the neck, the vessel remained occluded. The same happened in two cases with neck resection and suture. In these three occluded vessels a non-reflow phenomenon could be observed and the absence of local narrowings in the region of the suture or of the clipped neck; extensive manipulation under temporary clipping had to be performed in these three vessels.

In no case did the intraoperative Doppler reveal normal flow patterns when postoperative transcranial Doppler or control angiography revealed stenosis or occlusion.

Despite satisfactory Doppler findings, 4 patients presented new neurological deficits postoperatively. In these patients, neither the postoperative control angiography nor the transcranial Doppler showed any occlusions or severe stenosis. A re-evaluation of the operative reports and the CT scan revealed that these patients suffered from ischaemic deficits after prolonged temporary clipping.

The patients in whom effective intraoperative corrections were possible had no additional postoperative deficits.

Discussion

Intraoperative Doppler sonography has proved to be a simple, atraumatic, and reliable tool in aneurysm surgery for checking the patency after clipping[4, 5, 6, 13]. Local narrowings and occlusions can be detected with a high degree of accuracy[4, 6, 13]. The Doppler technique picks up local flow disturbances better than intraoperative angiography, which only provides better morphological information on residual aneurysmal necks, for example, and on the collateral circulation. The angiogram offers no physiological information on the haemodynamic effect of a local narrowing. Furthermore velocity is a more reliable indicator of vessel narrowing, because it reflects to the second power the lumen diameter reduction. Local narrowings from a 40% cross-sectional reduction onwards can be detected by increased flow velocities[6].

Operation on giant and large aneurysms bears a (high) risk of inadvertent occlusion of parent arteries[1, 2, 3, 10, 20, 21, 23]. In only 12 out of our 24 aneurysms did a normal outer aspect and the surgeon's impression correspond to a normal haemodynamic situation de-

tected by the Doppler method. In eight cases the satisfactory outer appearance turned out to be deceptive, as the Doppler revealed a disturbed patency. In four cases the Doppler results caused corrections to be made which were effective and in one case they eliminated an uncertainty. In six patients the Doppler indicated disturbed patency which could not be corrected for technical reasons. The possibility of identifying an aneurysm by its typical flow pattern played no major role during operation[5, 6].

Our results are not better than those of other large series of operations on giant and large aneurysms published in the literature[1, 2, 3, 8, 10, 14, 17, 18, 19, 20, 21, 23]. We can therefore suppose that personal skill and experience play an important role in achieving undistributed patency after aneurysm surgery. However, intraoperative Doppler is able to increase the safety of the procedure, especially if less experienced surgeons are operating on large and giant aneurysms. It replaces traumatic tests of patency such as the double forceps test, and reduces the surprises associated with the puncture of the aneurysmal sack. It also makes it possible to test different clip positions, rules out haemodynamic disturbances – sometimes when the position does not look perfect, and eliminates uncertainties. Optimal clip positioning should also reduce the risk of stenoses which decompensate secondarily, for instance, after hypotension or due to additional vasospasm.

We cannot prove that any individual patient has benefited from a clip repositioning or a Doppler induced suture, but evidence of undisturbed flow is more reassuring than merely trusting in a good collateral circulation.

References

1. Clark K (1976) Complications of aneurysm surgery. Clin Neurosurg 23: 342–356
2. Creissard P (1980) Table ronde de la réunion de la société française de neurochirurgie. Paris, Octobre 1979. Les anévrysmes géants. Neurochirurgie 26: 309–331
3. Drake CG (1979) Giant intracranial aneurysms: Experience with surgical treatment in 174 patients. Clin Neurosurg 26: 12–95
4. Gilsbach JM, Harders A, Mohadjer M (1987) The microvascular Doppler – an intraoperative tool for aneurysm surgery. In: Advances in Neurosurgery, vol 16 (in press)
5. Gilsbach JM, Harders AG (1987) Microvascular and transcranial Doppler sonographic evcaluation of cerebral aneurysm flow pattern. Neurol Res (in press)
6. Gilsbach J (1983) Intraoperative Doppler sonography in neurosurgery. Springer, Wien New York
7. Harders A (1986) Neurosurgical applications of transcranial Doppler sonography. Springer, Wien New York

8. Heros RC, Nelson PB, Ojemann RG, Crowell RM, DeBrun G (1983) Large and giant paraclinoid aneurysms: surgical techniques, complications, and results. Neurosurgery 12: 153–163

9. Hillmann J, Johansson I (1987) Peroperative angiography – a useful tool in cerebral aneurysm surgery? Acta Neurochir (Wien) 84: 39–42

10. Hosobuchi Y (1979) Direct surgical treatment of giant intracranial aneurysms. J Neurosurg 51: 743–756

11. Lazar ML, Watts CC, Kilgore B, Clark K (1971) Cerebral angiography during operation for intracranial aneurysms and arteriovenous malformations. J Neurosurg 34: 706–708

12. Loop JW, Foltz EL (1966) Applications of angiography during intracranial operations. Acta Radiol (Diagn) 5: 363–367

13. Nornes H, Grip A, Wikeby P (1979) Intraoperative evaluation of cerebral hemodynamics using directional Doppler technique. J Neurosurg 50: 570–577

14. Onuma T, Suzuki J (1979) Surgical treatment of giant intracranial aneurysms. J Neurosurg 51: 33–36

15. Parkinson D, Legal J, Holloway AF *et al* (1978) A new combined neurosurgical head holder and cassette changer for intraoperative serial angiography. Technical note. J Neurosurg 48: 1038–1041

16. Parkinson D (1969) Rapid serial simultaneous biplane stereoscopic angiography: an aid in the surgical management of cerebral arteriovenous malformations. Clin Neurosurg 16: 179–184

17. Silverberg GD (1984) Giant aneurysms: surgical treatment. Neurol Res 6: 57–63

18. Sindou M, Keravel Y (1984) Intracranial giant aneurysms. Therapeutic approaches. Neurochirurgie 30

19. Spetzler RF, Selman W, Carter LP (1984) Elective EC-IC bypass for unclippable intracranial aneurysms. Neurol Res 6: 64–68

20. Sundt TM Jr, Piepgras DG (1979) Surgical approach to giant intracranial aneurysms. Operative experience with 80 cases. J Neurosurg 51: 731–742

21. Symon L, Vajda J (1984) Surgical experiences with giant intracranial aneurysms. J Neurosurg 61: 1009–1028

22. Turner JW, Crossant KW (1972) Angiography during surgery for aneurysm. In: Fusch I, Kunc Z (eds) Present limits in neurological surgery. Elsevier, Amsterdam, pp 281–286

23. Yaşargil MG (1984) Microneurosurgery, vol II. Thieme, Stuttgart New York

Address for correspondence: Dr. G. Laborde, Department of Neurosurgery, University of Freiburg, D-7800 Freiburg, Federal Republic of Germany.

Acta Neurochirurgica, Suppl. 42, 81–84 (1988)
© by Springer-Verlag 1988

Cerebral Vasospasm After Subarachnoid Haemorrhage Investigated by Means of Transcranial Doppler Ultrasound

K.-F. Lindegaard, H. Nornes, S. J. Bakke, W. Sorteberg, and P. Nakstad

Department of Neurosurgery and Department of Radiology (Section of Neuroradiology), Rikshospitalet, The National Hospital, University of Oslo, Oslo, Norway

Summary

Measurements of flow velocity in defined segments of the basal cerebral arteries can be obtained through the intact adult skull using 2 MHz pulsed Doppler ultrasound. We compared flow velocity in these vessels with findings from 56 cerebral angiographies obtained in 51 patients at from day 1 to day 21 after subarachnoid haemorrhage (SAH). The diameter of the proximal segment of the middle cerebral, anterior cerebral, and posterior cerebral arteries (MCA, ACA, and PCA, respectively) were measured from anteroposterior films produced in one angiographic laboratory. In patients investigated on day 1–2, the median MCA diameter was 2.8 mm with range 2.3–3.4 mm. The median flow velocity was 56 cm/s, range 36–88 cm/s (within normal limits). There was a clear inverse relationship between the MCA diameter and MCA flow velocity. Eleven of the 13 MCA's having diameter 1.5 mm or less showed flow velocity in excess of 140 cm/s. This seems a useful limit to diagnose pronounced MCA spasm (50% diameter reduction) with this method. Further clues to the severity of MCA spasm were obtained from the ratio calculated dividing the MCA flow velocity by the flow velocity in the ipsilateral, extracranial internal carotid artery (ICA), since spasm probably does not involve the neck vessels. This ratio was from 1.1 to 2.3, median 1.7 at day 1–2, but rose to over 10 in patients with the most severe MCA lumen narrowing.

The PCA flow velocity was inversely related to the PCA diameter.

Assessment of ACA spasm requires considering findings from both hemispheres combined, since the two proximal ACA's usually anastomose through the anterior communicating artery. Therefore, and ACA spasm or developmental anomaly notwithstanding, the larger ACA showed the higher flow velocity in 17 of the 22 patients with a clear asymmetry of the circle of Willis.

Keywords: Cerebral arteries; cerebral vasospasm; subarachnoid haemorrhage; transcranial Doppler ultrasound.

Introduction

The lack of an objective and clinically acceptable method to follow the course of vasospasm explains some of the uncertainty concerning the clinical significance of cerebral vasospasm in patients who survive the acute subarachnoid haemorrhage (SAH). The introduction of the noninvasive transcranial Doppler (TCD) technique[3, 9, 10, 11] may change this situation. In the first report on patients with SAH investigated by means of this method, Aaslid *et al.* demonstrated blood velocities increased to over two times the normal in vessels with angiographical evidence of vasospasm[1]. Being able to detect asymptomatic vasospasm and to predict if and when these patients will go on to developing symptomatic ischemia would be of great clinical value. The present study further analyses angiographic and blood velocity data in order to assess the TCD method for evaluating patients with SAH.

Material and Methods

Patients

A total of 51 patients with confirmed SAH, and aged between 14 and 68 years, were investigated. Angiography was performed in the same angiography laboratory at from 12 hours (day 1) to 13 days (day 14) after the bleed; and the TCD investigation was performed not more than one hour before angiography. Repeat angiography was performed at from one to 12 days after the first in five patients. Thus, a total of 112 carotid and 34 vertebral angiography series were available for analysis.

Blood Velocity Measurements

The 2 MHz pulsed wave Doppler instrument had acoustic focusing and real-time spectrum analysis (laboratory prototype and production model TC 64-2, made by Eden Medizinische Elektronik GmbH, Überlingen, Federal Republic of Germany). The working principles and the procedure for clinical investigation of blood velocity in the middle, anterior, and posterior cerebral arteries, and in the distal extracranial segment of the internal carotid artery (denoted MCA, ACA, PCA, and ICA, respectively) have been outlined elsewhere[3, 9]. Following the identification of each specific basal cerebral artery, the probe position and the sampling distance was adjusted to record

Table 1. *MCA Diameter and Blood Velocity Correlated with Day After SAH**

	No. vessels	Vmca (cm/s) median (range)	Dmca (mm) median (range)	Vmca/Vica median (range)
Day 1–2	28	58 (38–88)	2.8 (2.3–3.4)	1.7 (1.1–2.3)
Day 3–4	22	64 (44–164)	2.6 (1.7–3.2)	1.8 (1.0–4.3)
Day 5–7	28	109 (34–218)	2.0 (1.2–3.0)	2.7 (1.2–5.9)
Day 8–14	28	108 (40–202)	2.1 (0.8–2.9)	3.1 (1.2–10.2)
Day 15–21	6	93 (68–206)	2.2 (0.6–3.0)	2.6 (1.8–11.4)

* MCA: Middle cerebral artery; SAH: Subarachnoid haemorrhage.
Vmca: MCA blood velocity.
Dmca: Diameter of proximal MCA.
Vmca/Vica: Vmca divided by the blood velocity in the distal extracranial internal carotid artery (ICA) on the same side.

from the vessel segment showing the highest blood velocity. The annotated blood velocity was the time-mean value of the velocity spectrum outline from 10 consecutive cardiac cycles.

Cerebral Angiography

Using the femoral artery route and local anaesthesia, a standard bolus of 10 ml iohexol with 300 mg iodine/ml (Omnipaque 300, made by Nycomed, Oslo, Norway) was injected at a rate of 7 ml/s for carotid artery studies. The standard bolus for vertebral artery studies was 7 ml injected at a 5 ml/s. The exposure rate was about one frame per second. A magnification ratio of 1.3 to 1 for the circle of Willis in the antero-posterior projection was determined from phantom studies. The diameter of the proximal portions of the MCA, ACA, and PCA was measured at the point showing the narrowest lumen, using a 0.1 mm reticle and 4 × magnification. The results were divided by 1.3.

Results

Middle Cerebral Artery

When considering data from all patients combined there was a highly significant inverse correlation between the MCA blood velocity Vmca) and the MCA diameter ($r = -0.905$, $p \ll 0.01$). In the 14 patients investigated on day 1 or day 2, the MCA diameter was from 2.3 to 3.4 mm, median 2.8 mm. The Vmca was between 38 and 88 cm/s, median 58 cm/s (Table 1). These findings are within previously established normal limits[3, 5, 9], and were regarded as representative samples from a patient population with recent SAH and no vasospasm. The blood velocities were higher by the end of the first week and throughout the second week after SAH. In the same period, decreased MCA diameter indicating MCA spasm was present in the majority of patients (Table 1).

The Vmca was ⩾ 110 cm/s in 30 of the 35 MCA's showing a diameter ⩽ 2.1 mm (MCA lumen diameter supposedly narrowed by 25% or more); hence, sensitivity was 0.85 with specificity 0.98 using

Vmca = 110 cm/s as the lower cut-off limit to predict the presence of MCA spasm (⩾ 25% MCA diameter narrowing). With a prevalence of MCA spasm of 35/112, the positive predictive value (probability of disease given a positive test) was 0.98. Changing the lower cut-off limit to 100 cm/s increased sensitivity to 0.94, while reducing specificity and positive predictive value to 0.90 and 0.86, respectively.

The Vmca was ⩾ 140 cm/s in 11 of the 13 MCA's having a diameter ⩽ 1.5 mm (MCA lumen diameter supposedly narrowed by 50% or more); nonetheless, Vmca was ⩾ 140 cm/s also in 10 of the 22 MCA's with diameter between 1.5 and 2.1 mm (from 25 to 50% MCA diameter narrowing or mild spasm). This cut-off limit had sensitivity 0.84, while the specificity in differentiating between mild and pronounced MCA spasm was only 0.54.

The ratio calculated by dividing Vmca with the blood velocity in the ipsilateral extracranial ICA provided further clues to the severity of MCA spasm. This dimensionless index, denoted Vmca/Vica, was between 1.1 and 2.3, median 1.7, at day 1–2 (Table 1). The Vmca/Vica index showed a peak in the second week after SAH (Table 1). The Vmca/Vica ratio was ⩾ 3.0 in 31 of the 35 hemispheres showing MCA ⩽ 2.1 mm. Thus, this index showed a sensitivity of 0.88 for predicting MCA spasm. The specificity and the positive predictive value were both 0.98, since the Vmca/Vica index was greater than 3.0 in just one hemisphere with MCA diameter > 2.1 mm.

The Vmca/Vica index exceeded 10 in the two patients with MCA diameter reduction to less than 1.0 mm; being ⩾ 5.0 in 10 of the 13 hemisphers showing an MCA diameter ⩽ 1.5 mm, and in two hemispheres with MCA diameter between 1.5 and 2.1 mm. Thus, with regard to differentiating between mild and pronounced MCA spasm, sensitivity was 0.77 with specificity 0.90. Changing the cut-off limit to 5.5 eliminated the false positive findings at the cost of increasing false negative findings from three to five, thereby giving a 1.0 specificity but reducing sensitivity to 0.61.

The Circle of Willis

The blood velocity in the PCA was inversely related to the PCA diameter ($r = -0.671$, $p < 0.01$). Analysis of the relationship between ACA diameter and ACA blood velocity showed $r = -0.247$ ($0.02 < p < 0.05$). With regard to the ACA combinding findings from both hemispheres is required, since the proximal ACA's usually anastomose via the anterior communicating artery. Spasm or developmental anomaly in this unique artery system notwithstanding, the larger of the two ACA's showed the higher blood velocity in 17 of the 22 patients angiographically demonstrating a clear asymmetry of the anterior circle of Willis.

Discussion

Interpreting the data obtained from investigating SAH patients by means of the TCD method implies considering the impact on cerebral artery blood velocity from several factors, including the intracranial pressure and the state of the brain metabolism in the individual patient after the bleed, as well as the degree and the extension of vasospasm. The diameter of the spastic artery can be measured objectively and compared with the TCD data; however, blood velocity is not always inversely proportional to the vessel diameter, since a reduction in volume flow often follows after SAH[6, 15]. These are important considerations when chosing parameters and cut-off limits to diagnose vasospasm from absolute blood velocity measurements. The introduction of the Vmca/Vica index alleviates some of this problem by compensating for changes in volume flow. This is in keeping with previous experience that the best estimate of extracranial carotid stenosis is obtained from comparing the blood velocity in the narrowed segment with the blood velocity in the adjacent normal artery segment[8]. Furthermore, in the anterior circle of Willis blood velocity is usually higher in the largest proximal ACA, due to the unique possibilities for collateral flow. As is well known, an ACA developmental asymmetry is not uncommon in patients with intracranial aneurysms.

Specific premonitory clinical symptoms and signs that herald symptomatic vasospasm have not been established[4]. Radioactive tracer methods can demonstrate local reduction of cerebral blood flow, yet these investigations give no warning of escalating vasospasm and provide no clue as to the anatomical nature of the lesion[12]. Angiography, while indispensable to reveal the source of the bleed, is unacceptable for monitoring of cerebral vasospasm in clinical situations. Therefore, the controversy surrounding the clinical significance of cerebral vasospasm stems at least partly from the lack of an objective and clinically acceptable method to follow the development and the resolution of vasospasm. This situation may change as investigations with TCD become increasingly employed. The present study shows that investigations by means of the TCD method permit ruling out vasospasm using either the absolute blood velocity or the Vmca/Vica index. When vasospasm does occur, the Vmca/Vica index seems sufficiently accurate to permit assessing the degree of lumen narrowing. Thus, obtaining longitudinal and frequently updated information on the vasospasm status in individual patients can be integrated in the day to day clinical routine. Studies reported so far suggest that TCD will be an important tool in the management of patients with SAH[2, 7, 10, 13, 14]. Moreover, the improved insight in the cerebral hemodynamics in patients at risk for cerebral vasospasm may help unveiling new and clinically important aspects of this enigmatic condition. Now that we have the method, the results should be forthcoming.

References

1. Aaslid R, Huber P, Nornes H (1984) Evaluation of cerebrovascular spasm with transcranial Doppler ultrasound. J Neurosurg 60: 37–41

2. Aaslid R, Huber P, Nornes H (1986) A transcranial Doppler method in the evaluation of cerebrovascular spasm. Neuroradiology 28: 11–16

3. Aaslid R, Markwalder T-M, Nornes H (1982) Noninvasive transcranial Doppler ultrasound recording of flow velocity in basal cerebral arteries. J Neurosurg 57: 769–774

4. Fisher CM, Roberson GH, Ojemann RG (1977) Cerebral vasospasm with ruptured saccular aneurysm – The clinical manifestations. Neurosurgery 1: 245–248

5. Gabrielsen TO, Greitz T (1970) Normal size of the internal carotid, middle cerebral and anterior cerebral arteries. Acta Radiol (Diagn) 10: 1–10

6. Grubb RL, Raichle ME, Eichling JO, Gado MH (1977) Effects of subarachnoid haemorrhage on cerebral blood volume, blood flow and oxygen utilization in humans. J Neurosurg 46: 446–453

7 a. Harders AG, Gilsbach JM (1987) Time course of blood velocity changes related to vasospasm in the circle of Willis measured by transcranial Doppler ultrasound. J Neurosurg 66: 718–728

7 b. Harders A (1986) Neurosurgical applications of transcranial Doppler sonography. Springer, Wien New York

8. Lindegaard KF, Bakke SJ, Grip A, Nornes H (1984) Pulsed Doppler techniques for measuring instantaneous mean and maximum flow velocities in carotid arteries. Ultrasound Med Biol 10: 419–426

9. Lindegaard KF, Bakke SJ, Grolimund P, Aaslid R, Huber P, Nornes H (1985) Assessment of intracranial hemodynamics in carotid artery disease by transcranial Doppler ultrasound. J Neurosurg 63: 890–898

10. Lindegaard KF, Grolimund P, Aaslid R, Nornes H (1986) Evaluation of cerebral AVM's using transcranial Doppler ultrasound. J Neurosurg 65: 335–344

11. Lindegaard KF, Bakke SJ, Aaslid R, Nornes H (1986) Doppler diagnosis of intracranial artery occlusive disorders. J Neurol Neurosurg Psychiatry 49: 510–518

12. Mickey B, Vorstrup S, Lindewald H, Harmsen A, Lassen NA (1984) Serial measurement of regional cerebral blood flow in patients with SAH using ^{133}Xe inhalation and emission computerized tomography. J Neurosurg 60: 916–922

13. Seiler RW, Grolimund P, Aaslid R, Huber P, Nornes H (1986) Cerebral vasospasm evaluated by transcranial ultrasound correlated with clinical grade and CT-visualized subarachnoid haemorrhage. J Neurosurg 64: 594–600

14. Seiler RW, Grolimund P, Zurbruegg HR (1987) Evaluation of the calcium-antagonist Nimodipine for the prevention of vasospasm after aneurysmal subarachnoid haemorrhage. Acta Neurochir (Wien) 65: 7–16

15. Voldby B, Enevoldsen EM, Jensen FT (1985) Regional CBF, intraventricular pressure and cerebral metabolism in patients with ruptured intracranial aneurysms. J Neurosurg 62: 48–58

Address for correspondence: K.-F. Lindegaard, M.D., Department of Neurosurgery, Rikshospitalet, The National Hospital, N-0027 Oslo 1, Norway.

Acta Neurochirurgica, Suppl. 42, 85–87 (1988)

Prostacyclin: a New Treatment for Vasospasm Associated with Subarachnoid Haemorrhage

P. A. Stanworth, J. Dutton, K. S. Paul, R. Fawcett, and E. Whalley

University Departments of Neurosurgery, Neuroradiology and Pharmacology, Manchester Royal Infirmary, Manchester, U.K.

Summary

One of the major problems associated with the treatment of ruptured intracranial aneurysms is the syndrome of late onset ischaemia. Patients so affected deteriorate neurologically and cerebral angiography often shows narrowing of the intracranial arteries, commonly known as vasospasm.

Many drugs have been used to treat the condition but with little success. A new group of compounds have come into clinical use recently, the Prostaglandins. One member, Prostacyclin (PGI$_2$ or Epoprostenol) is claimed to be one of the most potent vasodilators known. It was used at Manchester first on an experimental model. An isolated piece of human basilar artery was caused to contract using various agents. Prostacyclin was then used in an attempt to relax the contracted segment of artery. It, suprisingly, caused profound relaxation at very low concentrations of Prostacyclin, yet at higher concentrations it again caused the artery to contract. A literature search suggested this also was seen in the living subject and the crossover occurred at a dosage of 5 ng/kg/min.

A limited pilot trial was therefore devised using Prostacyclin at the low concentration of 1 ng/kg/min and used on six patients. Patients were assessed, in the main, for clinical improvement and change in radiological spasm. Clinically, the results exceeded our expectations in that all patients improved, some back to normal. Radiologically, the vasospasm changed but did not revert completely and also unusual extracranial-intracranial anastomoses appeared in the angiograms. In addition, in one patient, cerebral blood flow showed a more than threefold increase. No generalised cardiovascular collapse occurred and no bleeding tendency was observed.

Keywords: Vasospasm; subarachnoid haemorrhage; prostacyclin; PGI$_2$; Epoprostenol.

Introduction

The majority of patients presenting with a subarachnoid haemorrhage will have a ruptured intracranial aneurysm as the cause. Surgical technique has improved enormously over the last few years and the outlook, after successful surgery, for a patient without complications is extremely good. However, the associated syndrome of late onset ischaemia, or vasospasm, is still a major problem. Patients with this syndrome gradually deteriorate neurologically at some time after the haemorrhage. An angiogram shows "spasm" of the intracranial arteries and blood flow studies that the brain is severly ischaemic (Kwak *et al.* 1979).

The syndrome has proved very resistant to treatment and many drug regimes have been proposed including, for example, the use of sympathomimetic amines (Sundt *et al.* 1973), alpha-adrenergic blocking drugs (Takamatsu *et al.* 1972), and a variety of vasodilator agents such a nitroprusside (Allen 1976), phosphodiesterase inhibitors (Flamm 1976) and the more recently developed calcium blocking drugs (Allen *et al.* 1983). The results from most of these studies have been inconclusive.

A logical approach is to reverse the vasospasm by giving a vasodilator, thereby improving cerebral perfusion and with it brain function. The vasodilator substance should theoretically (a) act selectively on cerebral blood vessels and have no effect on peripheral vascular beds, (b) reverse contractions of cerebral vessels induced by a variety of substances believed to be involved in the aetiology of vasospasm, (c) be orally active.

Taking these factors into consideration we decided, using the isolated human basilar artery as a model (Forster *et al.* 1980), to evaluate the effect of a variety of vasodilator substances to reverse contractions of the basilar artery to a range of contractile agents including the unidentified spasmogenic factor which is present in the CSF of patients with vasospasm (Boullin *et al.* 1976).

Evidence from the literature indicated that prostacyclin, a product of the arachindonic acid pathway, may function as a physiological cerebral vasodilator

(Boullin *et al.* 1979) and possessed potent relaxing actions on cerebral arteries from various species (Boullin *et al.* 1979, Chapleau and White 1979, Jarman *et al.* 1979). We therefore analysed the effect of prostacyclin (PGI$_2$ Epoprostenol) in detail using the in vitro model. Prostacyclin was found to produce a concentration dependent relaxation of tissues contracted with a variety of spasmogens thought to be involved in the aetiology of vasospasm, including contractions to vasospastic CSF (Paul *et al.* 1982). The reversal of the "spasm" induced in the basilar arteries was greater than 100%, thus prostacyclin relaxed the tissues beyond their normal resting levels. The effect of prostacyclin was compared on a human peripheral tissue in vitro (mesenteric artery) and it was found that the peripheral artery was less sensitive to the relaxant action compared to the cerebral vessel (Paul 1983). At very high concentrations prostacyclin contracted basilar and mesenteric arteries (Lye *et al.* 1983).

Thus, from the in vitro studies prostacyclin possessed 2 of the important criteria required theoretically for treatment of vasospasm. On the basis of these results it seemed very reasonable to use prostacyclin clinically at a very low concentration. From the available literature it was felt that the crossover from vasoconstriction to dilatation occured at a dosage of 5 nanograms/kg/min, and it was therefore decided to use a dose of 1 nanogram/kg/min for the trial.

Other aspects of the clinical use of prostacyclin had to be considered. First, because of its apparently unrelated property of reduction of platelet aggregation, it might cause spontaneous bleeding. Secondly, it might cause generalised vasodilatation and cardiovascular collapse.

A *limited pilot study* was therefore devised. Patients had to have a ruptured aneurysm as a cause of their subarachnoid haemorrhage and to have also suffered a neurological deterioration after a baseline had been established. That deterioration had to be due to vasospasm alone. Our primary aim was to assess their clinical state but, in addition, cerebral vasospasm would be monitored. Attention would also be paid to their cardiovascular state and a watchful eye kept for any bleeding tendency.

Six patients were treated. All patients initially improved. The majority went on to make a full recovery and virtually returned to normal, some from being in coma at the start of treatment. The first patient was complicated by a staphylococcal pneumonia and eventually died. Post-mortem examination showed death was due to herniation secondary to a pre-existing hae-matoma and hydrocephalus. Apart from this patient, all improved to grade 4 (moderately disabled) or grade 5 (good recovery) on the Glasgow Outcome Scale.

Cerebral arterial vasospasm was assessed radiologically. The pattern of the spasm changed, some arterial segments dilated but the vessel calibre did not revert to normal. However, unexpected extracranial-intracranial natural anastomoses became prominent.

Apart from the blood pressure dropping by 10–20 mmHg, the cardiovascular state remained stable. No evidence of any bleeding tendency was observed in any of the patients. Because the first patient eventually died and was shown to have raised intracranial pressure (I.C.P.), it was decided to monitor I.C.P. in the subsequent five. In all these patients, the pressure showed a steady fall which parallelled their clinical improvement.

Opportunity arose for the measurement of cerebral blood flow in one patient. It showed an initially ischaemic brain but a remarkably greater than threefold increase subsequently.

There were two unexpected problems. First, because such a minute dose was used, there was difficulty with dilution of the prostacyclin. Secondly, one patient had her aneurysm clipped during a course of prostacyclin. The Anaethetist remarked during the operation on having great difficulty dropping the blood pressure electively. "It was as though the patient was fully vasodilated".

This preliminary study was very encouraging; it far exceeded our expectations. We may be accused of bias but independent and very objective arbiters, the nursing staff, were so impressed with the results that they pressed us very strongly to continue the trial.

Perhaps the one problem with the use of prostacyclin is its instability, thus the need for a more stable prostacyclin-like compound; indeed recent studies using the in vitro human basilar artery model have demonstrated an equivalent profile of action with Iloprost, a stable prostacyclin analogue (Lye *et al.* 1986).

In conclusion, from the results from the above study, prostacyclin (or a more stable analogue) appears to offer potential as a treatment for vasospasm associated with subarachnoid haemorrhage.

Ackowledgements

We wish to thank Mr. F. A. Strong, F.R.C.S., for allowing us to treat some of his patients and the Wellcome Foundation for the supply of prostacyclin.

References

1. Allen GS (1976) Cerebral arterial spasm, part 8: The treatment of delayed cerebral arterial spasm in human beings. Surg Neurol 6: 71–80
2. Allen GS, Ahn HS, Preziosi TJ *et al* (1983) Cerebral arterial spasm – a controlled trial of nimodipine in patients with subarachnoid haemorrhage. New Eng J Med 308: 619–624
3. Bouillin DJ, Bunting S, Blaso WP *et al* (1979) Responses of human and baboon arteries to prostacyclin: their relevance to cerebral arterial spasm in man. Br J Clin Pharmacol 7: 139–149
4. Bouillin DJ, Mohan J, Grahame-Smith DG (1976) Evidence for the presence of a vasoactive substance (possibly involved in the aetiology of cerebral arterial spasm) in cerebrospinal fluid from patients with subarachnoid haemorrhage. J Neurol Neurosurg Psychiatry 39: 756–766
5. Chapleau CE, White RP (1979) Effects of prostacyclin on the isolated canine basilar arteries. Prostaglandins 17: 573–580
6. Flamm ES (1976) Treatment of cerebral vasospasm by control of cyclic adenine monophosphate. Surg Neurol 6: 223–226
7. Forster C, Whalley ET, Mohan J, Dutton J (1980) Vascular smooth muscle response to fibrin degradation products and 5-hydroxytryptamine: possible role in cerebral vasospasm in man. Br J Clin Pharmacol 10: 231–236
8. Jarman DA, du Boulay GH, Kendall B *et al* (1979) Responses of baboon cerebral and extracranial arteries to prostacyclin and prostaglandin endoperoxide in vitro and in vivo. J Neurol Neurosurg Psychiatry 42: 677–686
9. Kwak R, Nizuma H, Ohi T *et al* (1979) Angiographic study of cerebral vasospasm following rupture of intracranial aneurysms. Part 1. Time of appearance. Surg Neurol II: 257–262
10. Lye RH, Parsons AA, Whalley ET (1986) Effect of iloprost (ZK 36374) and prostacyclin and in vitro human cerebral arteries. Br J Pharmacol 89: 691
11. Lye RH, Paul KS, Whalley ET (1983) Analysis of the contractile action of prostacyclin on the human basilar artery in vitro. Br J Pharmacol 78: 164
12. Paul KS (1983) Prostanoids and the in vitro human basilar artery: relevance to cerebral arterial spasm. Univ of Manchester, Ch.M. (Neurosurgery) Thesis
13. Paul KS, Whalley ET, Forster C, Lye R, Dutton J (1982) Prostacyclin and cerebral vessel relaxation. J Neurosurgery 157: 334–340
14. Sundt TM, Onofro BM, Meredith J (1973) Treatment of cerebral vasospasm from subarachnoid haemorrhage with isoproterenol and lidocaine hydrochloride. J Neurosurg 38: 557–560
15. Takamatsu M, Miyazaki Y, Hottait Y *et al* (1972) Cerebral angiogram and adrenergic blocking agents. In: The cerebral vasospasm: 5th conference on specific topics in neurosurgery. Tokyo, pp 165–177

Address for correspondence: P. A. Stanworth, M.D., Consultant Neurosurgeon, Walsgrave Hospital, Department of Neurosurgery, Clifford Bridge Road, Walsgrave, Coventry CV2 2DX, U.K.

Acta Neurochirurgica, Suppl. 42, 88–92 (1988)

Treatment of Carotid-cavernous Fistulas by Embolization of the Cavernous Sinus Through Venous Affluents or Direct Puncture

P. Albert, M. Polaina, F. Trujillo, J. Romero, and F. Durand

Regional Department of Neurosurgery of Spanish Social Security, Virgen del Rocio General Hospital, Seville, Spain

Summary

Eight patients with nine carotid-cavernous fistulas (CCF), one of them bilateral, have been treated in our Department by embolization of the venous components of the cavernous sinus, either through one of its venous effluents or by direct puncture of the sinus, in an effort to preserve the internal carotid artery (ICA). Five of the fistulas were of traumatic origin and four were spontaneous. In two cases the cavernous sinus was embolized through bridging temporal veins, and in another two cases embolization was carried out by catheterization of the cavernous sinus via the superior ophthalmic vein, while in the remaining five cases it was performed by direct puncture of the lateral wall of the sinus. The embolizing materials employed were bone wax mixed with an oily iodised contrast media, small pieces of oxidized cellulose or gelfoam vehicled by saline solution, and fibrin sealant. Obliteration of the CCF was complete in all but one case, in which this was accomplished only partially. On six occasions carotid patency was maintained, while in two patients failure of the technique made trapping of the ICA necessary. In one patient the ICA was ligated prior to the embolization of the cavernous sinus. The surgical techniques employed and the results achieved are exposed.

Keywords: Carotid-cavernous fistula; cavernous sinus; therapeutic embolization.

Introduction

Treatment of traumatic or spontaneous CCF has always been a difficult problem. As many other neurosurgeons, we believe that simple ligation of the carotid artery at the neck does not lead to the obliteration of the fistula in many cases. Although efficiency of trapping is greater it does not achieve CCF occlusion in all instances. However, embolization of the CCF with different types of materials through the carotid artery after its intracranial ligation did produce effective occlusion of the fistula. Introduction of a Fogarty catheter through the cervical carotid artery was also a good procedure, and obviated the craniotomy necessary to place a supraclinoid clip. All these techniques imply the sacrifice of the carotid artery and in some few cases may lead to blindness of the affected eye when the supraclinoid carotid artery is ligated beyond the emergence of the ophthalmic artery, and obviously, they could not be applied in bilateral CCF. Parkinson's method, the direct approach to the fistula with hypothermia, circulatory by-pass and arrest, requires a difficult and dangerous operation. Therefore, since 1972, in certain cases we have tried to occlude the fistula by embolizing its venous portion through different pathways, in an attempt to save carotid patency.

Material and Methods

From January 1972 to June 1986 in our Department we have treated eight patients with nine CCF, one of them being bilateral, by embolization of the venous portion of the cavernous sinus pursuing carotid patency. The cases are summarized in Table 1.

Common symptoms and signs were chemosis, intracranial bruit and exophthalmos. The frequency of visual loss and oculomotor palsies was also high. In those of traumatic etiology ocular symptoms were apparent between two days and two months following head injury.

In all patients the fistual was demonstrated by carotid angiography and, in two cases of spontaneous CCF, filling of the fistula was noticed to be by the external carotid artery as well as the ICA.

Surgical Techniques Employed

Nine CCF were treated surgically. Two different surgical techniques have been used. In seven patients (cases 1, 2, 3, 4, 5, 7, and 9) we decided on a direct approach to the cavernous sinus and the fistulas were obliterated by embolization with different materials (Table 1): on five occasions after direct puncture of the cavernous sinus (cases 1, 2, 3, 5, and 9), and through temporal venous affluents to the sinus in two other instances (cases 4 and 7). In the remaining two patients (cases 6 and 8) embolization was carried out by catheterization of the sinus via the superior ophthalmic vein.

Due to different technical reasons, on two occasions (cases 2 and 5), after direct puncture of the cavernous sinus we were unable to maintain carotid patency and trapping of the carotid artery was

Table 1. *Summary of Treatment and Results*

Case	Age/sex	Etiology	Type	Date of operation	Surgical approach	Materials	Results
1	19/f	traumatic	unilateral	1. 14. 1972	trapping + direct embolization of sinus	bone wax	occlusion of CCF previously blind
2	66/f	spontaneous	unilateral	6. 6. 1980	direct embolization of sinus + trapping	bone wax bone wax	occlusion of CCF blind after surgery
3	64/f	spontaneous	bilateral	2. 24. 1981	direct embolization of sinus	oxidized cellulose	occlusion of CCF preservation of ICA asymptomatic
4	64/f	spontaneous	bilateral	3. 4. 1981	embolization through bridging temporal veins	oxidized cellulose	occlusion of CCF preservation of ICA asymptomatic
5	32/m	traumatic	unilateral	5. 11. 1981	direct embolization of sinus + trapping	oxidized cellulose	occlusion of CCF blind after surgery
6	27/m	traumatic	unilateral	2. 8. 1983	embolization via sup. ophthalmic vein	oxidized cellulose	occlusion as CCF preservation of ICA asymptomatic
7	61/m	spontaneous	unilateral	4. 22. '85	embolization through bridging temporal veins	oxidized cellulose	partial occlusion of CCF preservation of ICA symptomatic
8	6/m	traumatic	unilateral	1. 31. 1986	embolization via sup. ophthalmic vein	fibrin sealant	occlusion of CCF preservation of ICA asymptomatic
9	20/f	traumatic	unilateral	6. 12. 1986	direct embolization of sinus	fibrin sealant gelfoam	occlusion as CCF preservation of ICA asymptomatic

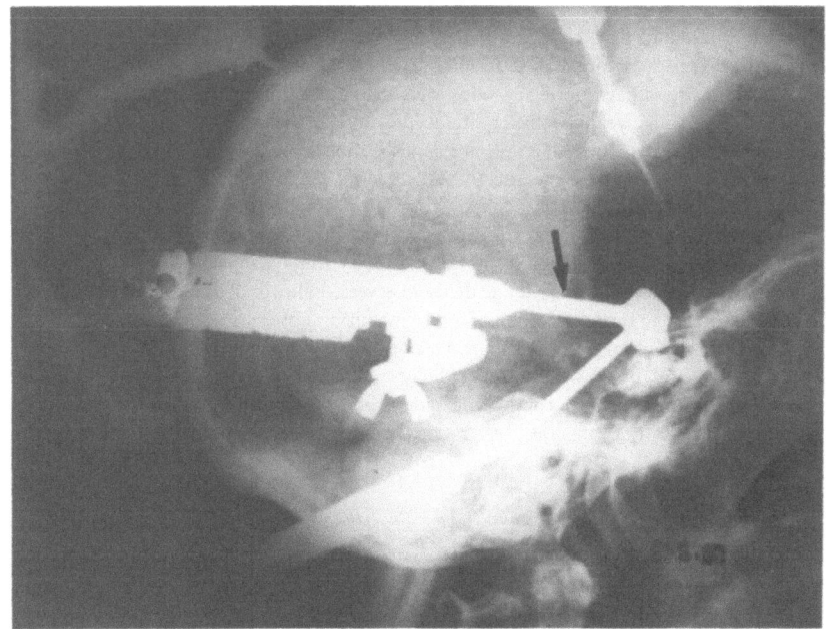

Fig. 1. Intraoperatory control. Puncture of the cavernous sinus at the emergence of the superior ophthalmic vein shows its repletion with contrast media prior to embolization. Arrow pointing to orthostatic separator retracting the temporal lobe

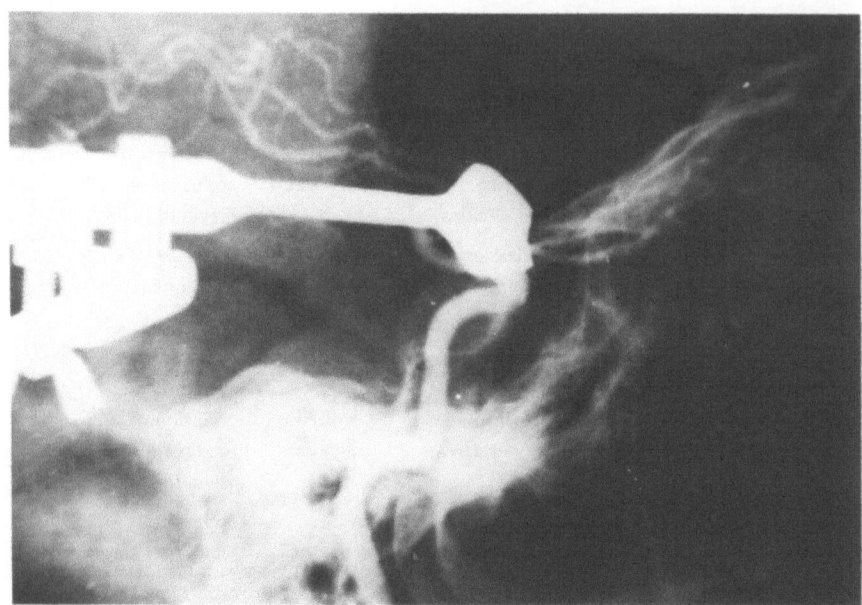

Fig. 2. Control arteriogram after embolization, with occlusion of the fistula and carotid patency

necessary. Another patient (case 1), had trapping of the carotid artery as the initial technique and because of failure to occlude the fistula, the cavernous sinus was embolized by direct puncture.

Direct Approach to the Cavernous Sinus

The seven patients who underwent a direct approach to the cavernous sinus were operated under general anaesthesia. First we dissect the comma carotid artery in the neck and a band is passed around it to allow its temporary occlusion. An arteriographic catheter is left in place for intraoperative control. With the patient's head placed in the lateral position we perform a pterional craniotomy, thus exposing the lateral wall of the sinus. In all instances this wall protruded showing a convex surface, to which led dilated veins bearing arterialized blood to the undersurface of the temporal lobe, so indicating the reversal of blood flow. These were constant findings except for patient 9, in whom the drainage of the sinus was preferrentially through the ophthalmic veins and in patient 5, who had sustained a severe head injury that required two previous operations and in whom these veins were found to be thrombosed. Since case 4, when we have evidence of patent bridging veins, we attempt canalization of one of these with a venous catheter after its previous section, and hereby reach the cavernous sinus. We accomplished this in cases 4 and 7. In the remaining cases we directly punctured the lateral wall of the sinus, also with a venous catheter (Fig. 1) to visualize the cavernous sinus and its drainages. Once its optimal placement is verified, as well as lack of filling of the carotid artery, embolizing substances are injected through the catheter in several bolus until we note resistence to further injection of more embolizing substance and blood stops flowing from the sinus. Once embolization is accomplished the wall of the sinus no longer bulges and it recovers its normal aspect. Intraoperative control angiograms prove disappearance of the fistula (Fig. 2) and demonstrate patency of the carotid artery.

Embolization of the Cavernous Sinus via the Superior Ophthalmic Vein

On two occasions (cases 6 and 8) we embolized the cavernous sinus through the canalized superior ophthalmic vein. In these cases we

had evidence by previous angiograms that most of the drainage of the sinus was carried out by this vein, as often occurs.

As in the previous method we first dissect the common carotid artery in the neck and pass a band around it to permit its temporary occlusion. A small incision of about 2 cms long is made over the supraorbital rim, centered at the supraorbital foramen. The superior ophthalmic vein is exposed as it leaves the foramen. In both cases it was dilated, becoming stenotic as it enters the bony channel. It is catheterized towards the orbit with a polyethylene catheter of adequate width to allow its passage through the foramen, and malleable to permit its adaptability to the tortuous vein within the orbit. We gently move forward towards the sinus simultaneously injecting saline solution in order to dilate and straighten, as much as possible, the sinuousities of the vein, and taking angiographic controls of its situation until the sinus is reached. Once in the cavernous sinus the embolizing substance is introduced while the carotid artery is temporarily occluded. When embolization is concluded the results are documented by arteriography and embolization may be repeated if necessary until total occlusion of the fistula (Fig. 3).

Embolizing Materials Employed

In all instances we have used materials whose plasticity make them suitable for adjustment to the cavernous sinus and that are capable of being introduced through catheters that are sometimes of quite small gauge. In our first two cases we used a mixture of 30% bone wax and 70% oily iodised contrast media*, but it solidified quickly and obstructed the catheter readily. Therefore, in the following cases we have used other embolizing substances. Finely cut oxidized cellulose** mixed with saline solution was employed in cases 3, 4, 5, 6, and 7 but it presents the drawback of forming a pasty blend which is difficult to inject into thin catheters. In our next patient we used fibrin sealant***, which is introduced easily and induces fast thrombosis of the sinus. In our last patient we combined two substances: small elongated pieces of gelfoam****, vehicled by saline solution,

---

   * Lipiodol Ultrafluide (Ethiodol).
  ** Surgicel.
 *** Tissucol.
**** Espongostan.

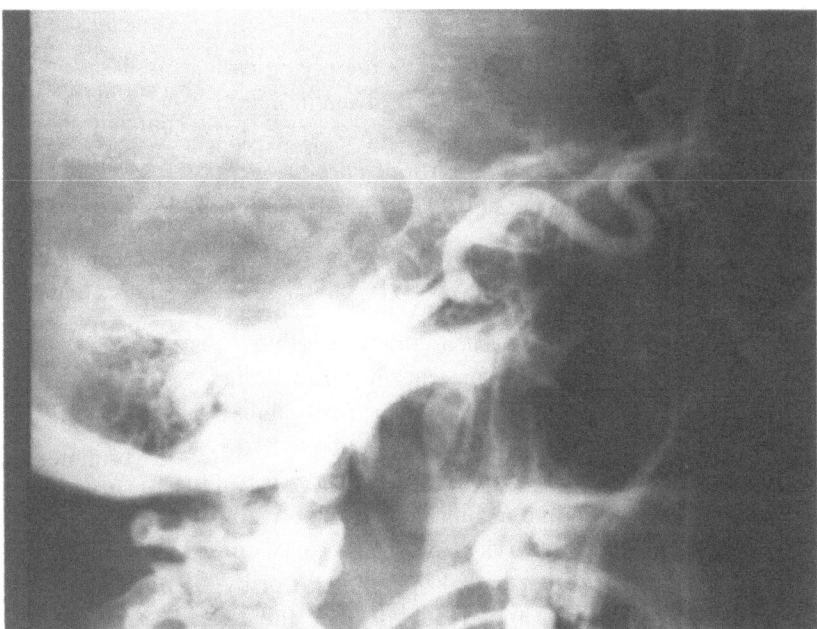

Fig. 3. Intraoperatory control. Canalization of the superior ophthalmic vein until the sinus is reached

injected in several bolus, and that offer no difficulty in passage through very thin catheters, followed by embolization with fibrin sealant. We believe that the combination of these two substances is the most adecuate embolizing material, as they pass in fluid form through the finest catheters without obstructing them and they induce a rapid thrombosis of the cavernous sinus thus occluding the fistula.

Results

Control angiograms performed after the operations demonstrated complete occlusion of the fistula in all patients but one (case 7), in whom an angiographic leak persisted in the basal portion of the cavernous sinus.

In two cases it was not possible to maintain carotid patency and trapping was necessary. In one of these, case 2, we punctured the lateral wall of the sinus twice and each time radiological control showed that the catheter was placed in the intracavernous carotid artery. Fearing the risk of creating a greater fistula rest we opted for carotid artery trapping and successive embolization of the intracavernous artery with a mixture of bone wax and oily iodised contrast media, thereby achieving the occlusion of the fistula. In the other patient (case 5), after direct puncture and embolization of the sinus with oxidized cellulose we found on intraoperative control angiograms that the carotid artery was obstructed at its intracavernous portion, possibly because of embolization of the material into the carotid artery or of spasm secondary to its compres-

sion. Upon this finding we again opted for trapping of the carotid artery. Case 1, a patient with severe head injury, blindness and ophthalmoplegia of the affected eye, underwent surgery because of marked chemosis and progressive exophthalmos. We performed trapping and due to persistence of the fistula the cavernous sinus was embolized with bone wax and oily iodised contrast media by direct puncture.

In the remaining six cases patency of the carotid artery was preserved and control angiograms did not show irregularities of its lumen, except in case 8, in which a false aneurysm persisted, with total occlusion of the venous portion of the fistula. Another control angiogram taken 9 months later showed that the aneurysm had not grown.

Ocular symptoms subsided in all patients with gradual disappearance of chemosis and exophthalmos. Intracranial noise and bruit ceased immediately except in case 6, in whom part of the fistula persisted and the patient complained of the same preoperative noise. One patient suffered transient third nerve palsy followed by complete recovery. Two patients in whom trapping was performed suffered unilateral blindness either due to the placement of the supraclinoid carotid artery clip beyond the emergence of the ophthalmic artery or due to embolisation of the injected materials into the artery. One patient was blind prior to the operation. Two patients that had moderate visual loss improved after surgery. There were no other neurologic complications.

Discussion

The occlusion of high-flow CCF while preserving carotid patency continues an unsolved problem in many instances. Multiplicity in techniques employed at the present moment support this assertion. Parkinson[1, 2] demonstrated that it was possible to occlude the fistula without necessarily sacrificing the carotid artery, by directly approaching the lateral wall of the cavernous sinus through a triangular space which he outlined, but the technique he used is of such complexity that it makes it difficult to apply. At present time endarterial catheterisation with detachable balloons, as designed by Serbinenko and Debrun[3, 4] are being applied frequently, but these methods require experience in intravascular navigation and technical means not available to all Neurosurgery Services. They also carry important disadvantages that have been pointed out by different authors[4, 5, 6, 7], and frequently they fail to preserve carotid blood flow. Other authors also obliterate the venous portion of the fistula with different techniques in order to achieve the thrombosis of the cavernous sinus and solve the fistula without acting directly on the carotid artery. Hosobuchi[8] uses different approaches, always through venous pathways, which vary depending on the location of the fistula and induces thrombosis with electric current. Complexity of the technical equipment necessary is responsible for its infrequent application. Mullan[9] also follows venous drainage pathways to reach the cavernous sinus and introduce thrombogenic wires as well as other plastic materials. Recently Isamat[10] has used Parkinson's approach to obliterate CCF by introducing muscle and/ or fibrin sealant directly into the cavernous sinus.

We have also embolized the venous portion of the fistula, whenever possible, through venous effluents of the cavernous sinus, either via the superior ophthalmic vein as Hosobuchi or Peterson[11]; or through bridging temporal veins, a cavernous sinus pathway which has not been reported previously in papers on this matter. This venous pathway obviates the direct puncture of the cavernous sinus, which in some case may lead to damage of the carotid artery, as so happened in one of our patients, or of the cranial nerves. When this route was not possible we have embolized the sinus by direct puncture of its lateral wall, in the triangular space described by Parkinson as free of nervous or vascular elements, always using malleable substances whose plasticity allow their adaptability to the sinus without

causing compression or damage to the elements that occupy it. Therefore, we believe that in the setting of a spontaneous or traumatic CCF, occlusion and carotid patency should always be sought, using different techniques adopted to the situation in the special patients. If drainage is preferentially via the superior ophthalmic vein, embolization through it should be attempted. If this is not achieved or the fistula drains to the posterior sinuses embolization may be done via the jugular vein or petrosal sinus (we do not have any experience with this pathway). These methods elegantly avoid craniotomy. When these pathways fail, a direct approach to the sinus and its embolization may be done through the bridging temporal veins, and if they are lacking or fail, it may be accomplished by direct puncture of the lateral wall of the sinus.

References

1. Parkinson D (1965) A surgical approach to the cavernous portion of the carotid artery. Anatomical studies and case report. J Neurosurg 23: 474–483
2. Parkinson D (1973) Carotid-cavernous fistula: direct repair with preservation of the carotid artery. Technical note. J Neurosurg 38: 99–106
3. Serbinenko FA (1974) Balloon catheterization and occlusion of major cerebral vessels. J Neurosurg 41: 125–145
4. Debrun G, Lacour P, Vinuela F, *et al* (1981) Treatment of 54 traumatic carotid-cavernous fistulas. J Neurosurg 55: 678–692
5. Vinuela F, Fox AJ, Debrun GM *et al* (1984) Spontaneous carotid-cavernous fistulas: clinical, radiological and therapeutic considerations. Experience with 20 cases. J Neurosurg 60: 976–984
6. Peters FLM, van der Werf AJM (1980) Detachable balloon technique in the treatment of direct carotid-cavernous fistulas. Surg Neurol 14: 11–19
7. Barrow DL, Fleischer AS, Hoffman JC (1982) Complications of detachable balloon catheter technique in the treatment of traumatic intracranial arteriovenous fistulas. J Neurosurg 56: 396–403
8. Hosobuchi Y (1975) Electrothrombosis of carotid-cavernous fistula. J Neurosurg 42: 76–85
9. Mullan S (1979) Treatment of carotid-cavernous fistulas by cavernous sinus occlusion. J Neurosurg 50: 131–144
10. Isamat F, Ferrer E, Twose J (1986) Direct intracavernous obliteration of high-flow carotid-cavernous fistulas. J Neurosurg 65: 770–775
11. Peterson EW, Valberg J, Whittingham DS (1970) Electrically induced thrombosis of the cavernous sinus in the treatment of carotid-cavernous fistula. In: Drake CG, Duvoisin R (eds) Fourth International Congress of Neurological Surgery, Ninth International Congress of Neurology, September 20–27, 1969, New York. Excerpta Medica, Amsterdam

Address for correspondence: Dr. P. Albert, Regional Department of Neurosurgery of Spanish Social Security, Virgen del Rocio General Hospital, Seville, Spain.

Acta Neurochirurgica, Suppl. 42, 93–97 (1988)

Surgical Management of Large AVMs

R. F. Spetzler and J. M. Zabramski

Division of Neurological Surgery, Barrow Neurological Institute, Phoenix, Arizona, U.S.A.

Summary

The surgical management of large AVMs (those greater than 6 cm in maximal diameter) should be based on a thorough understanding of the chronic hemodynamic changes produced by lesions and the acute stress placed on the cerebral vasculature by their removal. In addition to haemorrhage, seizure, and headache, these larger lesions often present with symptoms of cerebral vascular insufficiency[3, 4, 12]. Angiography frequently demonstrates a high-flow arteriovenous shunt with evidence of vascular steal from the surrounding brain. In many cases there is a virtual absence of normal hemispheric filling. When the steal is sufficient to produce an area of chronic ischaemia in the brain surrounding the AVM, there is an increased risk of swelling and haemorrhage associated with complete excision.

We have developed a strategy for the surgical management of these large lesions that involves a stepwise reduction of flow through the AVM using pre- and intraoperative embolization, followed by complete excision. The details of this management strategy are described, and results in 24 patients with exceptionally large AVMs are presented.

Keywords: Arteriovenous malformation (AVM); large AVM; treatment; embolization; excision; results.

Clinical Material

Twenty-four patients underwent staged embolization and resection of exceptionally large intracranial AVMs. There were 11 females (mean age: 37 years) and 13 males (mean age: 36 years). All 24 patients had large AVMs greater than 6 cm in size. The lesions ranged from 6 to 9 cm with a average diameter of approximately 7 cm. In addition, all lesions were located in or adjacent to eloquent cortex. The most common complaint was headache, followed by reports of fluctuating or progressive neurologic deficits in 75% and 67% of the patients respectively. Fifteen of 24 patients (62%) had a positive history for intracerebral haemorrhage, while slightly less than half (47%) had a history of seizures.

Angiographic evaluation demonstrated feeding vessels arising from two or more major arterial distributions. All patients had evidence of vascular steal by the AVM from the surrounding brain. Patients selected for surgery comprised about 25% of those with large AVMs: Surgery was recommended in this population only for those patients with evidence of previous haemorrhage or progressive neurologic deficit[17].

Methods

All patients in this series were managed using a multistaged approach to eliminate the major vascular supply to the AVM prior to resection. When collaterals from the external carotid artery supplied the AVM, the initial procedure usually consisted of transfemoral embolization to eliminate these feeding vessels. Ivalon particles (ranging from 150 to 290 microns) were injected as a slurry along with contrast material (Conray) under continuous angiographic monitoring. Aggressive transfemoral embolization to obliterate intracranial vessels was not performed. The rationale for this decision is discussed below.

Staged intraoperative embolization procedures were then performed, usually eliminating one major group of feeding vessels during each procedure. The typical large AVM with feeders from two or three vascular distributions would thus require a minimum of two operative stages for embolization prior to excision. All operative procedures were performed under barbiturate anesthesia, with extensive monitoring of EEG and evoked potential responses[11]. The scalp incision(s) were designed to allow access to the AVM and the feeding vessels around it without compromising blood flow to the edges of the flap. When possible, the nidus of the AVM was left covered by the dura. This assured that the AVM did not become adherent to the dura prior to excision.

Feeding vessels were isolated as close to the AVM as possible and cannulated with a 4- or 5-French catheter. Intraoperative digital subtraction angiography was performed to evaluate the extent of AVM filling. Ivalon particles were mixed in a slurry with Conray and injected under continuous fluoroscopic guidance until there was no further evidence of AVM filling. The feeding vessel was then permanently occluded after re-

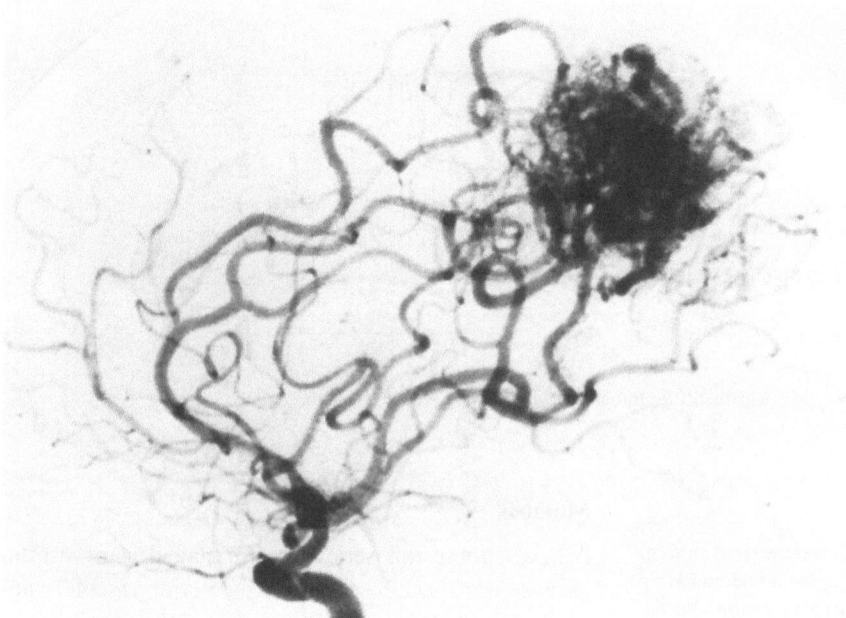

Fig. 1 a, b. Preoperative left internal carotid artery angiogram (AP and lateral views) demonstrating MCA and ACA feeders to a large parietal lobe AVM. Note the marked enlargement of the vessels supplying the AVM, and the decrease in flow to the surrounding normal brain

a

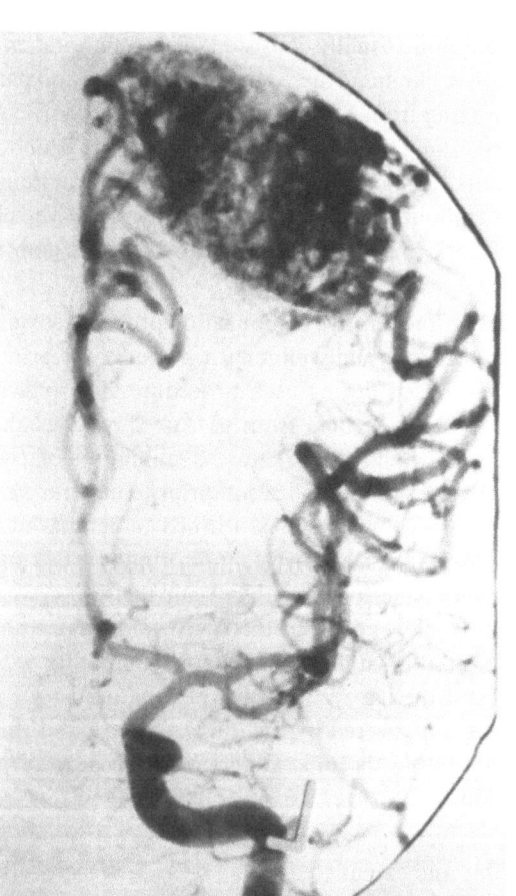

b

lization. The branches to normal cortex in these two cases were occluded with temporary aneurysm clips during embolization of the AVM.

The final operative approach was designed to completely excise the AVM. Following excision, all patients underwent angiography to determine whether complete resection had been obtained.

Results

Complications are listed in Table 1. There were no deaths in this series of 24 patients. Minor complications and temporary deficits occurred in six patients. Temporary deficits lasting less than 5 days were not considered complications.

Four patients had serious complications. Normal perfusion pressure breakthrough with extensive haemorrhagic infarction occurred in one patient after complete resection of his left parietal lobe AVM. He remains disabled, with a right hemiparesis and expressive aphasia. One patient developed a postoperative intracerebral and intraventricular haemorrhage following incomplete resection of a right parietal-occipital AVM. This complication was initially associated with a complete hemianopia. The patient made a protracted but complete recovery. Another patient developed a left hemiparesis following resection that has improved, though it has not resolved completely: The patient can ambulate well independently, but has significant residual deficit in the left upper extremity. Finally, one pa-

moval of the catheter in all but two cases. In these two cases the posterior cerebral artery was repaired after catheter withdrawal because branches to the normal occipital cortex originated distal to the site of embo-

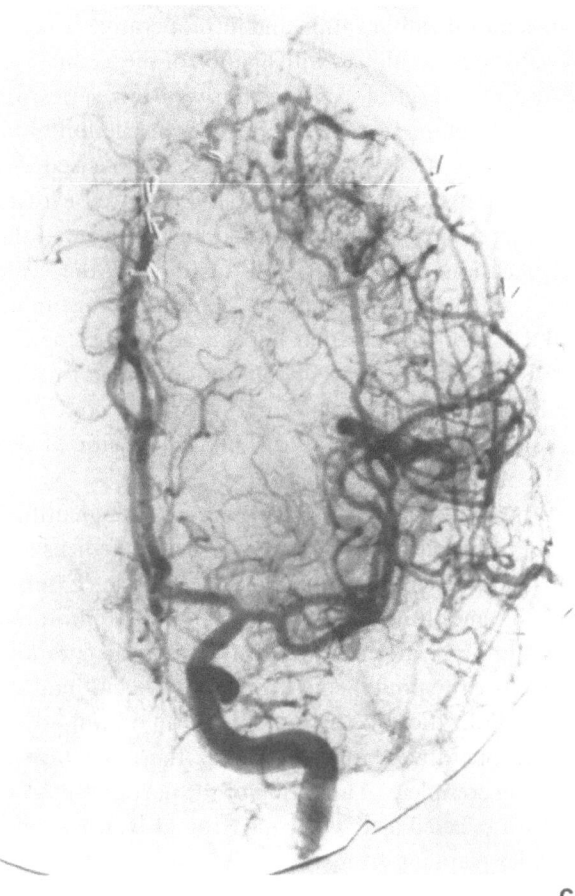

Fig. 1 c. Selective internal carotid artery angiogram (AP view) following second stage embolization. Both the MCA (first stage) and ACA (second stage) feeders have been embolized and occluded. There is only minimal residual flow through the AVM

Table 1. *Complications*

Minor complications

1. Temporary mild right sided hemiparesis and aphasia
2. Temporary mild ataxia and dysarthia
3. Temporary aphasia
4. Temporary aphasia and right hemiparesis
5. Slight increase in visual field defects with macular sparing
6. Mild increase in aphasia, detectable only with rapid speech

Major complications

1. Intracerebral hemorrhage with transient increase in preoperative visual field defect
2. Normal perfusion pressure breakthrough producing severe hemiparesis and aphasia
3. Osteomyelitis of craniotomy bone flap
4. Mild left upper extremity paresis

tient developed an infection of his bone flap, necessitating its removal and delayed cranioplasty.

Complete excision of the AVM was performed in 22 of 24 cases. Four patients in this series required two stages to complete excision. Incomplete resection occurred in two patients: One patient declined to undergo further surgery after three embolization procedures obliterated the majority of her AVM and relieved her symptoms. The second patient (who is described above) refused excision of her residual AVM after recovery from a postoperative intracerebral haemorrhage.

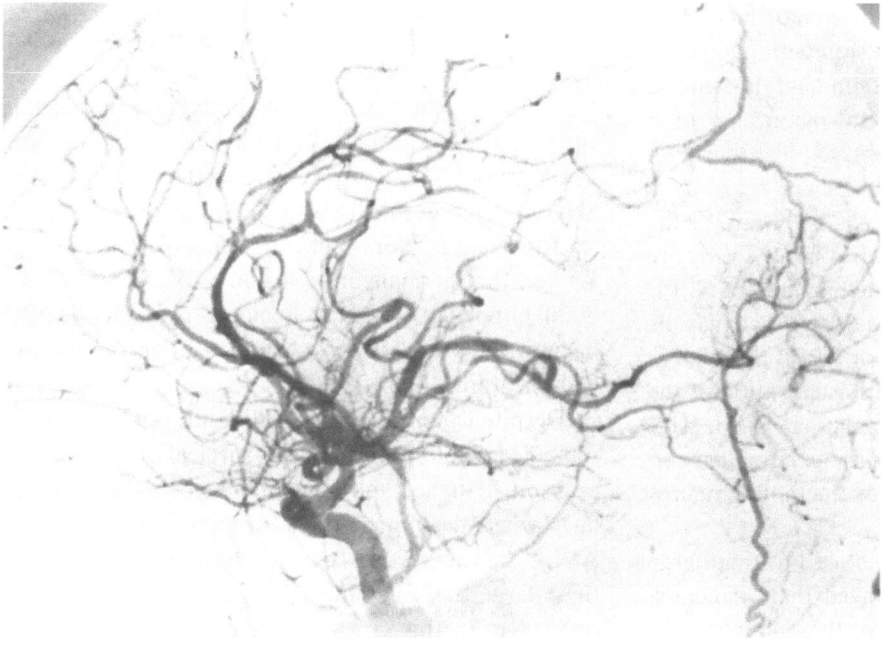

Fig. 1 d. Selective internal carotid artery angiogram (lateral view) one week following complete excision of AVM. Of interest, note that the vessels supplying the AVM are already returning to normal diameter

Discussion

The surgical management of large AVMs presents special problems. In addition to size, the risks of excision in these cases is frequently increased by the presence of drainage to deep venous structures and location within or adjacent to eloquent cortex[12, 17]. The majority of these large lesions also have angiographic evidence of vascular steal. In the more dramatic cases, there may be a virtual absence of normal hemispheric filling until a portion of the AVM is obliterated.

Vascular steal by the AVM can reduce cerebral perfusion pressure (CPP) in the region surrounding the malformation[9, 15]. Flow through the low resistance arteriovenous shunt of large AVMs results in the reduction of pressure in the feeding arteries and elevation of pressure in the draining veins. Our measurements and those by others have confirmed that feeding artery pressures in high-flow AVMs are uniformly lower than systemic arterial pressures by a range of 40 to 80 torr[8, 9, 13]. Because the feeding arteries and draining veins of the AVM are derived from the normal cerebral circulation, changes in pressure in AVM vessels are transmitted to the adjacent brain. In the region surrounding large, high-flow AVMs cerebral perfusion pressure may fall below the limit of autoregulation resulting in regional hypoperfusion[8, 9, 15]. The cerebral vasculature in these regions of ischaemia are chronically dilated and when exposed to sudden changes in pressure (such as those produced by complete excision of the AVM) may be unable to respond. If the autoregulatory capacity is exceeded, then local oedema and haemorrhage may occur in a pattern similar to that seen with malignant hypertension despite normal or even slightly reduced intravascular pressures. This clinical picture has been labeled normal perfusion pressure breakthrough (NPPB)[16].

Over all, this complication is relatively rare in the AVM literature[2, 6, 20]. However, when large AVMs are considered as a separate group, the frequency of NPPB is higher[6]. Clinical features of an AVM that may be associated with an increased risk of NPPB are 1. large size, 2. high-flow shunt with poor visualization of the surrounding normal cerebral vasculature, 3. extensive contribution of AVM feeders by the external carotid artery, 4. history of progressive or fluctuating neurologic deficits.

A variety of methods have been used to treat large AVMs[2, 18–21]. We have found staged management a valuable strategy in the therapy of these lesions[13]. In this series, large AVMs were managed by preoperative transfemoral embolization and intraoperative selective embolization combined with ligation of the feeding vessel prior to surgical excision. Early in this series one patient developed NPPB hours after completion of his surgery. His AVM had been aggressively excised. We believe that a more gradual obliteration (as evolved with increasing experience) would have prevented this complication. This approach of obliterating the AVM gradually in stages promotes a stepwise change in cerebral hemodynamics that appears to reduce the risk of NPPB. The extensive embolization integral to this method also facilitates the intraoperative control of bleeding – a limiting factor for surgical excision of very large AVMs.

When external carotid artery feeders are identified, the initial treatment of the AVM usually consists of transfemoral embolization of these vessels. External carotid embolization must be carefully monitored. During embolization extra- to intracranial anastomotic channels can spontaneously open producing inadvertent embolization to surrounding normal brain[5, 14].

Intraoperative embolization has the advantage of allowing complete obliteration of the part of the AVM fed by a selected artery. We have found Ivalon sponge particles (ranging from 150 to 290 microns) to be the material of choice for these procedures. The particles are small enough to pass through the feeding artery and lodge within the nidus of the AVM, reducing the possibility of collateral vessels resupplying the lesion. Bucrylate embolization of AVMs is not recommended when surgery is being considered. Bucrylate produces a brittle, noncompressible mass that makes operative dissection more difficult[1, 10].

Following intraoperative embolization the feeding vessel is ordinarily ligated at the site of cannulation. Embolization increases pressure in the feeding vessel, and ligation prevents transmission of this elevated pressure to residual nonoccluded segments of the AVM. We feel that permanent occlusion of the feeding vessel is an important factor in avoiding postembolization haemorrhage and is an advantage of intraoperative embolization[7].

Despite extensive embolization, all of these lesions remained viable, requiring long surgical procedures for excision. Patience and meticulous technique, particularly when dealing with the deepest portions of the AVM, are the keys to successful removal. Excision of these large lesions cannot always be accomplished in one stage. In this series 4 patients required a second procedure to complete removal.

Conclusion

Previously, the surgical management of large AVMs has been associated with an unacceptably high incidence of complications. In the literature, the combined mortality and serious morbidity for resection has approached 50%[6, 19]. By using a management strategy of staged intraoperative embolization combined with feeding vessel ligation prior to surgical resection, we have been able to completely excise 22 of 24 exceptionally large AVMs with no mortality and only one seriously disabling morbidity.

References

1. Debrun G, Fox A, Drake C, Drake CG (1981) Comments. In: Samson D, Ditmore QM, Beyer CW: Intravascular use of isobutyl-2-cyanoacrylate: Part 1. Treatment of intracranial arteriovenous malformations. Neurosurgery 8: 43–51
2. Drake CG (1979) Cerebral arteriovenous malformations: considerations for and experience with surgical treatment in 166 cases. Clin Neurosurg 26: 145–208
3. Feindel W, Yamamoto YL, Hodge CP (1971) Red cerebral veins and the cerebral steal syndrome. Evidence from fluorescein angiography and microregional blood flow by radioisotopes during excision of an angioma. J Neurosurg 35: 167–179
4. Hachinski V, Norris JW, Cooper PW, Marshall J (1977) Symptomatic intracranial steal. Arch Neurol 34: 149–153
5. Kerber CW, Ahn HS, Deeb ZL (1979) The production of stroke during therapeutic embolization via external carotid to intracerebral anastomoses. Presented at the annual meeting of the American Society of Neuroradiology, Toronto, May 1979
6. Luessenhop AJ, Rosa L (1984) Cerebral arteriovenous malformations. Indications for and results of surgery and the role of intravascular techniques. J Neurosurg 60: 14–22
7. Mullan S, Kawanaga H, Patronas NJ (1979) Microvascular embolization of cerebral arteriovenous malformations. A technical variation. J Neurosurg 51: 621–627
8. Nornes H (1984) Quantitation of altered haemodynamics. In: Wilson CB, Stein BM (eds) Intracranial arteriovenous malformations, vol 4. Williams and Wilkins, Baltimore, pp 32–43
9. Nornes H, Grip A (1980) Haemodynamic aspects of cerebral arteriovenous malformations. J Neurosurg 53: 456–464
10. Samson D, Ditmore QM, Beyer CW Jr (1981) Intravascular use of isobutyl 2-cyanocrylate: Part 1. Treatment of intracranial arteriovenous malformations. Neurosurgery 8: 43–51
11. Selman W, Spetzler R, Zabramski J (1985) Induced Barbiturate Coma. In: Wilkins RH, Rengachary SS (eds) Neurosurgery. McGraw-Hill Book Company, New York, pp 343–349
12. Spetzler RF, Martin NA (1986) A proposed grading system for arteriovenous malformation. J Neurosurg 65: 476–483
13. Spetzler RF, Martin NA, Carter LP, Flom RA, Raudzens PA, Wilkinson E (1987) Surgical management of large AVMs by staged embolization and operative excision. J Neurosurg 67: 17–28
14. Spetzler RF, Modic M, Bonstelle C (1980) Spontaneous opening of large occipital vertebral artery anastomosis during embolization. J Neurosurg 53: 849–850
15. Spetzler RF, Selman WR (1984) Pathophysiology of cerebral ischaemia accompanying arteriovenous malformations. In: Wilson CB, Stein BM (eds) Intracranial arteriovenous malformations, vol 3. Williams and Wilkins, Baltimore, pp 24–31
16. Spetzler RF, Wilson CB, Weinstein P, Mehdorn M, Townsend J (1978) Normal perfusion pressure breakthrough theory. Clin Neurosurg 25: 651–672
17. Spetzler RF, Zabramski JM (1987) Operative selection of patients with arteriovenous malformations. Proceedings of the International Symposium on Surgery for Cerebral Stroke, Sendai, in press
18. Stein BM, Wolpert SM (1980) Arteriovenous malformations of the brain. II. Current concepts and treatment. Arch Neurol 37: 69–75
19. U HS (1985) Microsurgical excision of paraventricular arteriovenous malformations. Neurosurgery 16: 293–303
20. Wilson CB, U HS, Domingue J (1979) Microsurgical treatment of intracranial vascular malformations. J Neurosurg 51: 446–454
21. Wolpert SM, Stein BM (1975) Catheter embolization of intracranial arteriovenous malformations as an aid to surgical excision. Neuroradiology 10: 73–85

Address for correspondence: Prof. R. F. Spetzler, M.D., FACS, Barrow Neurological Institute, 350 West Thomas Road, Phoenix, AZ 85013, U.S.A.

Acta Neurochirurgica, Suppl. 42, 98–102 (1988)
© by Springer-Verlag 1988

Non-invasive Follow-up of Patients with Intracranial Arteriovenous Malformations After Proton-beam Radiation Therapy

H. M. Mehdorn and **W. Grote**

Department of Neurosurgery, University of Essen Medical Center, Essen, Federal Republic of Germany

Summary

In order better to follow non-invasively the effect of proton beam therapy upon intracranial arteriovenous malformations, transcranial Doppler sonography (TCD) was used in 10 patients, in addition to conventional follow-up with angiography and computed tomography (CT); in six patients, pre- and postradiation TCD studies were obtained. The mean flow values obtained from the major arteries in the circle of Willis indicate an early flow reduction through the AVM of approximately 15% with further reduction between 2/3 and 1/2 of the initial values within 15 months of follow-up. CT showed a reduction of contrast enhancement in those patients in whom scans could be compared. In this limited experience, TCD was proven to yield additional information, in repeated non-invasive studies, on the haemodynamic influence of AVMs and the effect of proton beam therapy.

Keywords: Arteriovenous malformation; radiation therapy; transcranial Doppler sonography.

Introduction

Although surgical excision is the most reliable method to prevent (re-)bleeding from intracranial arteriovenous malformations (AVM), the risk of surgery seems too high in some patients compared with the benefit of surgery. Therefore, often a more conservative approach is favoured, particularly in patients presenting with a history of focal seizures who are otherwise neurologically intact. In these patients, alternative therapies should be considered in view of the ever present risk of intracranial haemorrhage from an AVM, which is estimated to be approximately 2–3% per year with an annual mortality rate of approx. 1%[11], while the annual risk of recurrent haemorrhage is estimated at 6% in the first year and 2–3% in the following years.

As an alternative treatment reduce the risk of haemorrhage in the future, particularly for young patients, radiation therapy has been suggested by sev-eral authors, either as gamma radiosurgery[10], or as conventional radiotherapy[5], or as proton beam therapy[6]. While the first kind of treatment has mainly been restricted to AVMs with a diameter of less than 25 mm, Kjellberg[6,7] suggested that proton beam radiation therapy applied to a large AVM would also reduce the risk of subsequent haemorrhage.

Follow-up investigations of patients harbouring an irradiated AVM are mainly concerned with the clinical course. It is recommended that they include (I) angiographic control of therapy after a time interval of between 1 and 2 years, and that (II) computed tomography (CT) scans be performed every year following radiation therapy in order to evaluate any possible changes of visualization of the abnormal vasculature. Information obtained by other non-invasive and easily repeatable methods may be helpful in the future to study the effects of radiation therapy on AVMs.

This paper reports, on a preliminary basis, our experience gained with the application of transcranial Doppler sonography (1) in the follow-up of haemodynamics after radiation therapy of AVMs.

Material and Methods

Between 1981 and 1986, fifty-six patients were referred to stereotactic Bragg-peak proton beam therapy in Boston. Twenty-eight of them had suffered one or up to seven intracranial haemorrhages, and in the other 28 patients, a variety of clinical symptoms including seizures, ischaemic symptoms and major headache had finally led to non-invasive and invasive diagnostic procedures revealing an AVM. All AVMs except 2 were larger than 2 cm in diameter, as determined by angiography.

Follow-up evaluation of these patients included CT scans every year after proton beam therapy and control angiography two years after radiation. CT scans were mainly performed on an outpatient basis in the patients' home towns. Angiography was usually performed in our centre using digital subtraction angiography with intra-arterial contrast injection (i.a. DSA), however, some patients underwent angiography in their primary referral centre.

In the last 20 patients, transcranial Doppler sonography (TCD) was used to study noninvasively the haemodynamic influence of the AVM on the intracranial circulation. Among them, six patients were examined pre- and postradiation, and in 4 patients, several TCD follow-up studies were performed only after proton beam therapy.

TCD studies were performed by one observer using the TC 2-64 (EME Überlingen, Federal Republic of Germany). Instrumentation and techniques to identify the intracranial arteries have been described elsewhere in detail[1, 3, 4].

If possible, all major arteries of the circle of Willis were studied. The middle cerebral arteries (MCA) were measured at an insonation depth of 50–60 mm, the anterior cerebral artery (ACA – A_1-segment) at 70 mm, and the vertebrobasilar system (VBA) at 65 or 70 mm, occasionally at 80 mm. It was not attempted, however, to define the feeding arteries to the AVMs, rather the overall flow in the trunk of the MCA or the other respective arteries was measured.

Among the 6 patients studied pre- and postradiation, one had an AVM located in the posterior fossa and fed exclusively through the vertebrobasilar system, while the others hat supratentorially located AVMs with feeders from two or more systems. One of them had suffered from a series of 6 intracranial haemorrhages (ICH).

Results

1. Clinical Course

One patient with a previous history of ICH died from another ICH with intraventricular extension, 10 months following radiation therapy, another patient with a previous history of 2 bleedings from a deep left temporal AVM suffered another series of 3 bleeds within 4 weeks, 15 months after radiation therapy, from which he recovered completely. All other patients did not show any signs of ICH during the follow-up period between 6 and 66 months (average 35 months).

One patient, with a minor progressive deficit from a large left parietooccipital AVM, continued to develop a major hemiparesis in a follow-up interval of 6 years after treatment. As far as seizures are concerned, no definite conclusion could be reached; in some patients, no further seizures or a reduction of frequency occurred after proton beam therapy, however, in some patients seizures developed after therapy. Patients presenting with headaches or trigeminal neuralgia experienced a reduction of their pain. In one patient, in whom a posterior chorioideal AVM had disappeared on angiography 2 years after therapy, an enlarging scotoma was observed by the ophthalmologists without subjective complaints, another 3 years later.

2. Angiography

Thirty patients have undergone control angiography until now. In only 2 patients could complete obliteration of the AVM be observed. In 5 patients, a small residual AVM could be identified, and in 1 patient a draining vein was observed to be present. In 21 patients, no reduction of AVM size was evidenced.

3. CT-findings

A major difficulty was encountered when the comparison of sequential CT-findings was attempted: Because most of the follow-up CT scans had been performed elsewhere or with higher resolution instrumentation than initially used, these scans could not properly be compared. (Additional problems were varying positioning of the patients and variations in contrast application.)

CT scans which nevertheless could reasonably be compared, showed a decrease of contrast enhancement one year (Fig. 1) after proton beam therapy, and in some patients, further decrease of enhancement could be observed to occur in the ensuing years.

In patients developing seizures in the postradiation course, CT scans did not reveal any changes of brain tissue adjacent to the AVM which would have seen suggestive of additional glial scar formation. Also, in the patient mentioned above, CT scan did not reveal any infarction or radiation damage which could have explained the enlarge scotoma.

4. Transcranial Doppler Studies

Among the 20 patients studied, there were 14 patients in whom a marked side differences in the flow velocities in the proximal MCA could be observed. In the others, no difference was evident. Particularly, in the 6 patients studied preradiation and in the follow-up, the mean flow velocities in the proximal part of the MCA differed significantly in 5 of them from those of the contralateral side, and were above the normal age-related range. In patients in whom the VBA circulation contributed to the AVM, the mean flow velocities in the basilar artery were between 58 and 104 cm/sec.

Postradiation course of flow velocities in the feeding arteries was characterized, in the initial 14 months of follow-up, by a flow velocity reduction of 11–20% (average 15%) in the initial 3 months, another 4% in the next 3 months and 15% in the following 6 months. In a patient with a large right parietooccipital AVM fed through the VBA and the right MCA, flow velocity after 14 months was down to 58% (VBA) resp. 69% (MCA) of the preradiation values. In another patient with an AVM in the posterior fossa causing trigeminal neuralgia, TCD 13 months later showed a flow velocity reduction (Fig. 1) exclusively in the VBA down to 66% of the preradiation flow velocity.

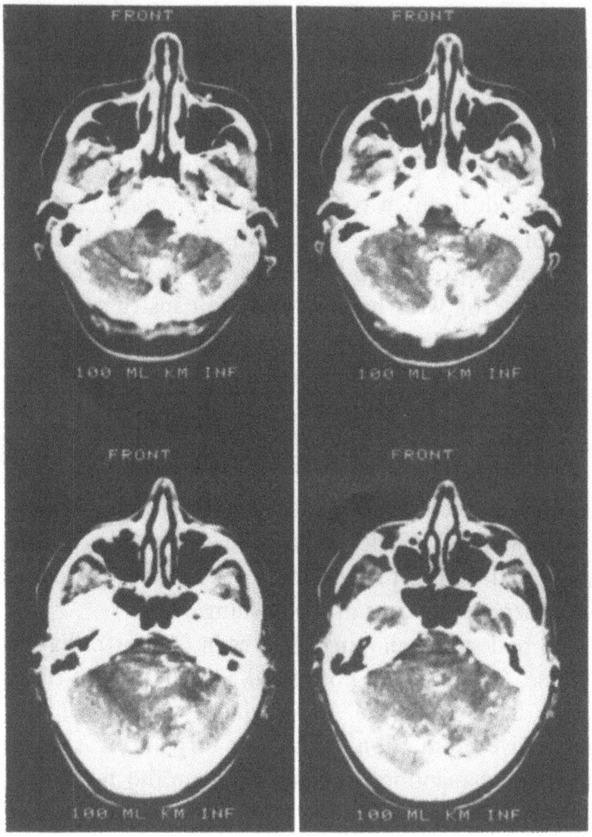

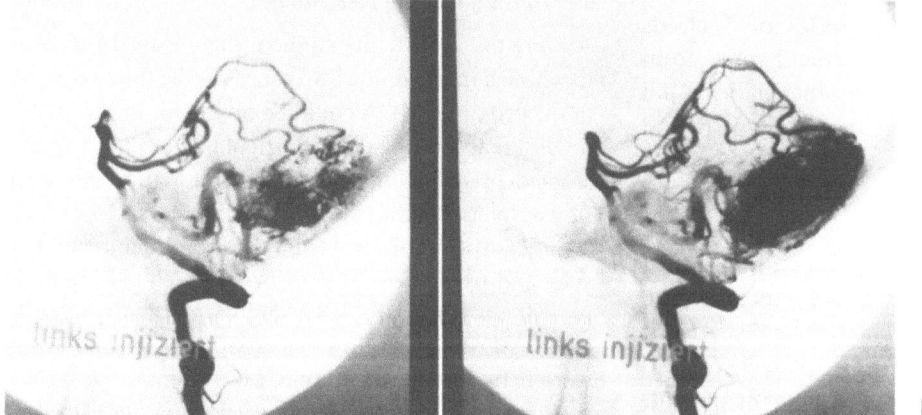

Fig. 1. Forty-six year old female patient presenting with history of trigeminal neuralgia. (a) Post-contrast CT scan shows hyperdense lesion in the right cerebellum suggestive of an AVM and draining veins. (b) Left vertebral angiography confirms the diagnosis of AVM fed through the basilar artery. (c) CT scan 13 months after proton beam therapy shows reduced contrast enhancement in the AVM. (d) Transcranial Doppler sonography study of the basilar artery. Numbers on the right side of the graphs indicate (from the top) insonation depth (mm), maximum systolic flow velocity (cm/sec) and mean flow velocity (cm/sec). In the later TCD studies, also the systolic/diastolic ratio is indicated. Proton beam radiation therapy was performed on April, 29, 1986

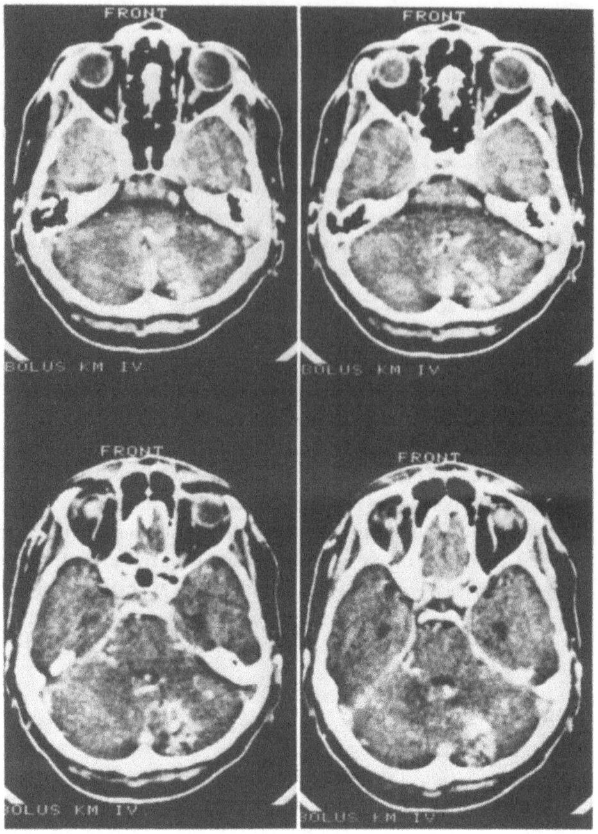

In patients in whom TCD follow-up was obtained only at later time during the follow-up, there was a different pattern of development: in 2 young patients with flow velocities of 106 resp. 130 cm/sec at intervals 6 and 11 months after radiation, no reduction of these values after additional intervals of 14 resp. 7 months was observed. In contrast, in another patient in whom the first TCD study was performed 4 years after ra-

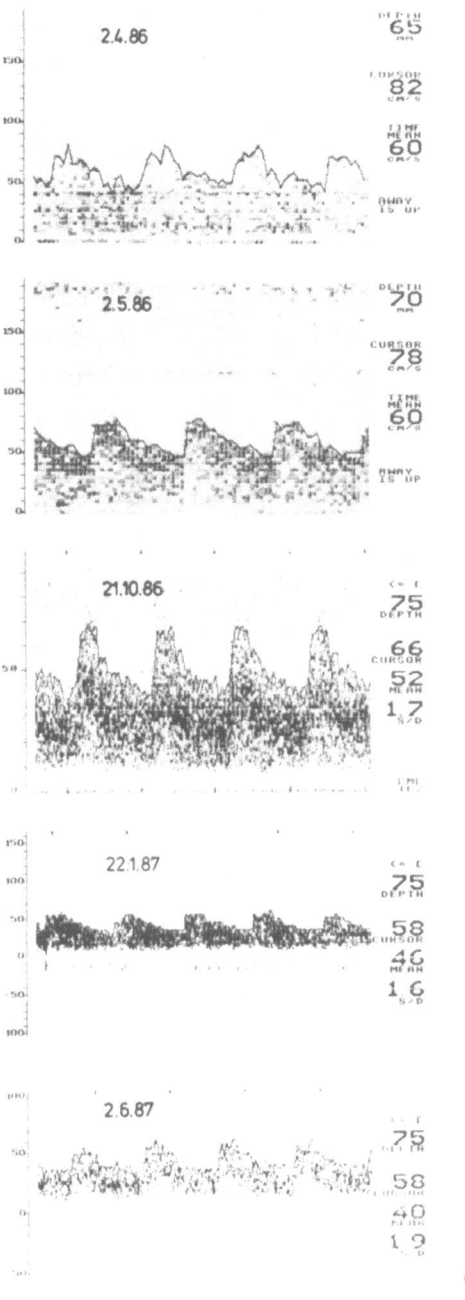

diation, the next study one year later showed a flow velocity reduction of 11% during this year on the affected side, while the contralateral MCA showed a flow increase of 50%.

Discussion

The value of proton beam therapy in the treatment of AVMs thought to carry an excessive surgical risk has been questioned. The "gold standard" by which to evaluate its benefit is the expected (re-)bleeding rate reduction. In this regard, out data do not allow any

definite conclusion because follow-up time is too short in view of the *longterm* bleeding risk of AVMs[9, 11]. Kjellberg *et al.*[7] state a significant risk reduction as analysed by life table analysis. On the other hand, Steiner[10] suggests that the rebleeding risk is present as long as a patient harbours an AVM that is visualized by angiography.

Our experience with control angiography in 30 patients was disappointing since wipe-out[2] of the AVM was seen only in 2 of them while there was reduction of the AVM in 6 and no obvious change in the remaining 21.

No additional information was given by CT scan. In the 2 patients mentioned who developed a neurological deficit in the later follow-up course, CT did not show any signs of infarction or other damage to the brain adjacent to or remote from the AVM. In view of the difficulties in obtaining comparable routine CT scans in a prolonged follow-up, we have concluded that CT is of limited value in estimating the longterm effect of proton beam radiation on reducing the AVM. However, it is still essential in order to evaluate any suspected brain damage caused by the AVM and/or radiation therapy, but it may well be that magnetic resonance imaging (MRI) will prove superior to CT in this period.

TCD studies were performed to evaluate flow velocity alterations following radiation. Initial flow velocities may deserve a comment: in the patients studied who presented with AVMs of a size more than 2 cm, TCD was able to detect the AVM by marked turbulence (bruits) in areas normally silent and/or by flow velocity increase. However, we found it difficult to identify the feeding arteries themselves close to the AVM because of superposition of the draining veins. Therefore, we preferred to use the flow velocities in the proximal MCA, and in ACA and VBA, to determine the alterations of haemodynamics caused by proton beam therapy. It was amazing to note, in 6 out of 20 patients, that the flow velocity in proximal MCA feeding the AVMs did not differ significantly from those in the MCA of the contralateral hemisphere; this has been described, previously[8] and may be explained by the well-known adaptation of the MCA diameter to the flow demand of the AVM.

In all 6 patients studied pre- and postradiation, there was a flow velocity reduction following radiation which began shortly after radiation and continued until 14 months postradiation at which time flow velocities reached 2/3 of the initial values. This reduction was observed in all the major feeding arteries studied, but was most striking in the VBA in patients presenting with either infratentorial AVMs or AVMs located in the parieto-occipital region that were fed also by the VBA system. It may well be that the VBA system is visualized best and without interference from draining veins, and therefore the follow-up is best achieved. The early flow velocity reduction with ensuing shrinking of the feeding arteries may well account for the twice observed time course of *early* postradiation relief of trigeminal neuralgia or headaches.

Among patients studied at later intervals no further significant reduction was seen. It may be speculated that after one year the feeding artery diameter will reduce in response to a reduced "demand" of a shrunken AVM nidus and therefore the flow velocity will remain constant. In only one patient was a further flow velocity reduction observed between years 4 and 5 postradiation; this was a patient with a large left temporal AVM which had already shown continuous reduction of enhancement during the preceding 4 years in well comparable CT scans.

TCD yields additional information concerning the haemodynamics situation of the main arteries of the circle of Willis feeding AVMs[4, 8]. The major advantage of TCD compared to i.a. DSA is that is can be repeated any time, and that it may therefore yield a time profile of radiation action upon the AVM nidus.

Further clinical experience and longterm follow-up of patients who have undergone proton beam therapy should show whether this flow velocity reduction is related positively to a lower rate of later intracranial haemorrhage. Furthermore, TCD should be used to classify AVMs according to their haemodynamic influence on the circle of Willis and to relate this to the beneficial effects of flow velocity reduction and possibly to the risk of later haemorrhages. It is hoped that this information may help to select patients who might benefit most from radiation therapy or combined surgery-radiation therapy.

References

1. Aaslid R (1986) Transcranial Doppler sonography. Springer, Wien New York
2. Drake C (1983) Arteriovenous malformations of the brain. The options for management. N Engl J Med 309: 308–331
3. Harders A (1986) Neurosurgical application of transcranial Doppler sonography. Springer, Wien New York
4. Hassler W (1986) Haemodynamic aspects of cerebral angiomas. Acta Neurochir (Wien) [Suppl] 37 Springer, Wien New York
5. Johnson RT (1975) Radiotherapy of cerebral angiomas with a note on some problems in diagnosis. In: Pia HW, Gleave JRW, Grote E, Zierski J (eds) Cerebral angioma: advances in diagnosis and treatment. Springer, Berlin Heidelberg New York, pp 256–259
6. Kjellberg RN, Poletti CE, Roberson GH, Adams RD (1977) Bragg peak proton beam therapy treatment of arteriovenous malformations of the brain. Excerpta Med Int Congr Ser 433: 181–187
7. Kjellberg RN, Hanamura T, Davis KR, Lyons SL, Adams RD (1983) Bragg peak proton beam therapy for arteriovenous malformations of the brain. N Engl J Med 309: 269–274
8. Lindegaard KF, Grolimund P, Aaslid R, Nornes H (1986) Evaluation of cerebral AVMs using transcranial Doppler ultrasound. J Neurosurg 65: 335–344
9. Mehdorn HM, Ebbing H, Grote W (1983) Late results of surgical vs. conservative therapy of intracranial angiomas. 7th Europ Congr Neurosurg, Brussels (Abstr no 1)
10. Steiner L, Leksell L, Forster DMC, Greitz T, Backlund ED (1974) Stereotactic radiosurgery in intracranial arteriovenous malformations. Acta Neurochir (Wien) [Suppl] 21. Springer, Wien New York, pp 195–209
11. Wilkins RH (1985) Natural history of intracranial vascular malformations: a review. Neurosurgery 16: 421–430

Address for correspondence: Priv. Doz. Dr. H. M. Mehdorn, Neurochirurgische Universitätsklinik, Hufelandstrasse 55, D-4300 Essen 1, Federal Republic of Germany.

Acta Neurochirurgica, Suppl. 42, 103–106 (1988)
© by Springer-Verlag 1988

774 Carotid Endarterectomies for Strokes and Transient Ischaemic Attacks: Comparison of Results of Early vs. Late Surgery

H. L. Khanna and **A. G. Garg**

Department of Neurological Surgery, Northeastern Ohio University College of Medicine, Western Reserve Care System, Youngstown, Ohio, U.S.A.

Summary

A total of 774 carotid endarterectomies were done; 363 (47%) for completed strokes and strokes with unstable neurological status and 411 (53%) for transient ischaemic attacks. One hundred eight emergency carotid endarterectomies were done in the stroke group for either recurrent strokes or where the lumen of the internal carotid was less than 0.5 mm. Eighty-four emergency carotid endarterectomies were done in the transient ischaemic attack group for either recurrent T.I.A.s or T.I.A.s with carotid lumen less than 0.5 mm. There were 14 (3.8%) deaths in the stroke group and 3 (0.7%) in the transient ischaemic attack group. In the stroke group, the lowest mortality of 1.6% (4 deaths) was in 246 carotid endarterectomies done in the first week, while 87 done in the second week had 3.4% mortality (3 deaths) and 30 done in the third week or later had 23.3% mortality (7 deaths). The highest mortality was in patients with recurrent strokes during the same admission before surgery and in whom surgery was delayed until the third week or later. Primary cause of death was myocardial infarction, pulmonary embolism but not neurological.

Keywords: Stroke; carotid endarterectomy.

Introduction

The present study was undertaken to determine if early carotid endarterectomy in completed strokes, strokes in evolution and strokes with unstable neurological status (C.S.) would offer any benefit in reducing mortality and morbidity. It was started after two patients died of recurrent strokes in four weeks while waiting for surgery in 1975. The transient ischaemic attack (T.I.A.) group was included to see if there is any difference in the mortality and morbidity between the two groups. It was also to determine the most common cause of death in both groups.

Material and Method

774 carotid endarterectomies (C.E.) were done for T.I.A. and C.S. All had complete history, physical, neurological and cardiological examination. In addition to the routine work-up, they also had CT brain scan and bilateral carotid, vertebral (both intracranial and extracranial) and arch arteriography within 48 hours of our initial examining the patient. In cases of stroke, an attempt was made to get arteriography and CT brain scan within 24 hours.

Patients were not recommended for carotid endarterectomy if they had (1) complete hemiplegia which showed no signs of improvement or recovery of function, (2) source of emboli proximal to the carotid bifurcation, (3) CT brain scan showing intracranial haemorrhage and oedema, (4) a recent myocardial infarction (M.I.), and (5) any medical condition which precluded surgery.

Age varied from 23 years to 92 years, and the mean age in both C.S. and T.I.A. group was 67 years. The size and degree of stenosis of the internal carotid artery (I.C.A.) was measured in millimeters at its narrowest point of stenosis. In the C.S. group of 363, 52 (14.3%) had more than 2 mm, 96 (26.4%) had 1 to 2 mm, and 215 (59.2%) had less than 1 mm lumen. In the T.I.A. group of 411, 73 (17.7%) had greater than 2 mm, 130 (31.6%) had 1 to 2 mm, and 208 (50.7%) had less than 1 mm lumen (see Table 1).

In the C.S. group, 363 C.E. were done in 345 patients. Eighteen bilateral C.E. were done for bilateral strokes, and of these, eight were done as one-stage emergency bilateral carotid endarterectomy for bilateral strokes before surgery during the same admission.

Table 1. *Showing Degree of I.C.A. Stenosis in Strokes and T.I.A.*

SIZE OF I.C.A. LUMEN	NO. OF C.E. FOR STROKES	NO. OF C.E. FOR T.I.A.
> 2 mm.	52 (14.3%)	73 (17.7%)
1 - 2 mm.	96 (26.4%)	130 (31.6%)
< 1 mm.	215 (59.2%)	208 (50.7%)
TOTAL	363	411

Table 2. *Showing Time Interval Between Onset of Symptoms and Carotid Endarterectomy*

WEEKS		NO. OF CAROTID ENDARTERECTOMIES	
	DAYS	STROKES	T.I.A.s
1ST WEEK		246	245
	0	4	3
	1	34	20
	2	46	46
	3	27	32
	4	34	29
	5	29	35
	6	29	32
	7	43	48
2ND WEEK		87	100
3RD WEEK OR LATER		30	66
TOTAL		363	411

Twenty-two patients in this group had complete occlusion of the internal carotid artery (I.C.A.). Since the duration of stroke was within less than 48 hours, these were explored as emergencies in an attempt to reopen the occluded arteries. An arbitrary cut-off limit of 48 hours was used in the hope that thrombus may still not be firmly adherent to the arterial wall. Seventy (19.2%) C.E. were done for mild motor and/or sensory neurological deficit, 93 (25.6%) for moderate C.S. (able to raise the affected limb against gravity but not hold it up) and 200 (55.1%) for severe neurological deficit (unable to raise the limb against gravity with or without severe dysphasia or homonymous hemianopsia). Forty-four (12.1%) had recurrent strokes or strokes preceded by T.I.A.s while in the hospital before surgery, and 25 (7.2%) had hemiplegia initially which improved before surgery. Seventy-eight patients (22.4%) showed a definite infarct on CT brain scan soon after admission. CT brain scan was not repeated after surgery unless there was increased neurological deficit. Carotid endarterectomies for C.S. and T.I.A.s were divided into three groups. In the C.S. group 1, 246 C.E. (67.7%) were done in the first week, group 2, 87 C.E. (23.9%) done in the second week, and group 3, 30 C.E. (8.2%) done in the third week or later after the stroke. In the T.I.A. group 1, 245 C.E. were done in the first week; group 2, 100 C.E. in the second week; and group 3, 66 in the third week or later. Table 2 shows a detailed breakdown and information.

Management

All diagnostic work was completed within 48 hours of our seeing the patient in most cases. Patients with strokes in evolution or fluctuating neurological status, recurrent strokes, recurrent T.I.A.s, patients with less than 0.5 mm lumen, and strokes with complete occlusion of the corresponding I.C.A. had emergency carotid endar-

terectomy (E.C.E.) within 12 hours of establishing the diagnosis. If there was a delay in surgery, then they were started on Heparin four hours after arteriography and were monitored in the intensive care unit (I.C.I.) until the time of surgery. One hundred eight E.C.E. within 12 hours of diagnosis were performed in C.S. group and 84 in the T.I.A. group. Blood pressure was controlled, and patients were maintained in a normotensive range. Early surgery within the first week was the goal in both T.I.A. and C.S. patients. The delay in surgery was primarily because of delayed referral of patients to us for surgery. The presence of an infarct on CT brain scan was not considered a contraindication to surgery.

All patients were evaluated by anesthesiologist, and an arterial line was started before anesthetic. Manipulation of the carotid artery was avoided. An intraluminal by-pass shunt to maintain cerebral circulation during surgery was used in all patients during C.E. Heparin, 7,500 units, was given intravenously two minutes before clamping of the carotid arteries. After opening the artery, the lumen was sucked out thoroughly, and retrograde blood flow was checked before installing the by-pass shunt. All shreds of loose tissue were meticulously removed with magnification and microsurgery. Patch graft was not used in any patient. After closure of arterotomy, circulation was first restored between the common and external carotid artery, and then the internal carotid artery was also opened. Heparin was only partially neutralized with 20 mg of Protamine ten minutes after the final closure of arterotomy and restoration of circulation. Blood pressure (B.P.) was maintained between 110 mmHg and 180 mmHg systolic. This figure was chosen in consultation with cardiologists. Patients were kept in I.C.U. and continuously monitored after carotid endarterectomy for at least 24 hours. Hypertension was controlled with the use of sodium nitroprusside, nitroglycerin, trimetaphan camsylate or a combination of these. Hypotension was controlled primarily with Dopamine (Intropin) or Neo-Synephrine (phenylephrine). The selection of medication and dose was determined according to individual need of each patient. In rare circumstances, Epinephrine, 0.2 to 0.5 mg bolus dose was given if Neo-Synephrine or Dopamine failed.

Results

Among 411 C.E. in T.I.A., there were three deaths (0.7%) mortality; two from M.I. and one from pulmonary embolus (P.E.). There was no relationship to the time of surgery. There was permanent increased neurological deficit after eight C.E. (1.9%).

Amongst 363 C.E. for C.S., 290 (80%) improved, 51 (14.0%) remained unchanged, 8 (2.2%) deteriorated and 14 (3.8%) died. Seven patients died of M.I., four of P.E., one from G.I. bleeding, one from associated head injury, and one from brain stem stroke. Five of the 14 patients who died had two or more recurrent strokes, and three had recurrent T.I.A.s before the stroke during the same admission and deteriorated before being referred for surgery. Seven of these patients were operated upon two to seven weeks after the initial stroke. There was a high incidence of death from M.I. and P.E. in patients with recurrent strokes. Five out of 44 patients with recurrent stroked died; two from

Table 3. *Showing Breakdown of Mortality for C.E. in Completed Strokes at Different Time Intervals*

WEEKS	NO. OF C.E.	MORTALITY
1ST WEEK	246	4 (1.6%)
2ND WEEK	87	3 (3.4%)
3RD WEEK OR LATER	30	7 (23.3%)
TOTAL	363	14

M.I. and three from P.E. There was no incidence of an ischaemic infarct changing into a haemorrhagic infarct in the 363 C.E. done for C.S. Of the 246 C.E. done in the first week, four died (1.6%); of the 87 C.E. in second week, three died (3.4%) and out of 30 C.E. in third week or later, seven died (23.3%) (see Table 3).

Discussion

Early C.E. for T.I.A.s is well accepted. There is, however, a controversy as to the recommended time for C.E. in C.S. because of a risk of haemorrhage in ischaemic infarcts following restoration of cerebral circulation. Delayed surgery has been recommended by Wylie *et al.*[17] and others[1, 2, 4, 6, 10, 11, 15, 16]. Early C.E. has been favored by Mentzer[12] and others[3, 7, 8, 9, 14] even though the number of cases studied was small. In the present study, the lowest mortality was in patients with only one stroke and operated in the first week. There were only four deaths (1.6%) in 246 C.E. done in the first week, 80% of the patients improved, there was 2.2% increased morbidity. The higher rate and better return of function following early C.E. may be from restoration of better or normal blood flow to ischaemic penumbra (ischaemic viable but nonfunctional neural tissue bordering an infarct). Surgical mortality of 1.6% in group 1, compared to 0.7% in T.I.A. group, is not significant statistically. However, when compared to mortality in group 3 of C.S. patients, the difference is significant ($p < 0.000001$).

Of the 78 patients operated in the second week, three (3.8%) died, while 30 patients in the third week or later had seven (23.3%) deaths. The difference is again significant ($p < 0.001$). High mortality was associated with patients with two or more recurrent strokes in the same admission before surgery. Five of the 14 patients who died had two or more recurrent strokes. The presence of cerebral infarct on the CT scan did not affect

the outcome. The low incidence of cerebral infarcts on CT scan in C.S. patients may be because most of these patients had CT brain scan in less than 24 hours of their admission or stroke. CT scan after surgery was only done in patients who had recurrent neurological deficit and was not done as a routine in all cases. None of the 78 patients who showed infarct on CT scan before developed haemorrhage in the ischaemic infarct after C.E. and restoration of normal blood circulation to the affected cerebral hemisphere. The mortality was primarily from M.I. and P.E. There were only two partially neurological deaths, one from associated head injury and one from brain stem stroke and pneumonitis. Neither of these were directly related to C.E. Fisher[6] suggested hypertension as the main cause of haemorrhage in ischaemic infarcts. In the present series, attention was paid to maintain the systolic pressure between 110 mmHg and 180 mmHg during and after surgery. This figure, as mentioned earlier, was taken arbitrarily in consultation with cardiologists. This may have helped in preventing haemorrhage in ischaemic infarcts. Harrison and Marshall[9] found presence of thrombus and thromboembolic material in 66% of patients in carotid arteries at C.E. for four weeks. In the present series, an attempt was made to surgically eliminate the cause of recurrent embolization by early C.E. in both T.I.A. and C.S. with gratifying results.

Conclusion

From the present study it appears that early C.E. within one week of C.S. gives much better results with very low mortality. It is not significantly different statistically from similar surgery for T.I.A. A risk of ischaemic infarct changing to haemorrhagic infarct can be greatly reduced by strict control of blood pressure during and after surgery. Recurrent strokes during the same admission before surgery increased mortality and morbidity significantly, irrespective of the timing of surgery.

References

1. Blaisdell WF, Clauss RH, Galbraith JG *et al* (1969) Joint study of extracranial arterial occlusion I.V. A review of surgical consideration. JAMA 209: 1889–1895
2. Bruetman MF, Fields WS, Crawford ES, DeBakey ME (1963) Cerebral haemorrhage in carotid artery surgery. Arch Neurol 9. 458–467
3. Callow AD (1980) An overview of the stroke problem in the carotid territory. Am J Surg 140: 181–190

4. Caplan LR, Ojemann R (1978) Intracerebral haemorrhage following carotid endarterectomy: A hypertensive complication. Stroke 9: 457–460

5. Ciriello J (1985) Associate Professor of Physiology, University of Western Ontario, London, Ontario, Canada: A paper read in "An Update on Cerebrovascular Disease" in September

6. Fisher CM (1951) Occlusion of the carotid arteries. Arch Neurol Psychiatry (Chicago) 65: 346–377

7. Goldstone J, Moore WS (1978) A new look at emergency carotid artery operations for the treatment of cerebrovascular insufficiency. Current concepts of cerebrovascular disease. Stroke 9: 599–602

8. Goldstone J, Moore WS (1976) Emergency carotid artery surgery in neurologically unstable patients. Arch Surg 111: 1284–1291

9. Harrison MJG, Marshall J (1977) The finding of thrombus at carotid endarterectomy and its relationship to the timing of surgery. Br J Surg 64: 511–512

10. Hass WK, Clauss RH, Goldberg AF, Johnson AL, Imparito AM, Rausohoff J (1966) Special problems associated with surgical and thrombolytic treatment of strokes. Arch Surg 92: 27–31

11. Lehv MS, Salzman EW, Silen W (1970) Hypertension complicating carotid endarterectomy. Stroke 1: 307–313

12. Mentzer RM, Finkelmeir BA, Crosby IK, Wellon HA Jr (1981) Emergency carotid endarterectomy for fluctuating neurological deficit. Surgery 89: 60–66

13. Najafi H, Javid H, Dye WS, Hunter JA, Widemann FE, Julian OC (1971) Emergency carotid thromboendarterectomy: Surgical indications and results. Arch Surg 103: 610–614

14. Ojemann RG, Croweel RM, Robertson GH, Fisher CM (1975) Surgical treatment of extracranial carotid occlusive disease. Clin Neurosurg 22: 214–262

15. Sundt TM, Sharbrough FW, Piepgras DG, Kearns TP, Messick JM, O'Fallon WM (1981) Correlation of cerebral blood flow and electroencephalographic changes during carotid endarterectomy: With results of surgery and hemodynamics of cerebral ischaemia. Mayo Clinic Proc, Sept 56 (9): 533–543

16. Thompson JE, Talkington CM (1979) Carotid surgery for cerebral ischaemia. Surg Clin North Am 59: 539–553

17. Wylie EJ, Hein MF, Adams JE (1964) Intracranial haemorrhage following surgical revascularization for treatment of acute stroke. J Neurosurg 21: 212–215

Address for correspondence: Hira L. Khanna, M.D., Assistant Professor in Neurological Surgery, 3020 Fredrick Drive, Youngstown, OH 44505-2098, U.S.A.

III. Intracranial Tumours

Acta Neurochirurgica, Suppl. 42, 109–112 (1988)

Cystic Craniopharyngioma Treated by [90]Yttrium Silicate Colloid

J. A. Guevara, H. J. Bunge, J. J. Heinrich, G. Weller, A. Villasante, and **A. B. Chinela**

Centro de Radiocirugía Neurológica, Clínica del Sol, Buenos Aires, Argentina

Summary

Fourteen children and 3 adults with cystic craniopharyngiomas were treated with intracavitary ^{90}Y, by the procedure described by Backlund. Their ages ranged from 2 to 65 years and postoperative follow-up ranged from 6 to 40 months. Leksell's stereotactic technique was employed to determine coordinates by CAT. Cyst volume was quantified both geometrically and isotopically with ^{99}Tc, values differing by 7%. Dosimetry was determined by applying the formula developed by Loevinger et al., and 20,000 rads were administered throughout to the cystic wall. All 17 patients, except for 4 children, had previously received surgery, shunts or radiotherapy, alone or combined. In 4 cases, ^{90}Y injection was the only treatment, while in 6, the cyst was evacuated at 10 days following radiocolloid injection. Skull and spinal column gamma chamber studies were carried out on all patients at 24, 48, and 72 hours post injection, but no isotope leakage could be detected. The patients returned to normal activities except one with multiple cysts who died. There were no changes in the endocrinologic profile. In one case, a decrease in visual acuity 18 months after treatment, improved following corticoid administration.

Conclusions: Though preliminary, these results are encouraging since it seems that the severe neuro-endocrinologic sequelae of open surgery may be avoided.

Keywords: Craniopharyngioma; cyst; treatment; [90]Yttrium silicate colloid.

Introduction

Craniopharyngiomas make up 9% of cerebral neoplasms in infancy and 4.5% of all brain tumours[19, 25]. In spite of being histologically benign tumours, their clinical behaviour is commonly aggressive in paediatric patients[20]. Owing to the fact that the cystic component often makes up the larger portion of the tumour mass, it may be possible to concentrate treatment on this part in many cases.

During the last 50 years, with the advent of hormonal replacement therapy, the use of focused radiotherapy and the introduction of microsurgery, the prognosis for these patients has been significantly improved. However, despite this progress controversies over the treatment of these neoplasms have not yet been settled. Hoffman and coworkers supported surgical excision as the treatment of choice, given the expectations of recovery, the prevention of recurrences and the avoidance of undesirable radiation effects[8]. An entirely different approach has been advanced by Mori et al., who contend that craniopharyngioma should be considered a malignant tumour, warning against any radical operation which might damage neighbouring structures[18]. For their part, Thomsett and coworkers demonstrated that partial tumour removal associated with radiotherapy led to a better neurological and endocrinological prognosis, in spite of the potential risk due to radioactive exposure[24].

Craniopharyngioma symptomatology is commonly due to recurrence of the tumour's cystic portion, while the solid part presents a slower growth, providing a substrate for the development of new cysts. At the Neurosurgical Department, Karolinska Hospital, Sweden, a methodology has been developed to treat cysts by means of stereotaxic injection of a colloidal beta emitter ([90]Yttrium). The aim of the present report is to describe the results of this recent technique in a series of paediatric and adult patients.

Clinical Material and Methods

A total of 17 patients, 14 children and 3 adults, were treated from December 1983 to December 1986. Children's ages ranged from 1.6 to 14 years, with a mean of 8 years; adults' ages ranged from 53 to 68, with a mean of 63 years. There were 10 females and 4 males in the paediatric group, while the 3 adults were females.

As regards prior treatment, three groups of patients were discerned:

- primary, with [90]Yttrium as the initial procedure (5 cases),
- previous surgery, with prior subtotal removal of the tumour (8 cases) and
- previous surgery plus radiotherapy (4 cases).

All patients presented cystic-solic craniopharyngiomas. In 12 patients partial tumour resection had been previously carried out,

Table 1

Patients	Sex	Age/years	Neurological symptoms	Visual defects	Endocrine status
G.P.	f	10	headache—nausea vomiting—hippus	hemianopia papilloedema	short in stature
A.M.	f	10	headache—vomiting anorexia—somnolence	hemianopia	panhypopituitarism diabetes insipidus
M.S.G.	m	5	headache—vomiting andrexia—seizures	normal	panhypopituitarism diabetes insipidus
R.A.	f	9	headachew—nausea vomiting	III par palsy	hypothyroidism short in stature
G.A.	f	14	headache	hemianopia	short in stature
M.C.	f	7	vomiting	blindness	panhypopituitarism diabetes insipidus
T.G.	f	14	seizures—hemiparesis hydrocephalus		panhypopituitarism diabetes insipidus
O.D.	f	7	headache loss weight	strabismus	panhypopituitarism diabetes insipidus
F.F.	m	2	headache—vomiting ataxia—hydrocephalus	not performed	normal
M.M.	m	8	headache	hemianopia	panhypopituitarism diabetes insipidus
L.M.C.	f	7	headache—vomiting	campimetric. defect	short in stature
B.N.B.	f	6	coma	optic atrophy	panhypopituitarism diabetes insipidus
D.L.	f	10	headache—vomiting somnolence—seizures	normal	panhypopituitarism
N.P.	m	$1^1/_2$	seizures	not performed	panhypopituitarism short in stature
C.G.	f	68	headache—frontal lobe disorder	decreased visual acuity	panhypopituitarism diabetes insipidus amenorrhoea
C.A.	f	53	hemiparesis	decreased visual acuity	normal
S.P.L.	f	67	headache—vomiting stiffneck	hemianopia	normal

Table 2

Treatment	No. patients	Evolution		Deaths
		Good	Poor	
Primary	5	4	1*	0
Previous surgery	8	5*	3	0
Previous surgery Plus radiotherapy	4	3		1
Total	17	12	4	1

* One case with double irradiation.

some also required shunting. Time elapsing from surgery to recurrence ranged from 3 to 38 months, with a mean of 15.6 months. In addition, 4 of these patients received conventional radiotherapy, recurrence taking place from 12 to 48 months, with a mean of 30 months. The remaining 5 cases, out of the original 17, received [90]Yttrium as primary treatment.

All patients exhibited neurological and visual symptoms (Table 1), and some presented endocrinological symptomatology, as outlined in Table 2.

After general anaesthesia in paediatric cases, patients were fitted with Leksell's stereotaxic frame, setting blank coordinates by means of a stereotaxic CT scan. Cyst volume was determined: a) tomographically; and b) by the dilution method, injecting ^{99}Tc into the cyst. A 7.5% difference was found between the two determinations. Data thus obtained were employed to calculate the radioisotope dose to be injected, by means of Loevinger's formula[16]. A sufficient amount of [90]Yttrium was given so that the cystic wall should receive a 20 Krad dose. Following the procedure, all patients were monitored in the Gamma Chamber in order to check for radiocolloid loss, but no leakage could be detected in any case. Postradiation evacuation of the cyst was carried out in 9 patients and deemed unnecessary in the remainder.

Follow-up ranged from 6 to 40 months, and patients received neurological, endocrinological and ophthalmological control periodically, although 2 had incomplete eye tests due to limited age. Cyst size was evaluated by CT scan.

Results

Evolution was considered satisfactory when there was improvement or remission of neurological and visual symptoms with the disappearance of CT scan evidence of cysts, children were able to perform their school tasks and adults were capable of caring for themselves (Table 2).

Twelve patients had a favourable evolution, of whom 4 were primary cases, 5 had received previous surgery and the remaining 3 had already received surgery plus radiotherapy. In 4 cases the evolution was poor, due to cyst recurrence leading to severe visual defects in 3 children (M.C., B.N.B., and D.L.). The remaining patient was an adult (C.G.) who was lost to later follow-up. In 2 children (M.S.G. and M.C.) treatment was repeated 12 and 24 months after the first procedure, since they presented cyst recurrence. In one of them (M.S.G.) evolution was favourable, while in the other it was poor. One girl (G.P.) presented rapid progressive visual deterioration at 20 months posttreatment. CT scan showed no evidence of cyst recurrence. Sight was totally recovered after the patient received oral corticoids during 3 months. Two children (F.F. and M.M.) with a ventricular shunt developed ascites, so that the peritoneal derivation was switched to the circulatory system, with satisfactory disappearance of symptoms.

Out of the 17 patients treated, there was a single death (A.M.) 10 months post-treatment. This was a patient who had previously received surgery and radiation twice and presented multiple hypothalamic cysts with severe endocrinological involvement.

Discussion

It is evident that there is no single treatment for craniopharyngioma. Even in the most successful series, conventional surgery leads to less than 40% of total removal. In the 48 patients reported by Hoffman and coworkers, total tumour resection was only achieved in 17, with 2 deaths[8]. Likewise, Shillito *et al.* performed total removal in only 7 out of 20 patients[21, 22]. Besides, it is known that radiotherapy, as a complement to surgery, significantly lowers the incidence of recurrence and improves both neurological and endocrinological evolution[6, 20].

In paediatric patients treated by non-radical surgery and radiotherapy, Cavazzuti *et al.* demonstrated more satisfactory neurological evolution than in those in whom an attempt had been made to resect the tumour completely[5]. However, complications following radiotherapy include radionecrosis, the appearance of new tumours, vascular alterations, deafness and late hypopituitarism[15, 23]. Therefore, the use of GAMMA UNIT radiosurgery seems a logical alternative for the treatment of the solid portion, and the employment of Beta emitters for cyst treatment. The GAMMA UNIT III was applied on one occasion in the case of a boy whose solid tumour had been partially resected 18 months before. This case will be described in a separate communication.

Internal irradiation for cystic craniopharyngioma therapy was initiated by Leksell in 1952[13], employing radioactive chromic phosphate, and later adopted by other authors[1, 2, 3, 4, 11, 14]. The [90]Yttrium currently employed provides the ideal conditions for internal radiation: a short half-life, easy application and the fact that it is a pure beta emitter. In our cases no isotope loss was found through the injection orifice nor were signs of damage to neighbouring structures observed. As regards dosage, 20 Krad were used regardless of cyst volume, although this dose seemed insufficient in some case. There was hardly any difference in volumetric determinations according to the method applied, suggesting that no multiple cysts were found. As a rule CT scan data were taken into account, as reported by Lunsdorf *et al.*[17].

Best results were achieved in the group of primary cases. Four out of the 5 treated patients follow a normal life. One of them, who presented a quadrantopsia prior to treatment (L.M.C.), improved within 15 days postinjection. One year later this visual defect reappeared; at that time the CT scan was normal. One of the girls belonging to the primary group had an unfavourable evolution. This patient presented a cyst having a thick wall and very viscous contents. She was treated on two occasions and on the second a quantity of liquid, estimated to be less than required to collapse the cyst, was evacuated with difficulty. This failed to prevent further visual deterioration, so that a new subfrontal approach was performed to resect the cyst and a small solid portion. It would seem that in this case the thickness of the capsule for the dose administered was mainly responsible for therapy failure. There were no significant differences between results obtained in children and adults, in spite of findings reported by other authors[7, 10].

Although postoperative follow-up is admittedly brief, we believe that data presented here demonstrate the efficacy of this method as regards the quality of life achieved and the low morbidity and mortality.

References

1. Backlund EO (1972) Studies on craniopharyngiomas. I. Treatment past and present. Acta Chir Scand 138: 743–747
2. Backlund EO, Johansson L, Sarby B (1972) Studies on craniopharyngiomas. II. Treatment by stereotaxic and radiosurgery. Acta Chir Scand 138: 749–759
3. Backlund EO (1973) Studies on craniopharyngiomas. III. Stereotaxic treatment with intracystic [90]Yttrium. Acta Chir Scand 139: 237–247
4. Backlund EO (1973) Studies on craniopharyngiomas. IV. Stereotaxic treatment with radiosurgery. Acta Chir Scand 139: 344–351
5. Cavazzutti V, Fischer EG, Welch K, Belli JA, Winston KR (1983) Neurological and psychophysiological sequelae following different treatment of craniopharyngiomas in children. J Neurosurg 59: 409–417
6. Fischer EG, Welch K, Belli JA, Wallman J, Shillito J, Winston KR, Cassady R (1985) Treatment of craniopharyngiomas in children: 1972–1981. J Neurosurg 62: 496–501
7. Hoff JT, Patterson RH (1972) Craniopharyngiomas in children and adults. J Neurosurg 36: 299–302
8. Hoffman HJ, Hendrick EB, Humphreys RP *et al* (1977) Management of craniopharyngioma in childhood. J Neurosurg 47: 218–227
9. Julow J, Langi F, Hajda M, Simkovics M, Arany I, Toth SA, Pastor E (1985) The radiotherapy of cystic craniopharyngioma with intracystic installation of ^{90}Y silicate colloid. Acta Neurochir (Wien) 7: 94–99
10. Kahn EA, Gosch HH, Seeger JF *et al* (1973) Forty-five years experience with the craniopharyngiomas. Surg Neurol 1: 5–12
11. Kobayashi T, Kageyama N, Ohara K (1981) Internal irradiation for cystic craniopharyngioma. J Neurosurg 55: 896–903
12. Kodama T, Marsukado Y, Uemura S (1981) Intracapsular irradiation therapy of craniopharyngiomas with radioactive gold. Neurol Med Chir (Tokyo) 21: 39–58
13. Leksell L, Liden K (1951) A therapeutic trial with radioactive isotopes in cystic brain tumour. In: Radioisotope techniques, vol I. Proc. Isotope. Techniques Confer. Oxford. July 1957. H. M. Stationery Office, London
14. Leksell L, Backlund EO, Johansson L (1967) Treatment of craniopharyngiomas. Acta Chir Scand 133: 345–350
15. Liwnicz BH, Berger TS, Wiwnics RG, Aron BS (1985) Radiation associated gliomas: a reporte of four cases and analysis of postradiation tumours of the central nervous system. Neurosurgery 17: 436–445
16. Loevinger R, Japha EM, Brownell GL (1956) In: Hine and Brownell (eds) Radiation dosimetry. Academic Press, New York
17. Lunsford LD, Levine G, Gumeran L (1985) Comparison of computerized tomographic and radionucleide methods in determining intracranial cystic tumour volumes. J Neurosurg 63: 740–744
18. Mori K, Taketschi J, Ishikawa M, Handa H, Toyama M, Yamaki T (1978) Occlusive arteriopathy and brain tumour. J Neurosurg 49: 22–35
19. Russell DS, Rubinstein LJ (1977) Pathology of tumours of the nervous system, 4th edit. Arnold, London, 448 p
20. Shapiro K, Till K, Grant DN (1979) Craniopharyngiomas in childhood: A rational approach to treatment. J Neurosurg 50: 617–623
21. Shillito J (1976) Treatment of craniopharyngioma of childhood. In: Morley TP (ed) Current controversies in neurosurgery. WB Saunders, Philadelphia, pp 332–336
22. Shillito J Jr (1985) Treatment of craniopharyngioma. Clin Neurosurg 33: 533–546
23. Sundaresan N, Galicich JH, Deck MDF *et al* (1981) Radiation necrosis after treatment of solitary intracranial metastases. Neurosurgery 8: 329–333
24. Thomsett MJ, Conte FA, Kaplan SL, Grumbach MM (1980) Endocrine and neurologic outcome in childhood craniopharyngioma: Review of effect of treatment in 42 patients. J Pediatr 97: 728–735
25. Zülch KJ (1965) Brain tumours, their biology and pathology, 2nd Edit. Springer, Wien, pp 213–218

Address for correspondence: Dr. J. A. Guevara, Centro de Radiocirugía Neurológica, Clínica del Sol, Arenales 1468, (1061) Buenos Aires, Argentina.

Acta Neurochirurgica, Suppl. 42, 113–119 (1988)
© by Springer-Verlag 1988

Further Experiences in the Treatment of Cystic Craniopharyngeomas with Yttrium 90 Silicate Colloid

J. Julow, F. Lányi, M. Hajda, G. Szeifert, M. Simkovics, Sz. Tóth, and **E. Pásztor**

National Institute of Neurosurgery, Budapest, Hungary

Summary

20 patients suffering from cystic craniopharyngioma were treated with intracavitary irradiation on 25 occassions. The beta emitting radionuclide ^{90}Y silicate colloid was instilled into the cyst or cystic part of the tumour. 17 patients on 22 occasions underwent follow-up CT and ophthalmological examinations 1 to 144 (average: 34) months after the intracavitary radioisotope therapy.

An 23 occasions there was an average of 82 to 90% volume decrease of the craniopharyngioma cysts. On two occasions the volume has remained unchanged and the result of the ^{90}Y therapy cannot be evaluated yet.

The neuroophthalmological prognosis was good only when a relatively intact optic disc was seen; when the disc was atrophic the visual deterioration proved to be irreversible.

Pathologically, it is the fibrotic tissue that is responsible for the shrinkage of the cyst.

Introduction

It was in 1972 that Backlund developed intracavital irradiation of cystic craniopharyngeomas[1,2]. Since then the method has been spreading throughout the world. Of the different isotopes in current use ^{90}Yttrium silicate colloid* (Amersham, England) seems the most practicable, although no comparative study has come out in the literature yet. With the passage of time since the introduction of the method in Hungary in 1975 it has become possible to evaluate the results of the earlier procedures. In the present paper we wish to examine data concerning changes in the size of cystic craniopharyngeomas following irradiation, modifications in ophthalmological state and the nature of histological changes.

Patients and Method

In Table 1, we have listed the various diagnostic and therapeutic methods used in 20 patients. The vertical lines that come under the abbreviated forms of the symbols of procedures indicate the date of the intervention in a time-proportionale manner. Details of the pro-

cedures applied have been given in previous publications[7]. What is new about them is that we now do the calculations with a micro-computer on our own software program[6] and we are introducing ^{90}Y for primary therapy of cystic craniopharyngeomas in patients prior to operation. In the first half of 1987 the degree of shinkage of the cysts in the great majority of patients have been checked with a high-resolution Siemens DRH CT appartus. The cyst volumes were measured within an average of 34 months/minimum 1, maximum 144 months/following irradiation with ^{90}Y.

Neuroophthalmological Data of Our Patients

Visual field defects or an impairment of visual acuity have been observed in 19 out of the 20 patients studied. Deterioration of visual functions began within 2 years before isotope treatment in 6 patients (patient 2, 9, 10, 13, 14, and 19) and in the remaining 13 cases it began 2 to 22 years earlier. All but one of the patients had at least one intracranial operation before the isotope instillation treatment.

In 15 cases a lesion of the chiasm was seen before the instillation of the isotope; 2 of these patients were blind in both eyes, and in 2 further patients one eye was amaurotic. An homonymous hemianopia, indicating a lesion of the optic tract was observed in 4 cases.

Only one patient had a normal optic disc. Temporal pallor was seen in 5 cases, and all these patients had histories shorter than 2 years. The optic disc was atrophic in 14 cases; in all but one case the ophthalmological symptoms appeared more than 2 years before admission.

Histopathological analysis was obtained in 5 cases when the administration of ^{90}Y was followed by operation or autopsy.

Results

Table 2 shows the original volume of the cyst, the activity of administered ^{90}Y in MBq, the time that elapsed since the administration and changes in volume size during that period. To sum up, the cystic parts of craniopharyngeomas shrank to 10–18% of their original size. In 4 cases the cyst disappeared, in 2 cases it showed only slight shrinkage at first, and response to repeated treatment cannot yet be fully evaluated (Fig. 1) In-

Table 1. *Treatment of Our Craniopharyngeoma patients*

		1957	1970	1972	1975	1976	1977	1978	1979	1980	1981	1982	1983	1984	1985	1986	1987		Control st
1	BJ ♂ 1959			Rc	Cc	(Yc)						Sa	Rs	I					2
2	AS ♂ 1958				Rc	Cc (Yc)													?
3	TI ♂ 1958		Rc				Ac (Yc) Sc						Rs Ac	Cc (Yc)		(Yc)			1
4	KM ♂ 1943	Rc							Cs Rc (Yc)										1
5	AE ♀ 1946									Rc	Ac Cc (Yc)								1
6	HGy ♂ 1963					Rc					Rc Rc Cs (Yc) Rs								2
7	VA ♂ 1972								Sp Cc Rc	(Yc)									1
8	BZ ♀ 1926					Rc					Cs (Yc) Cs +								+
9	LL ♀ 1934										Rs	Cs (Ys)(Ys)+ Cs							+
10	TB ♂ 1957											Rs I Cs (Ys) As As							1
11	LI ♂ 1949										Rc Rc Ac Rc Ac Rs (Yc) Sa								3
12	ZI ♂ 1956			Rc									Rs I (Ys) As					1
13	DSz ♂ 1980													Rc Cc (Yc)					1
14	GO ♂ 1941													Rc Ac Sa (Yc)					1
15	SzZ ♂ 1964													Rc	(Yc) Rs				1
16	NJ ♂ 1947								Rs						(Ys)	(Yc)			3
17	KG ♂ 1936													Rs Rc		(Ys)			1
18	FJ ♀ 1921														(Yc) Rc +				+
19	ME ♀ 1959															(Yc)			1
20	FGy ♂ 1971											Rc	Rc			Sa (Yc)			3

Key to the signs used

Therapeutic interventions

Partial resection of the cyst or tumour
  Rc Transcranial approach
  Rs Transsphenoidal puncture

Aspiration of the cyst
  Ac Transcranial puncture
  As Transsphenoidal puncture

Cystography
  Cc Transcranial puncture
  Cs Transsphenoidal puncture

I- External irradiation

Shunt operations

  Sa Ventriculoatrial shunt
  Sp Ventriculopertioneal shunt
  Sc Ventriculocisternostomy

Yttrium-90 silicate instillation via
  Yc Transcranial puncture
  Ys Transsphenoidal puncture

Results of control state

1 Good recovery—patient can lead a full an independent life with or without minimal neurological deficit

2 Moderately disabled-patient having neurological or intellectual impairment but is independent

3 Severely disabled-conscious patient but totally dependent on others to get through the activities of the day

4 Vegetative survival

+ Dead

? Present state is unknown

Table 2. *Changes in Cyst Volume After* ^{90} *Yttrium Silicate Therapy*

Names	Original volume of cyst in ml	Activity of 90 Y in MBq	Elapsed time (in months) after installation ^{90} Y	Volume (in ml) after 90 Y therapy	Volume (in percent) after 90 Y therapy	Percental decrease of original volume
1. B.J.	25	203.5	144	1	4	96
2. A.S.	25	547.6	?	?	?	?
3. T.I.	25	199.8	57	2	8	92
	52	555.0	20	0	0	100
	8.8	92.5	6	1	10.2	89.8
4. K.M.	16	155.4	86	1.5	9.4	96.6
5. A.É.	62	925.0	61	3 + 2	8	92
6. H.Gy.	13	577.2	57	0	0	100
7. V.A.	30	666.0	?	?	?	?
8. B.Z.	14	518.0	1	1.2	9	91
9. L.L.	13	140.6	5	2.1	16.1	83.9
	3.6	296.0	1	0.3	8.3	92.7
10. T.B.	2.8	37.0	36	0	0	100
11. L.T.	24	222.0	33	2.5 + 4.5	104.1	+ 4!!
	× 25	277.5	?	?	?	?
12. Z.I.	16	129.5	34	5	31.3	68.7
13. O.Sz.	22	192.4	26	2	9.09	90.9
14. G.O.	22	192.4	?	?	?	?
15. Sz.Z.	40	318.2	20	0	0	100
16. N.J.	38	255.3	11	35	92.1	7.9
	× 35	301.6	?	?	?	?
17. K.G.	9	88.8	10	1.4	15	85
18. F.J.	21.4	187.6	3	2.5	11.7	88.3
19. M.E.	× 34	355.2	?	?	?	?
20. F.Gy.	× 26	305.5	?	?	?	?

* Elapsed time is too short for evaluation of data.

vestigations undertaken after a longer period of time following ^{90}Y therapy confirmed a permanent shrinkage of the cyst. Neurological performance state of the patients is presented in Table 1. Mild to medium degree weakness of the upper limb – due to previous operations – is worth mentioning.

Improvement occurred in ophthalmological findings in 3 patients (numbers 10, 13, and 14). In the first case (no. 10) visual acuity improved in 3 months to 5/5 bilaterally from 5/15 on the right side and finger counting at 1 meter on the left; the visual fields became normal although there had been a bitemporal hemianopia before treatment. In patient number 13 visual acuity was 5/6 on the right and 5/35 on the before treatment with yttrium; five months later it improved to 5/5 on both sides and the hemianopia disappeared. In the third patient (no. 14) visual acuity was 5/35 before and became 5/5 on both sides after the treatment; bitemporal hemianopia became detectable only with the smallest isopter.

Treatment with ^{90}Y led to deterioration of vision in 2 patients. In one of them (no. 8) visual acuity of the left eye was 5/20 and in 3 weeks the eye lost light perception, while 5 months later the acuity of the right eye decreased from 5/8 to 5/50 and soon the patient died with the symptoms of a hypothalamic lesion. The other case (no. 15) had 5/5 acuity on the right and finger counting at 1 meter on the left; the visual acuity did not change shortly after instillation of the isotope but 6 months later bilateral amaurosis developed within a few days and this proved to be irreversible even after a transphenoidal removal of the cyst and the tumour.

Temporary paresis of the oculomotor nerve was observed in 3 patients 2 to 3 months after the isotope instillation.

Pathology: Biopsy samples taken before ^{90}Y treatment were composed of nests or trabeculae of epithelial cells and epithelial-lined cysts embedded in a loose connective tissue stroma (Fig. 2 a). There were either solid nests of squamous cells or stratified keratinization due

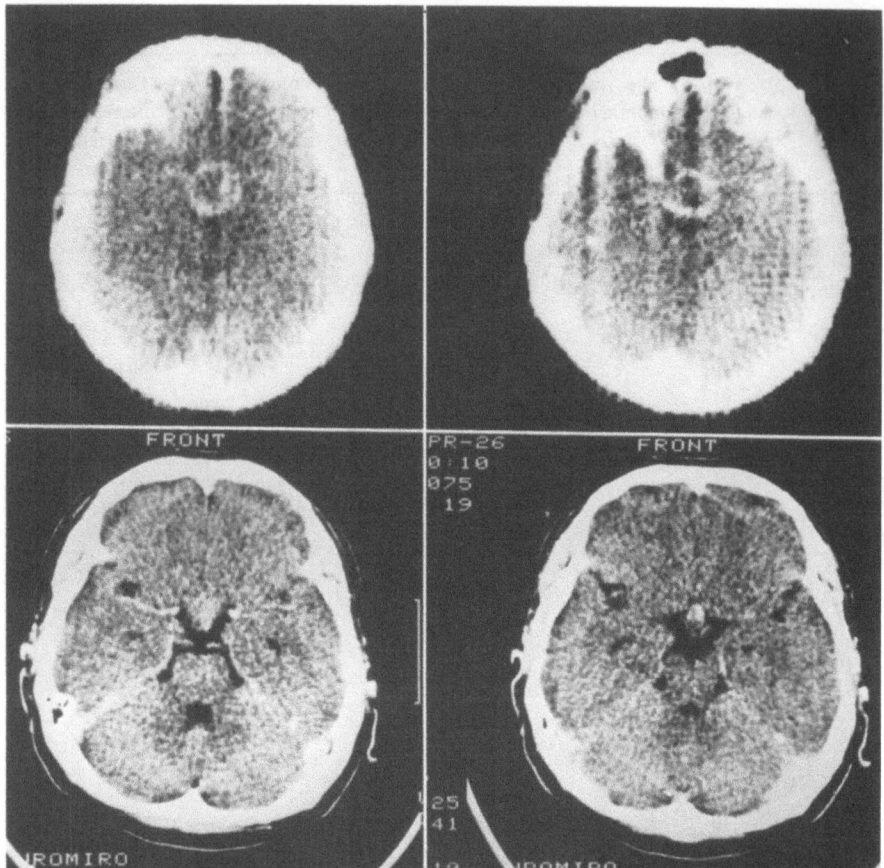

Fig. 1. Suprasellar cyst in the 12th case. Above: before ⁹⁰Y treatment. Below: 3 years after ⁹⁰Y instillation, the minimal remnants of the shrunken cyst can be seen

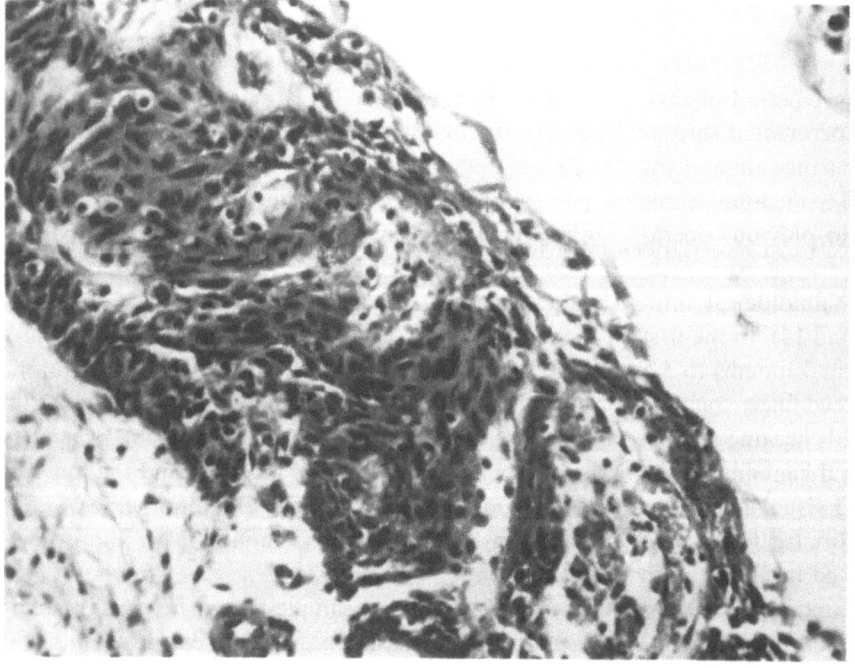

Fig. 2. (a) The lining epithelial cell layer of a cast wall. (b) Completely destroyed epithelium and thick collagen bundles in the cyst wall after ⁹⁰Y treatment. The lumen contains necrotic debris. (c) Subendothelial connective tissue proliferation in a small vessel with subsequent narrowing of lumen

a

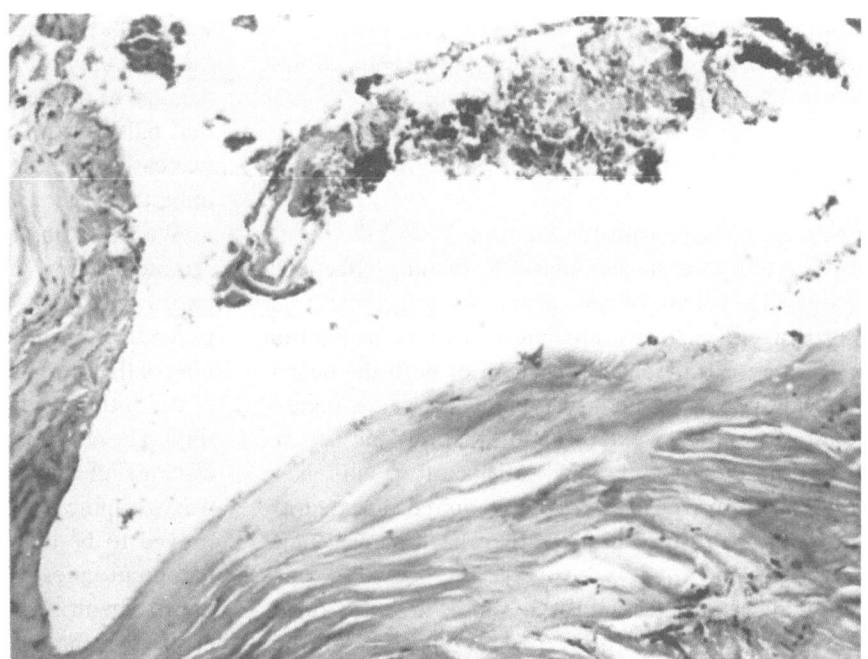

b

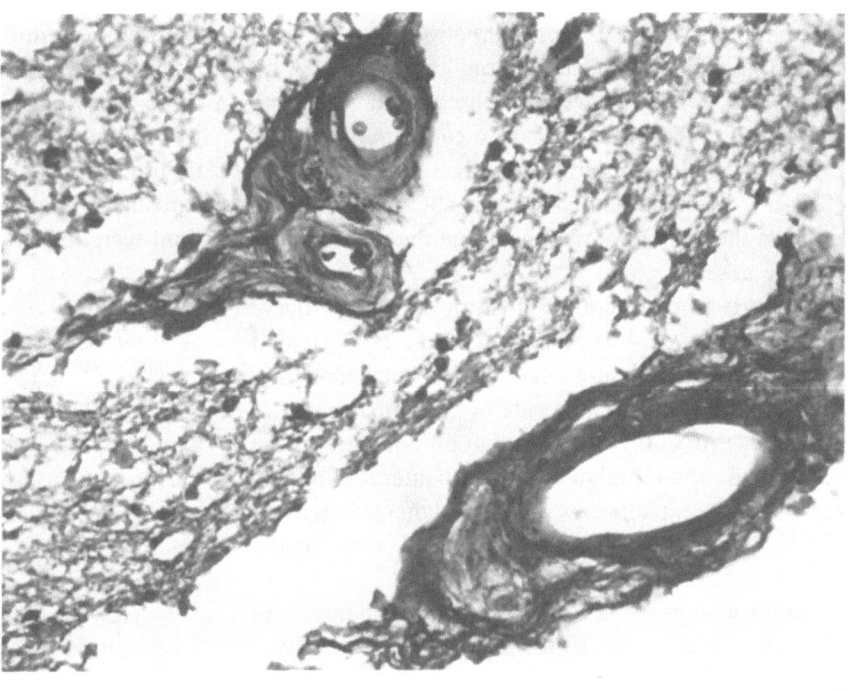

c

to maturation of squamous epithelium. The stroma contained thin-walled capillaries and small vessels. Cellular debris and sloughed-off squamous cells intermingled with foreign body giant cells accumulated in the lumen. Following ^{90}Y irradiation, histological examination of samples taken from the cyst wall revealed that the lining epithelial cell layer was damaged (Fig. 2 b) and the cyst wall was shrunken. It also showed thickened capillary walls, proliferation of endothelial cells, subendothelial connective tissue with focal calcification and narrowing of small vessel lumina (Fig. 2 c). A large amount of thick collagen bundles with hyaline degeneration was present. The cyst lumina contained haemorrhagic necrotic tissue with cholesterol clefts and foreign body giant cells.

Discussion

In accordance with the literature[3, 5, 7, 9, 10, 11, 13, 14] our results also suggest that ^{90}Y therapy effectively reduces the volume of cystic craniopharyngeomas. Cyst shrinkage by 82–90% after intra-cavitary irradiation usually results in an improvement of both the neurological and endocrinological deficits. It was a combination of surgery, external radiotherapy and ^{90}Y irradiation that gave the final satisfactory results. New cysts developed in 3 cases; they required repeated radioisotope therapy. By now they have shown some shrinkage.

In all cases which improved after isotope treatment the visual symptoms appeared within a year of starting treatment and on examination of the fundus only slight temporal pallor of the disc was seen. Similarly to the observations of Wycis and colleagues[16] the visual improvement in our cases occurred between 2 weeks and 6 months after treatment. It should be noted that all the patients who improved had thin walled cystic craniopharyngeomas. It is difficult to compare our results with the better success rate described in the literature (Szikla[15], Kodama and colleagues[8]) because of differences in the method of treatment and the nature of the isotope used.

Reports in the literature mention that optic nerve damage occurs in 4 to 5 per cent of the cases treated with isotope instillation (Backlund[2], Netzeband and colleagues[11]). In our first patient the impairment of vision was probably due to radiation damage. In the second case the rapidly developing bilateral blindness can hardly be explained by radiation damage; it is more likely that it was the consequence of optic nerve compression by the solid tumour.

In most cases the lack of opthalmological improvement may be explained by the fact that the isotope was administered at a rather late stage of the disease, several years after the first appearance of the symptoms and when the symptoms had progressed further in spite of one or more intracranial operations. The damage to the optic nerve was probably irreversible even before treatment. For this reason we think that in the case of a large cyst it is advisable to start treatment with the administration of isotope.

In a case of Huk and Mahlstedt[5] a third nerve palsy failed to improve. In our cases the damage to the nerve was probably due to mechanical displacement due to the shrinkage of the cyst or to temporary radiation damage to the nerve. This is also supported by the fact that palsy of the oculomotor nerve occurred in those two cases where laterally spreading tumour caused an optic tract lesion on the side of the nerve damage.

We have observed an improvement of the visual symptoms after ^{90}Y treatment of thin-walled cysts. The improvement was dependent on the time elapsed between the appearance of the symptoms and the beginning of the isotope treatment as well as on the condition of the optic disc before the treatment (Hajda and Pásztor[4]). The prognosis was good only when an intact optic disc of slight temporal pallor was seen. When the disc was atrophic at the time of treatment the process proved to be irreversible. The visual functions alone had no prognostic value even with regard to the visual improvement. We found no correlation between the changes in the ocular symptoms on the one hand and the amount of isotope administered on the degree of shrinkage of the cyst on the other. Any change of visual function due to treatment should be expected between the 2nd week and the 6th month after applying [90]Y.

Pathological changes indicate that fibrotic tissue is more likely to shrink and it is the fibrosis induced by irradiation together with the destruction of the squamous epithelium and the vascular changes that might explain the reduction of the cyst volume and diminished fluid secretion after ^{90}Y treatment[3, 14].

References

1. Backlund EO (1972) Stereotaktik behandling av kranio pharyngeom med intracystiskt Y[90] extern Co 60 Besträling. Stockholm: Doktor-thesis
2. Backlund EO, Axelson B, Noren G, Ribbesjö E, Rähn T, Schnell PO (1984) Stereotactic treatment of craniopharyngeomas. A 15-year material (manuscript)
3. Campbell JB, Hudson FM (1960) Craniobuccal origin, signs and treatment of craniopharyngiomas. Surg Gynec Obstet 111: 183–191
4. Hajda M, Pásztor E (1981) Prognostical importance of fundus in the improvement of visual functions after the operation of pituitary adenomas. (Hungarian) Szemészet 118: 209–217
5. Huk WJ, Mahlstedt J (1983) Intracystic radiotherapy (^{90}Y) of craniopharyngiomas: CT-guided stereotaxic implantation of indwelling drainage system. AJNR 4: 803–806
6. Julow J, Drasny G, Lányi F, Simkovics M, Hajda M, Szeifert G, Tóth Sz, Pásztor E (1987) Use of microcomputer in the treatment of cystic craniopharingeomas with ^{90}Y silicate colloid intracavitary irradiation. (Hungarian) Ideggyógy Szemle 40: 547–552.
7. Julow J, Lányi F, Hajda M, Simkovics M, Arany I, Tóth Sz, Pásztor E (1985) The radiotherapy of cystic craniopharyngioma with intracystic installation of ^{90}Y silicate colloid. Acta Neurochir (Wien) 74: 94–99

8. Kodama T, Matsukado Y, Uemura S (1981) Intracapsular irradiation therapy of craniopharyngiomas with radioactive gold. Neurol Med Chir (Tokyo) 21: 39–58

9. Lunsford LD, Gumerman L, Levine G (1985) Stereotactic intracavitary irradiation of cystic neoplasms of the brain. Appl Neurophysiol 48: 146–150

10. Musoline A, Munari C, Blond S, Betti O, Lajat Y, Schaub C, Askienázy S, Chodkiewicz JP (1985) Traitement stereotaxique des kystes expansifs de cranio-pharyngiomas par irradiation endocavitaire beta. Neurochirurgie 31: 169–178

11. Netzeband G, Sturm V, Georgi P, Sinn H, Schnabel K, Schlegel W, Schabbert S, Marin-Grez M, Gahbauer M (1984) Results of stereotactic intracavitary irradiation of cystic craniopharyngiomas comparison of the effects of Yttrium-90 and Rhenium-186. Acta Neurochir (Wien) [Suppl] 33: 341–344

12. Schaub C, Bluet-Pajot MT, Videau-Lornet C, Askienazy S, Szikla G (1979) Endocavitary beta irradiation of glioma cysts with colloidal Rhenium 186. In: Szikla G (ed) Stereotactic cerebral irradiation. North Holland, Amsterdam, pp 293–302

13. Strauss L, Sturm V, Georgi P, Ostertag H, Clorius FH, van Kaick G (1985) Radioisotope therapy of cystic craniopharyngeomas. Int J Radiat Oncol Biol Phys 8: 1581–1585

14. Szeifert G, Julow J, Slowik F, Bálint K, Lányi F, Pásztor E (1987) Pathological changes in cystic craniopharyngiomas following intracavital Yttrium-90 treatment. (Hungarian) Magyar Onkológia, 31: 223–229

15. Szikla G, Musolino A, Miyahara S, Schaub C, Askienázy S (1984) Colloidal Rhenium-186 in endocavitary beta irradiation of cystic craniopharyngiomas and active glioma cysts. Long term results, side effects and clinical dosimetry. Acta Neurochir (Wien) [Suppl] 33: 331–339

16. Wycis HT, Robbins R, Spiegel-Adolf M, Mészáros J, Spiegel EA (1954) Treatment of a cystic craniopharyngioma by injection of radioactive P 32. Confin Neurol 14: 193–202

Address for correspondence: Dr. J. Julow, National Institute of Neurosurgery, Amerikai ut 57, Budapest 1145, Hungary.

Acta Neurochirurgica, Suppl. 42, 120–123 (1988)

Treatment and Recurrences in 135 Pituitary Adenomas

B. Vlahovitch, C. Reynaud, J. Rhiati, H. Mansour, and **F. Hammoud**

Service de Neuro-Chirurgie A, Centre médical gui de Chauliac, Montpellier, France

Summary

In 135 operated pituitary adenomas of different histological nature, seen over 20 years, the initial treatment was surgical, by transsphenoidal or transcranial approach depending on the tumour extent, followed by radiotherapy if necessary.

In spite of modern investigation methods, diagnosis is still often made late, so that 118 of our cases already had visual symptoms.

After surgery alone (87 cases) made by the same team recurrences occurred in 16%. Reoperations (14 cases) were generally more difficult and could also be followed by new recurrences needing finally radiotherapy with still positive results.

Two cases had initial radiotherapy, the other 46 cases were treated by surgery and radiotherapy with a recurrence level of 8,3%.

These results indicate the need to use immediate postoperative radiotherapy in every pituitary adenoma showing even the slightest invasive potential because radical surgery is hypothetical and recurrences possible. The recurrence of pituitary adenomas remains a difficult problem of anticipation and of curative management. Our position in terms of these benign tumours is to adopt readily this therapeutic course: surgery followed by radiotherapy, which provides the best prognosis without major risks.

Keywords: Pituitary adenoma; treatment; operation; transcranial approach; transsphenoidal approach; radiotherapy; recurrence.

Introduction

Our series includes 135 hypophyseal adenomas of different histological types collected during the last twenty years. Within this period tremendous changes occured in neuroradiological and biological investigation techniques. The advent of CT scan and more recently of the MRI resulted in considerable advantages not only for the diagnosis of pituitary adenomas but also for the postoperative course of these tumours.

One is even allowed to speak now of two different investigation periods: pre- and postimaging eras. In spite of this improvement, the diagnosis of y hypophyseal tumour is still delayed not only by carelesness among patients but even among some physicians who omit to give any importance to some common symptoms such as headaches and slight visual symptoms.

For these reasons we have observed that 118 of our patients had visual impairment when the lesion was diagnosed.

Study of 135 Treated Hypophyseal Adenomas

In this presentation we are not attempting to undertake an epidemiological study but only to insist on the difficulties inherent in the surgical treatment of these benign tumours. 133 patients were operated by tumour exeresis, one patient has a transsphenoidal implantation of radioactive gold and another patient received external radiotherapy without surgery.

For the surgical cases the approach route was selected following the indications of the neuroradiological investigations.

84 pituitary adenomas were approached transsphenoidally (70 purely intrasellar tumours and 14 cases with a little suprasellar involvement.)

The 49 other cases were operated by subfrontal cranial approach generally when there was a suprasellar involvement. The majority of neurosurgeons agree that the best treatment for pituitary adenoma is the transsphenoidal approach which also is less risky[6, 7] 87 hypophyseal adenomas in our series were operated (71% by transsphenoidally approach and 29% by transcranial approach) with very satisfactory exeresis. But it is a well known fact that the more complete exeresis is not unfortunately always synonymous with a radical exeresis[3, 4, 8, 14, 17]. This explains why there was in our cases a recurrence rate of approximately 16%. The 46 other operated cases of pituitary adenomas (48% transsphenoidally and 52% transcranially) received postoperative radiotherapy. This complementary treatment was undertaken in spite of the controversial discussion in the literature on the value of X-rays in these cases[2, 13].

Nevertheless radiotherapy was necessary, when radical exeresis was difficult or even impossible, when the histology revealed atypcial tumour cells and also when postoperative scan controls showed residual tumour tissue insufficient to need a new operation[11].

The result of this therapeutic protocol resulted in a better score of only 8.3% recurrences[8, 9].

Recurrences

There was 16% recurrences (14 cases on 87) with surgical treatment alone and 8.3% recurrences (4 cases on 46) with combined surgical and X-ray treatment. The total recurrence rate for the 135 cases being 13.3% (two cases having radiotherapy without surgery).

Formerly, recurrences were usually recognised late, and entirely on the basis of clinical symptoms. Now, with neuro imaging, we are able to detect recurrences before the appearance of clinical signs, consequently earlier therapeutic intervention is possible[11]. All the postoperative recurrences happened almost within the first 10 years. From the analysis of the 87 pituitary adenomas operated without complementary radiotherapy it appears that the recurrence rate is two times higher. Even with radiotherapy the recurrence rate for pituitary adenomas remains important (8.3%) because of operative difficulties in the larger tumours or because the invasive lesions who have indistinct boundaries, which affects both the surgical approach and also radiotherapy planning.

It must be underlined that radiotherapy diminished only slightly the recurrence rate (16 to 13.6%) in transsphenoidal adenomas resection, while the transcranial operations with radiotherapy give a lower recurrence rate (16 to 4%) probably because of more accurate and complete resection[16]. On the contrary the recurrence levels are quite identical in pituitary adenomas with surgical treatment made alone by transsphenoidal as well as transcranial approaches.

Clinical Example

Franz P. Obs. No 11510. A 60-year-old man operated on twice at 1 year intervals for recurrence of a suprasellar chromophobe pituitary adenoma. After the second operation he was irradiated with 45 grays of Cobalt 60.

This man's illness began 12 years before the first operation with an important visual impairment which eventually progressed to a bitemporal hemianopsia. The first neuroradiological investigation in 1962 was a pneumo encephalogram which established the diagnosis of a sellar adenoma with suprasellar extension. His first operation was via a right transcranial approach and the patient was relieved of headaches but the bitemporal hemionopsia persisted. His condition worsened 1 year later and carotid angiography showed clearly

a suprasellar recurrence. A second operation was at once performed with complete exeresis of his suprasellar pituitary adenoma and it was immediately followed by radiotherapy. The patient recovered well but still with bitemporal hemianopsia. Seven years later he developed mental deterioration with amnesia and a low pressure hydrocephallus requiring a ventriculo-peritoneal shunt; he made an excellent recovery. 15 years after the first operation for pituitary adenoma a CT scan showed that there was no adenoma recurrence but only an empty sella. Radiotherapy in this case was very useful in preventing further growth of his invasive adenoma.

Those recurrences of pituitary adenoma which occurred after surgical treatment alone were all reoperated and irradiated; recurrences after surgery and radiotherapy underwent reoperation—in only one case of a large invasive tumour, and this patient was submitted to a second course of X-ray therapy.

CT scans made during a long follow-up (8, 12, and 15 years after the last operation) revealed tumour recurrence in three patients without any new clinical symptoms. These patients were neither reoperated nor treated with X-rays.

The Irradiation Protocol

Generally it consists of supervoltage and photons with cobalt 60 at 1 MEV. Irradiation with heavy particles with high acceleration by Cyclotron (5 to 20 MEV) allows deeper penetration and is more suitable and efficient biologically. The usual utilized dosage is about 40 to 50 grays by fractions of 2 or 2.5 grays twice per week for a month and a half. At least three irradiation fields are necessary (2 lateral and one anterior or superior); sometimes a rocking field can be used.

In our series of 135 cases of pituitary adenomas radiotherapy was utilized alone twice; once by interstitial transsphenoidal radioactive gold implantation, and once by external irradiation.

In 46 cases postoperative radiotherapy was performed to compensate incomplete resections or the finding of atypical tumour cells. Radiotherapy was also given in 15 recurrent postoperative adenomas. As mentioned the recurrence rate with radiotherapy diminished by approximatively 50%.

Radiotherapy Complications

61 patients received radiotherapy, 46 cases after the first surgical operation and 15 cases after the post recurrence operations. This gives for the series a high rate of radiotherapy, nearly 45%.

In spite of this high percentage in our cases no complication occured except for severe visual damage in one 58 year old woman who had been irradiated in another medical centre without operation for a scan enhanced hyperdense pituitary adenoma. This patient became blind with optic atrophy after only a usual 50 grays dosage. Later on a new scan showed an empty sella, proving the effectiveness of the X-ray therapy in this pituitary adenoma. (In this series a total of 3 cases of post radiotherapy empty sella were encountered.)

No radionecrosis were ever observed in our series, even for patient receiving radiotherapy twice for recurrent adenomas. This dreaded X-ray complication has been described as occuring rarely, even when using the commonly accepted dosage of 5,000 rad delivered slowly over a month and a half[1, 4, 5]. The other post-radiotherapy complications of sarcoma, fibro-sarcoma, glioma are very rare[10, 12, 15].

Discussion

It seems that one must accept that recurrences following the first operation for pituitary adenoma may occur up to 28 years later, as in one of our cases. None of the different surgical techniques (*i.e.* transcranial, transsphenoidal, and sophisticated microsurgical) can completely prevent recurrence which depends almost entirely on the invasive nature of these tumours whether they are microadenomas or larger lesions[14].

Radiation therapy reduces the recurrence rate but it is not yet a radical curative method. However it is obvious that radiation therapy is sometimes able to arrest the evolution of a pituitary adenoma. This is emphasised by those cases in which an empty sella was demonstrated after radiotherapy. While the CT scan is a remarkable tool for the postoperative follow-up, it can create difficult decisions concerning reoperation, *e.g.* confusion between tumour recurrence and fibrous tissue in a large sella.

Among different recent studies trying to bring out some correlation between functional classification of pituitary adenomas and their invasiveness[14] many unresolved problems of investigation remain. The estimated percentage of hypophyseal tumours of all types showing invasive features was calculated at 35% by Scheithauer and colleagues[14].

It is important to realize that radiation therapy is only recommended when radical removal has been difficult or impossible (because of cavernous sinus invasion, invasion of the dura, massive size of the adenoma, histology revealing atypical tumour cells) or the recognition of residual tumours on CT scan in the postoperative period.

Actually, radiation therapy has a beneficial effect,

since the difference in recurrence rate in our cases from 16 to 8.3% in the two groups was significant.

Conclusion

The analysis of 135 cases of surgical treatment of hypophyseal adenomas brings out the necessity and usefulness of complementary irradiation of these tumours after the first operation or after a recurrence. If pituitary adenoma are benign tumours, the modern recognized notion of their important rate of invasiveness explain in part the fairly frequent relapses after different surgical treatments. Radiotherapy as has been earlier suggested by Cushing still remains and is utilized by different authors with significant results like in our series (50% reduced recurrences). This useful treatment for pituitary tumours although not completely curative remains without major risks.

References

1. Aristizabal S, Caldwell WL, Avila J (1977) The relationship of time-dose fractionation factors to complications in the treatment of pituitary tumours by irradiation. Int J Radiat Oncol Biol Phys 2/7: 667–673
2. Baskin DS, Boggan JE, Wilson CB (1982) Transsphenoidal microsurgical removal of growth hormone secreting pituitary adenomas. A review of 137 cases. J Neurosurg 56/5: 634–641
3. Ciric IM, Mikhael T, Staford, Lawson L, Garces R (1983) Transsphenoidal microsurgery of pituitary macroadenomas with long term follow-up results. J Neurosurg 59: 395–401
4. Ebersold MJ, Quast LM, Laws ER, Scheithauer B, Randall RV (1986) Long-term results in transsphenoidal removal of non-functioning pituitary adenomas. J Neurosurg 64/5: 713–719
5. Fukamachi A, Tetsuo Wakado, Junichiro Akai (1982) Brain stem necrosis after irradiation of pituitary adenoma. Surg Neurol 18: 343–350
6. Kern EB, Laws ER, Randall RV, Westwood WB (1977) A transseptal, transsphenoidal approach to the pituitary: an old approach. A new technique in the management of pituitary tumours and related disorders. Trans Am Acad Ophthalmol Otolaryngol 84/6: 997–1010
7. Kern EB, Pearson BW, McDonaldt TJ, Laws ER (1979) The transseptal approach to lesions of the pituitary and parasellar regions. Laryngoscope 89/511 [Suppl] 15: 34 p
8. Kunc Z (1977) Surgical problems in chromophobe adenomas. J Neurosurg Sci 21/2–3: 143–149
9. Kutzner J (1983) Percutaneous radiotherapy in patients with pituitary tumours and recurrences. Therapiewoche 33/47: 6350–6352
10. Martin WH, Cail WS, Morris JL, Constable WC (1980) Fibrosarcoma after high energy radiation therapy for pituitary adenoma. Am J Neuroradiol 1/5: 469–474
11. Muhr C, Bergstrom K, Hugosson R, Lundberg PO (1980) Pituitary adenomas: computed tomography and clinical evaluation in a follow-up after surgical treatment. Neurol 19/3: 171–179

12. Powell HC, Marshall LF, Ignelzi RJ (1977) Post irradiation pituitary sarcoma. Acta Neuropathol 39/2: 165–167
13. Rubenstein R (1977) Treatment of pituitary microadenomas. Lancet 2/8035: 460
14. Scheithauer BW, Kovacs KT, Laws ER, Randall RV (1986) Pathology of invasive pituitary tumours with special reference to functional classification. J Neurosurg 65/6: 733–744
15. Shin H, Namba H, Ishigen N *et al* (1980) Pituitary fibrosarcoma secondary to radiation therapy for the treatment of chromophobe adenoma. Neurosurg 8/7: 605–614
16. Symon L, Logue V, Mohanty S (1982) Recurrence of pituitary adenomas after transcranial operation. J Neurol Neurosurg Psychiatry 45/9: 780–785
17. Wrightson P (1978) Conservative removal of small pituitary tumours: is it justified by the pathological findings. J Neurol Neurosurg Psychiatry 41/3: 283–289

Address for correspondence: Prof. Dr. B. Vlahovitch, Service de Neuro-Chirurgie A, Centre médical gui de Chauliac, Montpellier, France.

Acta Neurochirurgica, Suppl. 42, 124–129 (1988)
© by Springer-Verlag 1988

Management and Surgical Outcome of Suprasellar Meningiomas

J. Brihaye[1] and **M. Brihaye-Van Geertruyden**[2]

[1] Department of Neurosurgery, Institut Bordet, Free University of Brussels, [2] Department of Ophthalmology, Akademisch Ziekenhuis, Vrij Universiteit Brussel, Belgium

Summary

The authors analyse 22 cases of suprasellar meningiomas, drawing attention to factors influencing on the surgical outcome. In all but one case, symptomatology began with progressive visual failure in one eye. Bilateral anosmia was noted in 4 patients with large tumour. Mental disorders were conspicuous in 5 cases and 3 patients suffered from epilepsy. Headache was severe in 5 cases. Endrocrinological disorders were observed in 3 patients. The sella turcica was of normal shape in all cases. Marked hyperostosis of the planum or tuberculum existed in 7 cases. The tumour was heavy-calcified in 2 cases. CT scanning showed everytime a marked enhancement of the tumour and in 4 cases, a large hypodense area surrounded the tumour. The patients were operated on through a bifrontal approach or a unilateral frontal flap. A partial anterior frontal lobectomy was regularly performed on one side. While the tumour is piecemeal exacavated, the dural attachment at the base is reached as quickly as possible. Complications consisted in rhinorrhea of CSF in 2 cases, once in a transitory diabetes insipidus and in a secondary hydrocephalus. Postoperative mortality remains high. Among the eleven cases of large tumours, a direct postoperative death occurred, due to a severe arterial bleeding. Two other patients died 4 and 6 weeks respectively after operation. An other patient died 8 years after operation, from meningitis. Among the 5 cases of medium-sized tumours, one postoperative death occurred in a young female, 30 of age, following urinary infection by Klebsiella, complicated by toxicemia. No death and no morbidity occured in the cases of small-sized tumour. Recurrences were observed 3 and 7 years respectively after surgery in 2 patients. In conclusion: multiple factors influence on surgical outcome of suprasellar meningiomas: time interval between the onset of symptoms and the operation, tumour size, heavy calcification of the tumour, big zone of hypodensity, commonly interpreted as brain oedema, circumscribing the tumour. It has been postulated that the occurence of cerebral oedema in combination with meningioma is associated with growing factors of the tumour and results from a breakdown of the blood-brain barrier. The impairment of the blood supply to the surrounding brain (frontal lobes and hypothalamic region) is really aggravated by the intraoperative manipulations. Brain hypodensity associated with suprasellar meningioma appears to us a major factor influencing on the overall outcome.

Introduction

In 1929, reporting 15 cases of meningiomas arising from the tuberculum sellae, Cushing and Eisenhardt amplified their characteristic syndrome, namely bilateral primary optic atrophy in a middle-aged person, with progressive failure of vision and bitemporal field defects, combined with a normal sella turcica. From then, numerous studies have been devoted to suprasellar meningiomas[2, 10, 12, 13, 14, 15, 16, 17, 19, 21, 22, 26, 27, 28].

Material and Results

We are presenting an analysis of 22 cases of suprasellar meningiomas, drawing a particular attention to factors influencing the surgical outcome. Suprasellar meningiomas represent 7 to 10% out of intracranial meningiomas and 15 to 20% out of skull base meningiomas. Among 100 intracranial meningiomas found incidentally at necropsy, 12.7% were situated in the suprasellar region[32]. The preponderance of females (15 for 7 males) is noteworthy in all series. *The age of the patients* ranges from 15 to 84 years, 11 patients being between 50 and 60.

The site of attachment, determined on X-ray films and/or during operation, was the planum sphenoidale in 10 cases, the tuberculum sellae in 8 cases, the diaphragma sellae and one anterior clinoid process respectively in 2 patients; a large base of attachment was seen in 2 cases. Meningiomas arising from the diaphragma sellae are in a retrochiasmatic situation and set problems to the surgeon[11, 16]. Meningiomas confined to one anterior clinoid process are rare[31].

The tumour size was under 3 cms (small volume) in 3 cases, between 3 and 5 cms (median-sized mass) in 6 cases and over 5 or 6 cms in 13 patients.

In all but one case, *symptomatology* began with progressive visual failure in one eye, sometimes erroneously diagnosed as retrobulbar neuritis; 6 patients were blind or nearly blind at the time of operation; a syndrome of Foster-Kennedy was observed in one patient.

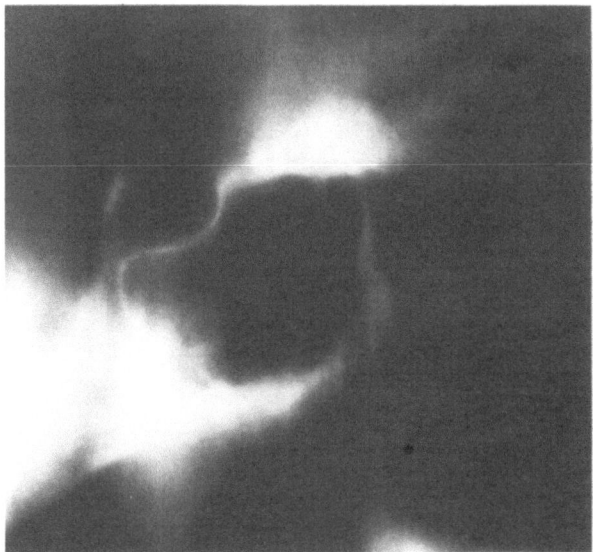
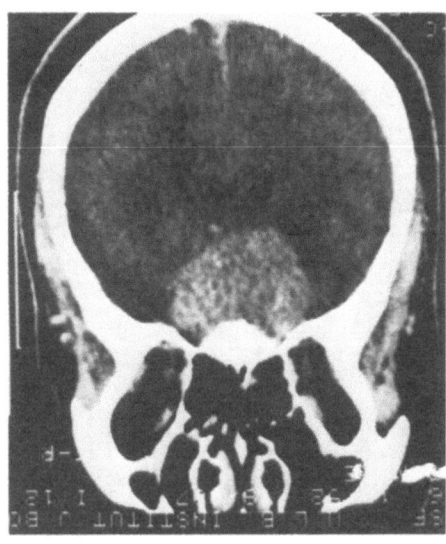

a   b

Fig. 1. Thickening and hyperostosis of the planum sphenoidale. (a) Lateral view on plain X-rays, (b) A.P. view on CT scan: the bone change is directly beneath the tumour

The field defects were on the form of an irregular bitemporal hemianopia. In the case of right anterior clinoid process meningioma, the onset was an *intra-tumoural haemorrhage* with subarachnoid bleeding which produced a sudden syndrome of the orbital apex: amaurosis, III, V, and VI cranial nerves impairment on the right side; no vascular anomaly was seen on the angiogram nor at the operation.

Spontaneous haemorrhage in patients with meningiomas has been reported several times[1, 6, 7, 23, 25].

Bilateral anosmia was noted in 4 patients with large tumour.

Mental disorders were conspicuous in 5 cases and 3 patients suffered from generalized epilepsy. Headache was severe in 5 cases. Endocrinological disorders were observed in 3 patients; two females weighing over 110 kg were amenorrheic but their endocrinological assessment was negative; in one male, existed an important deficit in 17 corticosteroids, 17 hydrocorticosteroids and in gonadotrophic hormones.

The time interval between the onset of symptoms and operation ranges from a few months to 20 years in one case; on average the duration was round about one year. But history was sometimes uncertain because of the low socio-economic level of some patients.

Regarding the *differential diagnosis*, it is important to note that the sella turcica was of normal shape in all our cases although some decalcification of the clinoid processes may be occasionally observed.

Marked hyperostosis of the planum sphenoidale or

tuberculum sellae existed in 7 cases (Fig. 1). Slight deposits of calcium within the tumour were seen in 4 cases and the tumour was heavy-calcified in 2 cases (Fig. 2).

These roentgenologic manifestations of suprasellar meningiomas (bone changes and calcification) have been attentively described by several authors[5, 9, 20, 29, 30].

Angiography only performed in half of the cases showed an upward displacement of the anterior cerebral artery on the AP view, resembling the dislocation observed in pituitary tumours. The meningioma appeared mainly but poorly stained, after injection of the internal carotid and ophthalmic arteries.

CT scanning is the most accurate imaging procedure for the diagnosis; performed in 13 patients it showed everytime a marked enhancement of the tumour on administration of contrast solution. However in our cases the degree of attenuation did not well correlate with the consistency of the tumour at operation, as it was postulated by Kendall and Pullicino. In 4 cases, a large hypodense area surrounded the tumour (Fig. 3); this image very probably is the radiological manifestation of parenchymal ischaemia with oedema and represented in our cases a factor of bad surgical outcome. In one of these cases, there was an important circulatory slowness on the angiogram, correlating well with the peritumoural hypodense image.

Cerebral oedema complicating meningiomas could be provoked by the vascular pericytic component of the tumour which is variable from one case to another[3,]

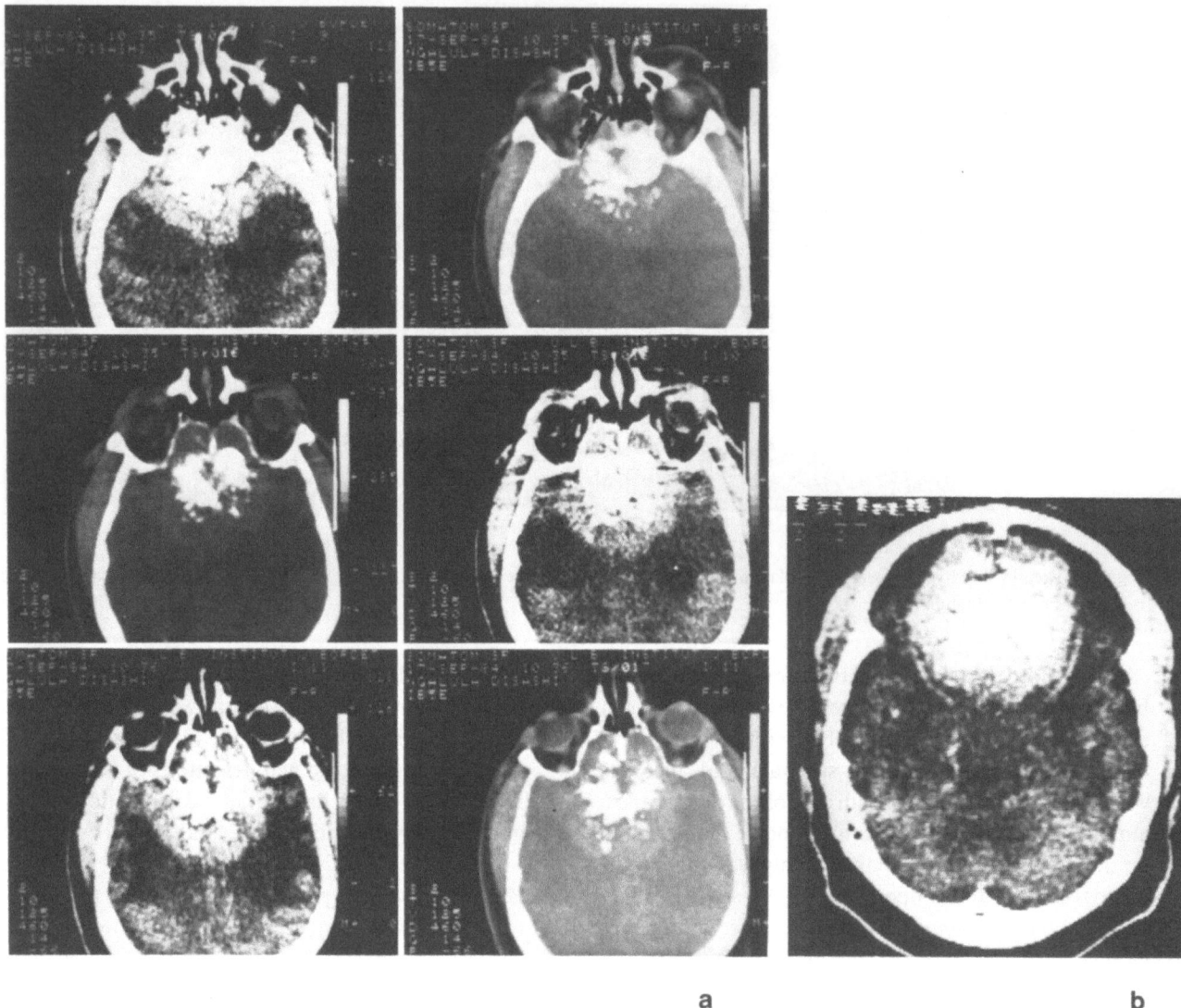

a   b

Fig. 2. CT scan without contrast. (a) Several slices demonstrating the heavy calcification of the tumour mass. (b) Around the calcified mass a large bilateral hypodense image is seen

[24]. In the discussion of Dr. Challa's paper, D.P. Becker thought that "the brain swelling itself may be a greater factor in patient morbidity than the direct brain compression caused by the tumour".

Surgery of suprasellar meningioma has benefited by modern operative technology although the surgical technique already described by Cushing[4] remains the key factor for reducing the morbidity and mortality.

A unilateral frontal flap may be used when the tumour size is under 3 or 4 cms. A bifrontal approach seems to us recommendable when the tumour is of large size. We have regularly performed a partial anterior frontal lobectomy on one side in case of medium or large-sized lesion in order to reduce to a minimum retraction of the frontal lobes.

Mandatory in the removal of the tumour is to spend all the time necessary for slow piecemeal excavation of the mass before trying to dissect and separate the tumour from the surrounding structures. The aim is to make from a large mass as small a tumour as possible in order to dissect under the microscope the optic pathways and arteries under and behind the lesion. When the piecemeal resection is too haemorrhagic, we recommend to reach as quickly as possible the zone of attachment to the skull base to obliterate the feeding vessels which arise from the ophthalmic and posterior ethmoidal arteries[14].

Dislodging the posterior surface of the tumour from the hypothalamic parenchyma and vessels is a difficult step; these vessels, the anterior cerebral arteries and

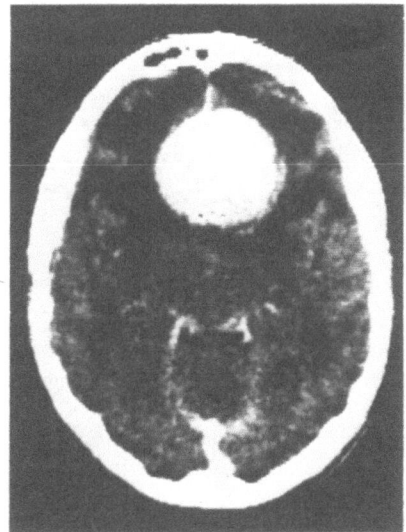 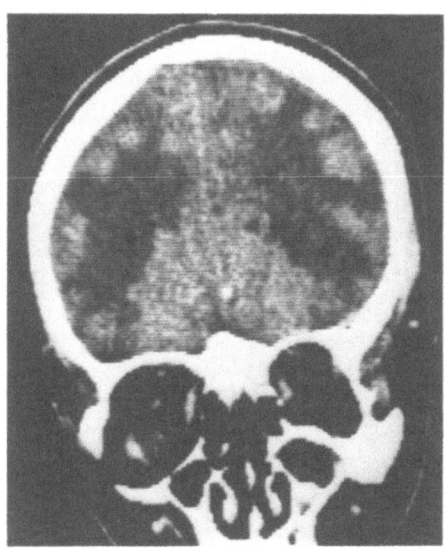

Fig. 3. Two different cases demonstrating the extensive bilateral hypodense appearance of the cerebral hemispheres. (a) Marked enhancement of the tumour. (b) Patchy oedema of both frontal lobes

a   b

their branches, are pushed upwards and backwards and sometimes are embedded in deep grooves in which they are concealed from view. We did not have recourse to presurgical embolization.

Surgery was performed on 19 patients. In one recent case, the tumour of small size was encountered while the patient was under care for a neck carcinoma; operation was not carried out in one female, 84 years of age, only complaining of unilateral visual failure; a third patient had a cerebral vascular accident with hemiplegia and speech disturbances two days before the planned operation. The other patients were operated on through a bifrontal approach in 10 cases and a unilateral frontal flap in the other 9 cases.

In one case of extensive involvement of the sphenoid and ethmoidal cells, ablation of the tumour necessitated resection of the invaded bone and reconstruction of the anterior skull base.

Complications consisted in rhinorrhea of CSF in 2 cases, in a transitory diabetes insipidus in one patient and in a secondary hydrocephalus in one other patient. Several patients operated on for a large suprasellar meningioma remained for a time varying from a few days to a few weeks in a state of poor and slow responsiveness such as we used to see long ago bilateral frontal lobectomy.

Postoperative mortality remains high. Among the thirteen cases of large tumours, a direct postoperative death occurred, due to a severe arterial bleeding while the posterior part of the tumour was dislodged. A second patient died six weeks after the tumour resection;

she was reoperated on first for a persistent rhinorrhea and secondly she received a ventriculo-peritoneal shunt because of a secondary hydrocephalus; her condition, reminiscent of reminding what Jefferson[16] called hypothalamic failure, continued to deteriorate until death. The third patient died suddenly from a pulmonary embolus four weeks after surgery while she was on the verge of leaving the hospital.

Among the six cases of medium-sized tumours, one postoperative death occurred in a young female, 30 years of age, who had a meningioma arising from the diaphragma sellae (Fig. 4). This woman was preoperatively dysmorphic, mentally defective and suffering from mitral valve disease. Within a few months, she developed behaviour disturbances and severe bilateral failure of visual acuity. At operation, the tumour was very soft, enveloping the optic nerves, chiasm and carotid arteries, penetrating the optic canals and extending in front of the brain stem. This kind of tumour looks very much like the third case reported by Guiot *et al.* The resection, partly made by aspiration, was easy and not haemorrhagic. Death occurred on the 14th day following urinary infection by Klebsiella, complicated by severe toxaemia.

No death and no morbidity occurred in the cases of small-sized tumours.

Recurrences were observed 3 and 7 years respectively after surgery in two patients. Postmortem examination could be made in one case, 4 months after operation: a minute fragment of tumour attached to the optic nerve could still be seen. In an other case, autopsy

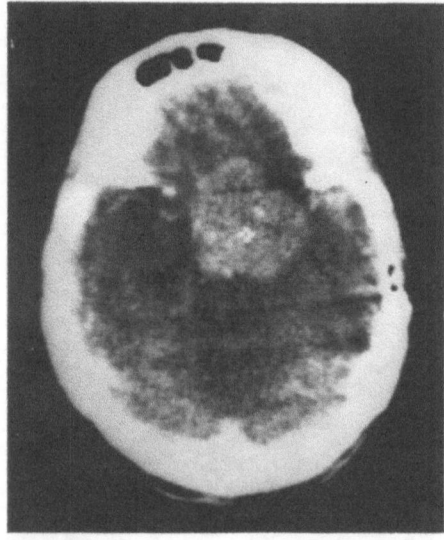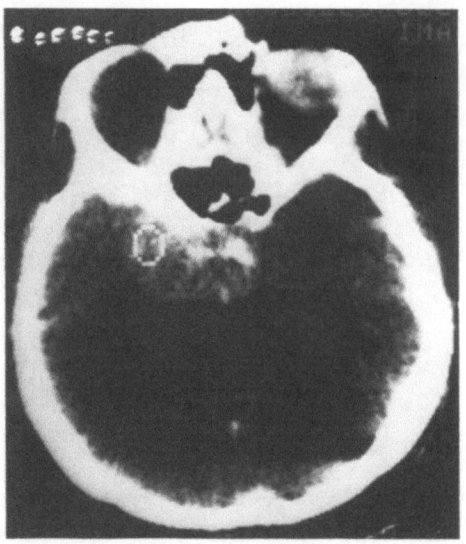

a   b

Fig. 4. CT scan without contrast material. (a) Medium sized soft meningioma with (b) a greater extension on the left side

revealed a tumour nodule within the sella turcica. In a third patient who died 12 years after operation, small quiescent remnants of tumour were seen at autopsy.

In *conclusion* of this analysis it is obvious that multiple factors influence the surgical outcome of suprasellar meningiomas.

The *time interval* between the onset of symptoms and the operation is of great consequence for visual improvement but it does not correlate with the tumour size and therefore with the eventual difficulty about surgery; biological determinants indeed make the tumour growth not uniform.

The *tumour size* as well as the predominantly retrochiasmatic situation of the mass certainly influence on the surgical outcome. However the technique of slow and progressive excavation of the tumour and if necessary of a prompt approach to the dural attachment at the base and its vascular component greatly facilitate the resection of huge lesions.

Heavy calcification of the tumour broadly anchored in the skull base makes the operation long and difficult; the resection of the invaded bone is mandatory for avoiding the recurrence of the tumour.

Large zones of hypodensity on CT, commonly interpreted as brain oedema, circumscribing the tumour, appeared to us a bad indicator with regard to the overall outcome. It has been postulated that the occurence of cerebral oedema in combination with meningioma is associated with growing factors of the tumour and results from a breakdown of the blood-brain barrier. The impairment of the blood supply to the surrounding brain (frontal lobes and hypothalamic region) being

on a par with the oedema, is easily aggravated by even gentle intraoperative manipulations.

Finally, the neurosurgeon greatly profits by personal experience and this factor is the most valuable for the patient.

References

1. Askenasy HM, Behmoaram AD (1960) Subarachnoid haemorrhage in meningiomas of the lateral ventricle. Neurology 10: 484–489
2. Cassinari V, Bernasconi V (1957) Considerazioni clinico-radiologiche su 35 casi di meningioma del tubercolo della sella. Acta Neurol XIII: 645–669
3. Challa VR, Moody DM, Marshall RB, Kelly DL (1980) The vascular component in meningiomas associated with severe cerebral oedema. Neurosurgery 7: 363–368
4. Cushing H, Eisenhardt L (1929) Meningiomas arising from the tuberculum sellae with the syndrome of primary optic atrophy and bitemporal field defects combined with a normal sella turcica in a middle-aged person. Arch Ophthalmol 1: 1–41 and 168–205
5. Di Chiro G, Lindgren E (1952) Bone changes in cases of suprasellar meningioma. Acta Radiol 38: 133–138
6. El-Banhawy A, Walter W (1962) Meningiomas with acute onset. Acta Neurochir (Wien) 10: 194–206
7. Goran A, Ciminello VJ, Fisher RG (1965) Haemorrhage into meningiomas. Arch Neurol 13: 65–69
8. Goutelle A, Allegre GE, Lapras Cl, Dechaume JP, Ravon R (1970) Méningiomes supra-sellaires antérieurs. A propos de douze observations. Neurochirurgie 16: 359–366
9. Gregorius FK, Bentson JR (1975) Comparison of radiological tests in the detection of presellar meningiomas. Neuroradiology 8: 267–274
10. Guillaumat LGJ (1937) Les méningiomes supra-sellaires (contribution à l'étude du syndrome chiasmatique) Thèse pour le Doctorat en Médecine. Ed Picavet, Paris

11. Guiot G, Montrieul B, Goutelle A, Comoy J, Langie S (1970) Méningiomes supra-sellaires rétro-chiasmatiques. Neurochirurgie 16: 273–285

12. Henderson WR (1938) The anterior basal meningiomas. Br J Surg 26: 124–165

13. Hullay J (1965) Planum (jugum) sphenoidale meningioma. Acta Neurochir (Wien) 12: 717–745

14. Hullay J, Gombi R, Velok Gy, Rozsa L, Borus F (1980) Planum sphenoidale meningioma. Attachment and blood supply. Acta Neurochir (Wien) 52: 9–12

15. Jane JA, McKissock W (1962) Importance of failing vision in early diagnosis of suprasellar meningiomas. Br Med J 5296: 5–7

16. Jefferson A, Azzam N (1979) The suprasellar meningiomas: a review of 19 years experience. Acta Neurochir (Wien) [Suppl] 28: 381–384

17. Kadis GN, Mount LA, Ganti SR (1979) The importance of early diagnosis and treatment of the meningiomas of the Planum sphenoidale and tuberculum sellae: a retrospective study of 105 cases. Surg Neurol 12: 367–371

18. Kendall B, Pullicino P (1979) Comparison of consistency of meningiomas and CT appearances. Neuroradiology 18: 173–176

19. Ley A, Gabas E (1979) Meningiomas of the tuberculum sellae. Acta Neurochir (Wien) [Suppl] 28: 402–404

20. Lindgren E, Di Chiro G (1951) Suprasellar tumours with calcification. Acta Radiol 36: 173–195

21. Passerini A, Cecchini A (1962) I meningiomi del tubercolo sellare (studio radiologico di 24 casi). Nuntius Radiol 28: 822–841

22. Rosenstein J, Symon L (1984) Surgical management of suprasellar meningioma. Part 2: prognosis for visual function following craniotomy. J Neurosurg 61: 642–648

23. Skultety FM (1968) Meningioma simulating ruptured aneurysm. Case report. J Neurosurg 28: 380–382

24. Smith HP, Challa VR, Moody DM, Kelly DL (1981) Biological features of meningiomas that determine the production of cerebral oedema. Neurosurgery 8: 428–433

25. Smith VR, Stein PS, McCarty CS (1975) Subarachnoid haemorrhage due to lateral ventricular meningiomas. Surg Neurol 4: 241–243

26. Solero CL, Giombini S, Morello G (1983) Suprasellar and olfactory meningiomas. Report on a series of 153 personal cases. Acta Neurochir (Wien) 67: 181–194

27. Symon L, Jakubowski J (1979) Meningiomas. Clinical features, technical problems, and results of treatment of anterior parasellar meningiomas. Acta Neurochir (Wien) [Suppl] 28: 367–370

28. Symon L, Rosenstein J (1984) Surgical management of suprasellar meningioma. Part 1: the influence of tumour size, duration of symptoms, and microsurgery on surgical outcome in 101 consecutive cases. J Neurosurg 61: 633–641

29. Tucker RL, Holman CB, McCarty CS, Dockerty MB (1959) The roentgenologic manifestations of meningiomas in the region of the tuberculum sellae. Radiology 72: 348–355

30. Weyand RD, Camp JD (1954) Roentgenographic examination in meningioma of the tuberculum sellae or olfactory groove. Radiology 71: 947–951

31. Wilson WB, Gordon M, Lehman RAW (1979) Meningiomas confined to the optic canal and foramina. Surg Neurol 12: 21–28

32. Wood MW, White RJ, Kernohan JW (1957) One hundred intracranial meningiomas found incidentally at necropsy. J Neuropathol Exp Neurol 16: 337–340

Address for correspondence: Prof. Dr. Jean Brihaye, 98. Av. des Franciscains, B-1150 Bruxelles, Belgium.

Acta Neurochirurgica, Suppl. 42, 130–136 (1988)

Microsurgery of Midbrain Lesions

G. Pendl and **P. Vorkapic**

Department of Neurosurgery, University of Vienna, Austria

Summary

Between 1973 and 1987 a total of 38 patients with midbrain lesions were encountered. In 15 cases surgical exploration was not warranted, 23 patients underwent definitive surgical exploration. The infratentorial supracerebellar approach proved to be the ideal route for the exposure of the 21 lesions which were located in the more dorsal aspect of the midbrain; in 7 cases the extension of the mass lesion reached from the cerebellum into the midbrain and, therefore, the exposure demanded a transcerebellar route. In one case with a hamartoma in the interpeduncular cistern and another case of a metastasis of the right cerebral peduncle the subtemporal approach was chosen. Three patients died as a consequence of the operation, but in the other 20 there was no increase in morbidity after surgery and the immediate postoperative course was favourable. Four cases with malignant tumours died from recurrence despite radiotherapy. The remaining 16 cases have been doing well up to 14 years after surgery.

Introduction

Since microneurosurgery was introduced more than a decade ago as a routine surgical procedure, it has enabled previously impossible results to be achieved in almost all fields of neurosurgery. Lesions which had been regarded as inoperable could be approached with more finesse and with much more optimism, especially tumours of the base of the brain, midline, and brain stem as well the latter apostrophized by Bailey et al.[1] as a "pessimistic chapter in the history of neurosurgery".

In the last decade lesions of the pineal region – since the stimulating publication by Stein[20] – have proved to be a less formidable surgical challenge as is confirmed by a number of favourable microsurgical studies, but this is not the case in lesions within the brain stem[9, 10]. The midbrain, in particular, has seemed to be too delicate a structure to allow surgical exploration on account of its important nuclei and the innumerable pathways incorporated within it. But sophisticated neuroimaging methods, CT and MRI, together with digital angiographic analysis, are of fundamental importance to localise and precisely define the lesion and its possible microtopographic relationships. Even with the possibility of preoperative diagnosis with tumour markers and CSF cytology, especially in cases of malignant tumours of the pineal region and midbrain, the need for surgical exploration in the light of the above diagnostic results before employing chemo- or/and radiotherapy has been stressed by Lassiter et al. 1971[7], who thereby achieved a longer survival rate especially in partly cystic lesions.

Case Material

In a period of 14 years, between 1973 and 1987, 23 cases of space-occupying lesions within the midbrain underwent surgical exploration and were histologically verified. The first 3 cases, as well as case 5 and case 10, underwent urgent surgical evacuation because the preoperative clinical course with decerebration left no other choice. Table 1 shows the clinical data of these 23 cases and Fig. 1 indicates the individual localisation of the lesions within the midbrain schematically. Localisation in the pre-CT era was primarily possible by scintigraphy and ventriculography, supported also by clinical signs. In the past 10 years all lesions have been localised primarily by CT scans and in the last 2 years also by MRI. Angiography has been of little help, but was able to rule out vascular malformations except in one case of a circumscribed harmartoma of high vascularity (case 7); also in cases of later proven haematomas the source of the haemorrhage could not be demonstrated (case 8 and case 10). Prior to the introduction of MRI, ventriculography in some cases of large masses helped to delineate extension of the tumour towards the 3rd ventricle.

Except for a hamartoma at the base of the midbrain in the interpeduncular cistern (case 4), a metastasis of the right peduncle (case 20) and a cavernoma with a haematoma of the left peduncle (case 23), all of which were approached by a subtemporal route, an infratentorial route was chosen. However, the supracerebellar approach was only possible in those cases, where the lesions were limited entirely to the midbrain. A transcerebellar resection of the tumour was necessary in case 5, 6, and 9, 13, 15, 18, and 22, since they extended towards the cerebellum and lower brain stem.

Preoperative neurological deficits improved in all cases except

Table 1. *Clinical Data in 20 Patients with Direct Microsurgical Removal of Lesions Within the Midbrain*

Case no.	Age	Sex	Histology	Localization	Signs and symptoms	Additional therapy	Survival or length of postoperative follow-up	Karnofsky Rating preoperative	postoperative
1	11 y	f	pilocytic astrocytoma	left cerebral peduncle	hydrocephalus, hemiplegia, Parinaud, acute decerebration	preoperative radio-therapy, VA-shunt	14 years, alive with mild deficite	40 (10)	90
2	9 y	m	ependymo-blastoma	midbrain-posterior 3rd ventricle	hydrocephalus, Parinaud	preoperative radio-therapy, Torkildsen	1½ years, death from disease	80	90 (1 y)
3	3 y	f	medullo-blastoma	midbrain-posterior 3rd ventricle	oculomotor disorder, acute decerebration	postoperative radio-therapy, VA-shunt	15 months, death from disease	(10)	(50)
4	16 y	m	hamartoma	interpedunuclar cistern	gelastic epilepsy, precocious puberty	none	5 years, alive with dificits, no further follow-up	40	60
5	2 y	m	medullo-blastoma	vermis-midbrain-posterior 3rd ventricle	hydrocephalus, hemiplegia	preoperative VA-shunt	2 months death post-operative related to surgery	—	—
6	5 y	m	astrocytoma II	4th ventricle-midbrain	Parinaud, cerebellar signs, VII nerve palsy	postoperative radiotherapy	2 years, alive with deficits, no further follow-up	40	60
7	57 y	f	hamartoma	right midbrain	hemiparesis, Parinaud, other oculomotor dis-orders, cerebellar signs, V + VII nerve signs, acute decerebration	preoperative VA-shunt	6 years, alive with mild deficits	40	90
8	31 y	f	microangioma	right midbrain	Parinaud, occulomotor disorder, VII nerve palsy	none	6 years, alive with mild deficits	70	90
9	17 m	m	ependymoma I	4th ventricle-left midbrain	hydrocephalus	postoperative radio-therapy, VA-shunt	2 years, death from disease	—	—
10	14 y	f	haematoma	right midbrain	subacute decerebration with occlusive hydro-cephalus	preoperative VA-shunt	4 years alive and well	20	100
11	52 y	f	glioblastoma	right midbrain	hemiparesis	postoperative radio-therapy	2 years and 9 months, death by recurrency	40	80 (2 y)
12	68 y	f	metastasis	left midbrain	chronic intracranial pressure, oculomotor and VIII nerve disorder	postoperative Torkildsen	11 days death post-operative related to surgery	40	—

Table 1 (continued)

Case no.	Age	Sex	Histology	Localization	Signs and symptoms	Additional therapy	Survival or length of postoperative follow-up	Karnofsky Rating preoperative	postoperative
13	16 y	m	pilocytic astrocytoma	midbrain	hydrocephalus, cerebellar signs	partial resection prior with VA-shunt	3 years and 2 months alive and well	40	100
14	13 y	m	glial tumour	quadrigeminal plate	hydrocephalus cerebellar signs	VA-shunt	3 months, alive with deficits, no further follow-up	90	90
15	5 y	m	astrocytoma II–III	left midbrain	hemiplegia, oculo-motor disorders	radiotherapy	2 months, alive with deficits, no further follow-up	—	—
16	17 y	m	pilocytic astrocytoma	midbrain	chronic intracranial pressure, oculomotor disorders	preoperative VA-shunt	2 years, alive with oculomotor disorder	50	70
17	68 y	m	cystic metastasis	right midbrain	extrapyramidal motor disorder	none	4 weeks, death by pneumonia	50	(70)
18	18 m	f	PNET	midbrain	cerebellar signs and symptoms	radiotherapy and chemotherapy	1 year and 2 months, well and alive	—	—
19	53 y	m	metastasis	right midbrain	oculomotor disorders, cephalalgia	radiotherapy	6 months, mild oculomotor disorder, death by disease	80	(100)
20	42 y	f	metastasis	right peduncle	hemiparesis	2 months previous frontal metastasis, lobectomy for lung tumour planned	10 months, well and alive	90	90
21	6 y	m	pilocytic astrocytoma	left midbrain	hemiparesis	preoperative radio-therapy, VA-shunt	8 months, well with mild deficits	70	90
22	9 y	m	pilocytic astrocytoma	right midbrain	hemiparesis	preoperative radio-therapy, VA-shunt	3 months, alive with progress	70	80
23	3 y	m	cavernoma with haematoma	left peduncle and upper pons	acute Hemiparesis	none	3 months, alive with progress	80	90

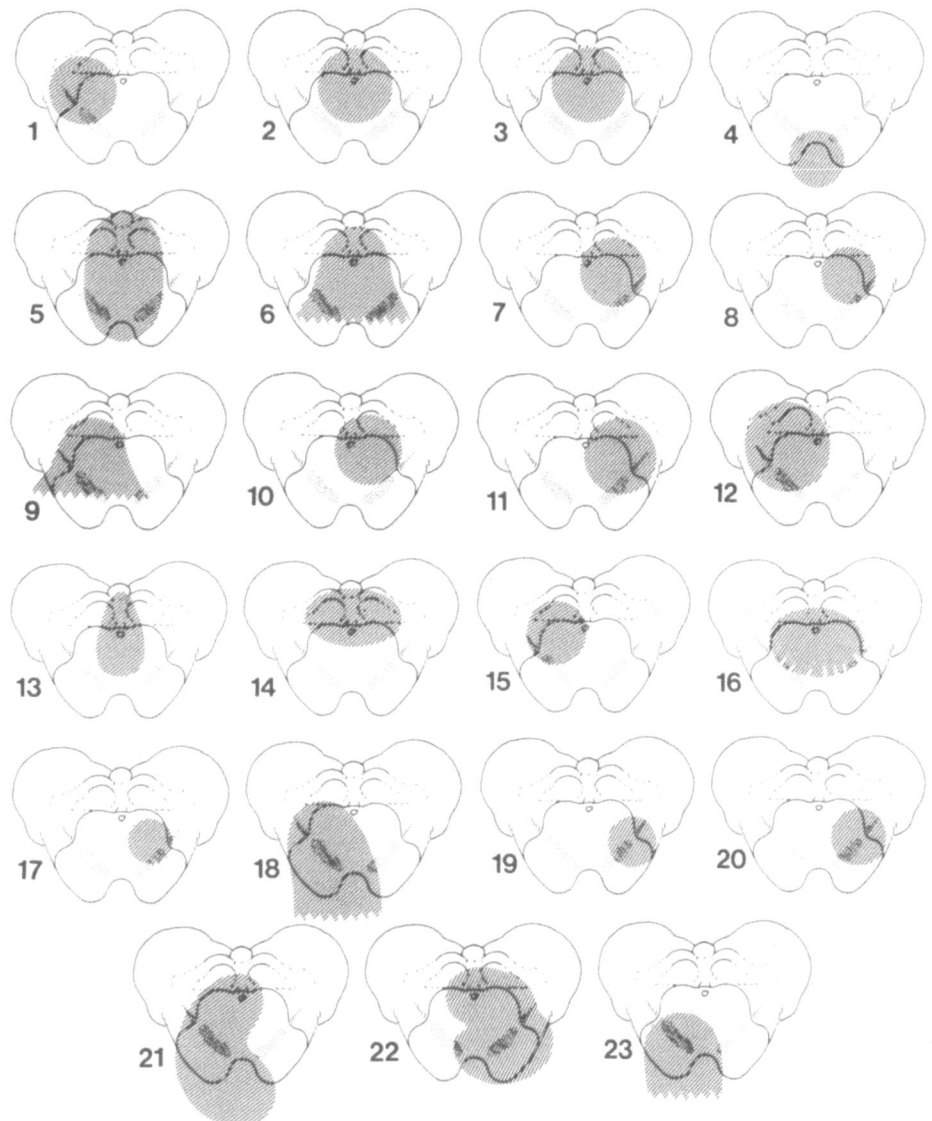

Fig. 1. Schematic drawing of the location of 23 histologically verified lesions within the midbrain according to Table 1

for two postoperative deaths; *i.e.* postoperative morbidity was short-term and in most cases it did not affect reintegration into normal life after discharge from the hospital. One child died two months after surgery as a result of secondary haemorrhage within the tumour cavity (case 5), and two other children with malignant tumours died from respectively a recurrence after 18 months (case 2), and from seeding to the spinal canal after 15 months (case 3) despite postoperative radiotherapy. In the adult group one postoperative death occurred when a patient with metastasis remained in a vegetative state with stable vital signs but died on day 11 after surgery from secondary haemorrhage into the tumour cavity (case 12). A 68-year-old patient with a cystic metastatic tumour and a coarse tremor of the left arm, who had urged us to relieve him of this movement disorder, died 4 weeks postoperatively as a result of pneumonia of his right lung – he had previously had a pneumonectomy on the left (case 17). Operative mortality rate, therefore, is 3 out of 23, *i.e.* 13%.

Another 15 cases were observed in this period, but either no surgical intervention was necessary or else no consent was given for major surgery or even for biopsy (Table 2). Thus those patients do not serve as a control group.

Discussion

Mass lesions within the midbrain originate from the glia of the midbrain, the ependyma of the aqueduct, the blood vessels (vascular malformations) and secondarily as haematomas, there are also calcified mass lesions, which in some cases proved to be tuberculomas[2, 3, 19]. Metastatic tumours may also be found in this region[21, 22]. Sanford *et al.*[16] report on low-grade astrocytomas presenting as pencil gliomas, lim-

Table 2. *Clinical Data in 15 Observed Non Surgical Cases with Lesions of the Midbrain Region*

Case no.	Age	Sex	Localization and tentative diagnosis	Signs and symptoms	Causes of conservative management and follow-up
1.	56 y	f	haematoma of the quadrigeminal plate	transient Weber's syndrome	no indication, 7 years follow-up with complete recovery
2.	6 y	f	calcified process of the left midbrain	hypaesthesia of left face, mild spasticity of the right leg since infancy	no indication, follow-up 6 years with gradual improvement
3.	33 y	f	large calcified lesions of the left midbrain with hydrocephalus	intracranial pressure, relieved by VA-shunt, no focal signs or symptoms	no indication, close follow-up for 4 years
4.	3 y	m	partially calcified lesion of the right midbrain	progressive ataxia	no consent, no follow-up
5.	5 y	m	mass lesion of the basal midbrain, not delineated	progressive ataxia and finally comatose	no indication, no follow-up
6.	19 y	m	hyperdense lesion of the left midbrain, well delineated	occulomotor disturbance	no consent, no follow-up
7.	44 y	f	haematoma of the midbrain	transient Parinaud and cephalea	no indication, 1 year follow-up with complete recovery
8.	36 y	m	glial mass lesion of the midbrain and left thalamus	personality change, slight hemiparesis of the right	no indication, radiotherapy with 1 year follow-up, unchanged
9.	4 y	m	glial mass lesion of the midbrain	personality change, slight hemiparesis on the right	no consent, no follow-up
10.	11 y	m	cystic tumour of the left midbrain	paraplegia after radiation of midbrain prior to neurosurgical consultation	spinal metastasis, no indication, no follow-up
11.	5 y	m	haematoma of the left midbrain	no deficit, spontaneous resorption	no indication 8 months follow-up
12.	20 y	f	calcified process of the quadrigeminal plate	hydrocephalus (VA-Shunt)	no indication, 4 years follow-up
13.	27 y	m	haematoma of the midbrain	apallic syndrome	no indication, no follow-up
14.	32 y	m	glial tumour of the quadrigeminal plate	personality change	no indication, no follow-up
15.	30 y	f	glial tumour of the left midbrain and pons	ataxia since 11 years	no indication

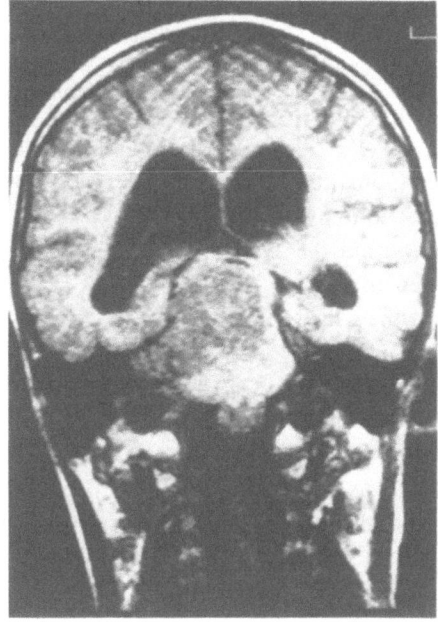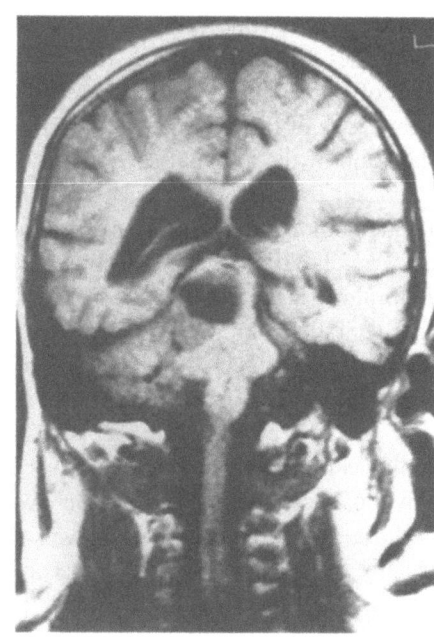

Fig. 2. MRI of a pilocytic astrocytoma within the right midbrain (case 22): (A) Preoperative MRI shows a well delineated mass lesion within the midbrain more to the right. (B) MRI of case 22 three weeks after an infratentorial supracerebellar approach and removal of the tumour. The tumour cavity shows residual tumour at its borders

A   **B**

ited to the aqueductal region of the tectum with fatal outcome. Finally, we have to expect medulloblastomas.

In 1971 Lassiter *et al.*[7] demonstrated the need for surgical exploration in this generally discouraging group of patients with brain stem gliomas either in the hope of finding a surgically amenable cyst or to obtain histological verification of a radiosensitive lesions. Despite allegations that mass lesions of the midbrain are not directly operable[20], there have been further reports on successful surgical interventions[3, 4, 5, 6, 8, 18, 21]. In most of the reported cases of successful operative treatment haematomas were encountered[4, 6, 8, 18]. Four out of these 6 reported haematomas were evacuated by a subtemporal route. In addition the total excision of a benign noncystic expansive tuberculoma of the rigth midbrain by a subtemporal transtentorial approach has been reported[3].

In our own series, 11 out of 23 cases with benign lesions justified surgical intervention in addition to the clinical necessity of an acute decompression after a shunting procedure. The primary nonsurgical management of a benign cystic astrocytoma of the left cerebral peduncle in a 12-year-old child almost led to a fatal outcome, since exact localisation had seemed impossible in the pre-CT era and radiotherapy was not successful (case 1). Therefore the decision to expose these lesions was not based on the need for histological material for planning appropriate postoperative treatment, but was the result of clinical deterioration which

left no other choice of treatment. Of course an exact histological diagnosis will prevent inappropriate further treatment, *e.g.* radiotherapy or chemotherapy. In case 7 and case 8 an erroneous diagnosis of multiple sclerosis led to frustrating cortisone therapy, which in the first case extended over a period of several years. In malignant tumours reduction of the tumour size by surgery may be of benefit for any subsequent radio- and/or chemotherapy.

Recent reports of microsurgically safely excised lesions in the pineal region and the microtopography of the pineal and midbrain region[11, 12, 14, 15, 23], which is currently better understood, have provided sufficient data with which suitable access to the midbrain itself. It is not only the deep-seated lesion to plan, which creates its own problems, but the surrounding vascular structures are also of great importance. In the pineal region the vascular architecture is often dislocated by the tumour and parts of the venous channels may even be obstructed; this is not the rule in most cases of lesions within the midbrain. Therefore the choice of approach is determined by the position of the normal anatomical structures overlying the lesion, given that the lesion has not extended beyond the midbrain area or is entirely limited to the base. In the latter situation a subtemporal approach is mandatory. When choosing an approach to the quadrigeminal plate to reach an underlying lesion, a suboccipital transtentorial, or an infratentorial supracerebellar approach will help to make possible a safe dissection concerning the vein of Galen and its tributaries. Equally, splitting the upper vermis or the edge of the cerebellar hemisphere will provide a sufficient view of the lateral aspects of the quadrigeminal plate and of the ambient cistern.

Except for tumours localised within the pons and medulla oblongata[13], we do not agree with Schönmayr and Agnoli[17] that in cases of mass lesions within the midbrain surgical treatment is only indicated and successful, if the tumour shows external growth, which allows decompression without harming the surrounding structures in the brain stem, as well as in tumours with marked cystic components. These pathomorphological circumstances will always help in the removal of these lesion[5, 7] but as our own cases, as well as those from other authors[3] prove solid and also especially well delineated lesions become accessible to surgery and total excision may even be possible (Fig. 2).

References

1. Bailey P, Buchanan DN, Bucy PC (1964) Intracranial tumours of infancy and childhood. The University of Chicago Press, Chicago and London, Third Impression
2. Bravo G, Vaquero J, Martinez R, *et al* (1981) Conservative management of a mesencephalic tuberculoma. J Neurosurg 55: 287–288
3. Dierssen G, Trigueros F, Sanz F *et al* (1978) Surgical treatment of a mesencephalic tuberculoma. J Neurosurg 49: 753–755
4. Durward QJ, Barnett HJM, Barr HWK (1982) Presentation and management of mesencephalic hematoma. J Neurosurg 56: 123–127
5. Enzian W (1983) Removal of intraponto-mesencephalic spongioblastoma. Neurosurg Rev 6: 67–70
6. Humphreys RP (1978) Computerized tomographic definition of mesencephalic hematoma with evacuation through pedunculotomy. J Neurosurg 49: 749–752
7. Lassiter KRL, Alexander E Jr, Davis CH Jr, *et al* (1971) Surgical treatment of brain stem gliomas. J Neurosurg 34: 719–725
8. La Torre E, Delitala A, Sorano V (1978) Haematoma of the quadrigeminal plate. J Neurosurg 49: 610–613
9. Pendl G (1975) Gelastic epilepsy in tumours of the hypothalamic region. Adv Neurosurg 3: 442–449
10. Pendl G (1976) Infratentorial approach to mesencephalic tumours. In: Koos WT, Böck F, Spetzler RF (eds) Clinical microneurosurgery. Thieme, Stuttgart, pp 143–150
11. Pendl G (1984) Microsurgical anatomy of the pineal region. In: Neuwelt EA (ed) Diagnosis and treatment of pineal region tumours. Williams and Wilkins, Baltimore London, pp 155–207
12. Pendl G (1985) Pineal and midbrain lesions. Springer, Wien New York
13. Pendl G, Koos WT (1980) Microsurgery of brain stem tumours in childhood and adolescence: A review of past experience. Adv Neurosurg 8: 403–408
14. Quest DO, Kleriga E (1980) Microsurgical anatomy of the pineal region. Neurosurgery 6: 385–390
15. Reid WS, Clark WK (1978) Comparison of the infratentorial and transtentorial approaches to the pineal region. Neurosurgery 3: 1–8
16. Sanford RA, Bebin J, Smith RW (1982) Pencil gliomas of the aqueduct of Sylvius. J Neurosurg 57: 690–696
17. Schönmayr R, Agnoli AL (1983) Brain stem tumours – diagnosis and surgical treatment. Neurosurg Rev 6: 57–65
18. Scoville WB, Poppen JL (1949) Intrapeduncular haemorrhage of the brain. Arch Neurol Psychiat 61: 688–694
19. Selekler K, Erbengi A, Saribas O *et al* (1983) Giant calcified and ossified midbrain tuberculoma. J Neurosurg 58: 133–135
20. Stein BM (1971) The infratentorial supracerebellar approach to pineal lesions. J Neurosurg 35: 197–202
21. Toberl WD, Sawaya R, Tew JM Jr (1986) Successful laser-assisted excision of a metastatic midbrain tumour. Neurosurgery 18: 795–797
22. Weber G (1974) Midbrain tumours In: Vinken PJ, Bruyn GW (eds) Handbook of clinical neurology. North Holland Publ Comp, Amsterdam, vol 17/II, pp 620–647
23. Yamato I, Kageyama N (1980) Microsurgical anatomy of the pineal region. J Neurosurg 53: 205–221

Address for correspondence: Prof. Dr. G. Pendl, Neurochirurgische Universitätsklinik, Währinger Gürtel 18-20, A-1090 Wien, Austria.

Acta Neurochirurgica, Suppl. 42, 137–141 (1988)

Diagnosis and Treatment of Pineal Tumours. Kyoto University Experience (1941–1984)

J. Yamashita and **H. Handa**

Department of Neurosurgery, Kyoto University Medical School, Kyoto, Japan

Summary

Our current policy for the treatment of pineal tumours is presented, based upon our experience of 139 cases of pineal tumours in a period of 44 years from 1941 through 1984. First of all, it should be emphasized that germinomas are extremely radiosensitive and are treated successfully by radiotherapy alone. Accordingly, the treatment of choice for germinomas would be radiotherapy without surgery, if the diagnosis could be made confidently by the modern armamentarium of investigations, including CT, cerebrospinal fluid cytology and serum level of tumour markers, such as alphafetoprotein and human chorionic gonadotropin.

When the diagnosis of germinoma is suspected although not straightforward, CT is reexamined after a trial of diagnostic radiation of 20 Gy. If the tumour size has not been reduced at all, it is unlikely that the tumour is a germinoma and surgery should be considered for removal as well as establishment of the histological diagnosis.

Keywords: Pineal tumour; germinoma; teratoma; treatment; operation; radiotherapy; outcome.

Introduction

Our attitudes toward the diagnosis and treatment of pineal tumours have been dramatically changed in recent years by several important factors:

1. Histological classification has been well established by introduction of the concept of germ cell tumours including germinomas and a variety of teratomas[5, 15].

2. The value of radiotherapy for germinomas has been firmly established[4, 11, 16].

3. Operative procedures have been refined by introduction of microsurgical techniques[12].

4. Preoperative diagnosis has been made easier and more precise by the development of CT and cerebrospinal fluid (CSF) cytology[17], and the discovery of tumour markers, such as alpha-fetoprotein (AFP) and human chorionic gonadotropin (HCG)[1, 7, 14].

5. Cisplatin (CDDP) is now available as a relatively specific chemotherapeutic agent for malignant teratomas, which are producing AFP and/or HCG[6, 9].

Clinical Materials

One hundred and thirty-nine cases of pineal tumours were encountered in our department over a period of 44 years form 1941 through 1984, which corresponded to 4.3% of all intracranial tumours treated in the same period. There were 60 cases of histologically verified pineal germ cell tumours, seven pineoblastomas, 15 other histologically verified tumours and 57 histologically non-verified pineal tumours (Table 1). The recent cases, whose diagnosis was made as germinomas by CSF cytology, were arbitrarily counted as germinomas. The majority of histologically non-verified pi-

Table 1. *Clinical Materials of Pineal Tumours* (Kyoto University, 1985)

	Pineal	Suprasellar	Other	Total
Germinoma	39	18	12	69
Teratoma				
Mature	16	3	4	23
Immature	5	7	6	18
	60	28	22	110
Pineoblastoma	7			
Others				
Verified*	15			
Non-verified	57			
Total	139			

* Verified Tumours: glioblastoma 7, astrocytoma 2, hemangloblastoma 2, ependymoma 1, ependymoblastoma 1, meningioma 1, epidermoid 1.

neal tumours were rather old in this series, diagnosed by ventriculography, and usually treated as germinomas. Germ cell tumours occur most frequently in the pineal region, followed by the suprasellar, basal ganglia and other regions. The germ cell tumours are grossly divided into germinomas and teratomas. The teratomas are further divided into mature and immature teratomas. The immature teratomas are composed of embryonal carcinomas, yolk sac tumours and choriocarcinomas. However, there are occasional cases which contain two or more different histological components of germ cell tumours even in a single tumour.

We have experienced 69 cases of histologically verified germinomas (39 pineal, 18 suprasellar and 12 in other regions), 23 cases of mature teratomas (16 pineal, 3 suprasellar and 4 in other regions) and 18 cases of immature teratomas (5 pineal, 7 suprasellar and 6 in other regions). Histologically verified germ cell tumours constituted 2.1% of all intracranial tumours in our institute. If histologically non-verified pineal tumours were included in germ cell tumours, germ cell tumours then amounted to 3.6% of all intracranial tumours in our institute. Among the 69 histologically verified germinomas, including the ones in the suprasellar and other regions, there were 56 males and 13 females, with a marked male preponderance. A male preponderance was also found for teratomas. Among the 110 histologically verified germ cell tumours, there were 84 males and 26 females. It is interesting to note that there is a female preponderance in the suprasellar location both for germinomas and teratomas.

The peak incidence was in the first half of the second decade for germinomas and immature teratomas. Mature teratomas occurred a little earlier with the peak incidence in the latter half of the first decade.

It has been reported that intracranial germ cell tumours occur much more commonly in Japan than in western countries[2]. According to the Brain Tumour Registry of Japan[3], germ cell tumours corresponded to 3.6% of all primary intracranial tumours, and 2.8% of all intracranial tumours including brain metastases.

Diagnosis

There has been a great confusion in the histological terminology of pineal tumours, because the origin of the epitheloid cells of the germinoma had been difficult to determine. Intracranial germ cell tumours are now understood as analogous to those in the male and female reproductive organs, according to the concept of germ cell tumours[5, 15]. It is believed that some of the

germ cells, arising from the yolk sac in an early embryonal stage, migrate along the ventral surface of the body towards the brain and give rise to the origin of intracranial germ cell tumours.

Histologically, the germinoma is composed of large epitheloid cells with clear cytoplasm and distinct nucleoli, and of lymphoid cells which are predominantly seen in perivascular distribution. The characteristic feature is the so-called "two cell pattern" of these two different types of cells.

In CT, the germinoma usually appears as a mass of normodensity or moderately high density, which is homogeneously enhanced in contrast studies. The border of the tumour is slightly ill-defined and sometimes extends along the ventricular walls.

The teratoma usually appears heterogeneous with areas of low density, suggestive of cyst formation. However, it is not always possible to differentiate between germinomas and teratomas by CT findings alone. In angiography, hypervascularity is one of the characteristics of malignant varieties of teratomas, such as embryonal carcinomas, yolk sac tumours and choriocarcinomas[8, 14].

In the case of germinomas, tumour cells tend to exfoliate into the CSF. We observed abnormal cells in the preoperative CSF in 62% of the cases with germinoma, whereas CSF cytology was always negative in teratomas[17]. Therefore, CSF cytology is one of the practically useful diagnostic tests for pineal tumours. In the typical cases, the two cell pattern corresponding to the histology is clearly observed in CSF cytology (Fig. 1).

The high serum level of AFP was observed in six out of seven cases of malignant varieties of teratoma but in none of the 14 cases of germinoma tested. On the other hand, the serum level of HCG was increased in three out of six teratomas with increased AFP and also in seven out of 14 germinomas. Serial determination of these tumour markers was found to be useful in evaluating the effect of treatment and in predicting the prognosis in some of the cases.

It was recently postulated that immunohistochemical staining of placental alkaline phosphatase (PALP) is useful for the diagnosis of germinomas[13]. In seven out of 8 cases of germinoma which we have studied, large polygonal epitheloid cells were postitively stained for PALP. Lymphocytes were persistently negative for PALP. The PALP may be another useful tumour marker for the germinoma, although a further study is required as to its specificity.

Immunohistochemical staining of AFP and HCG is

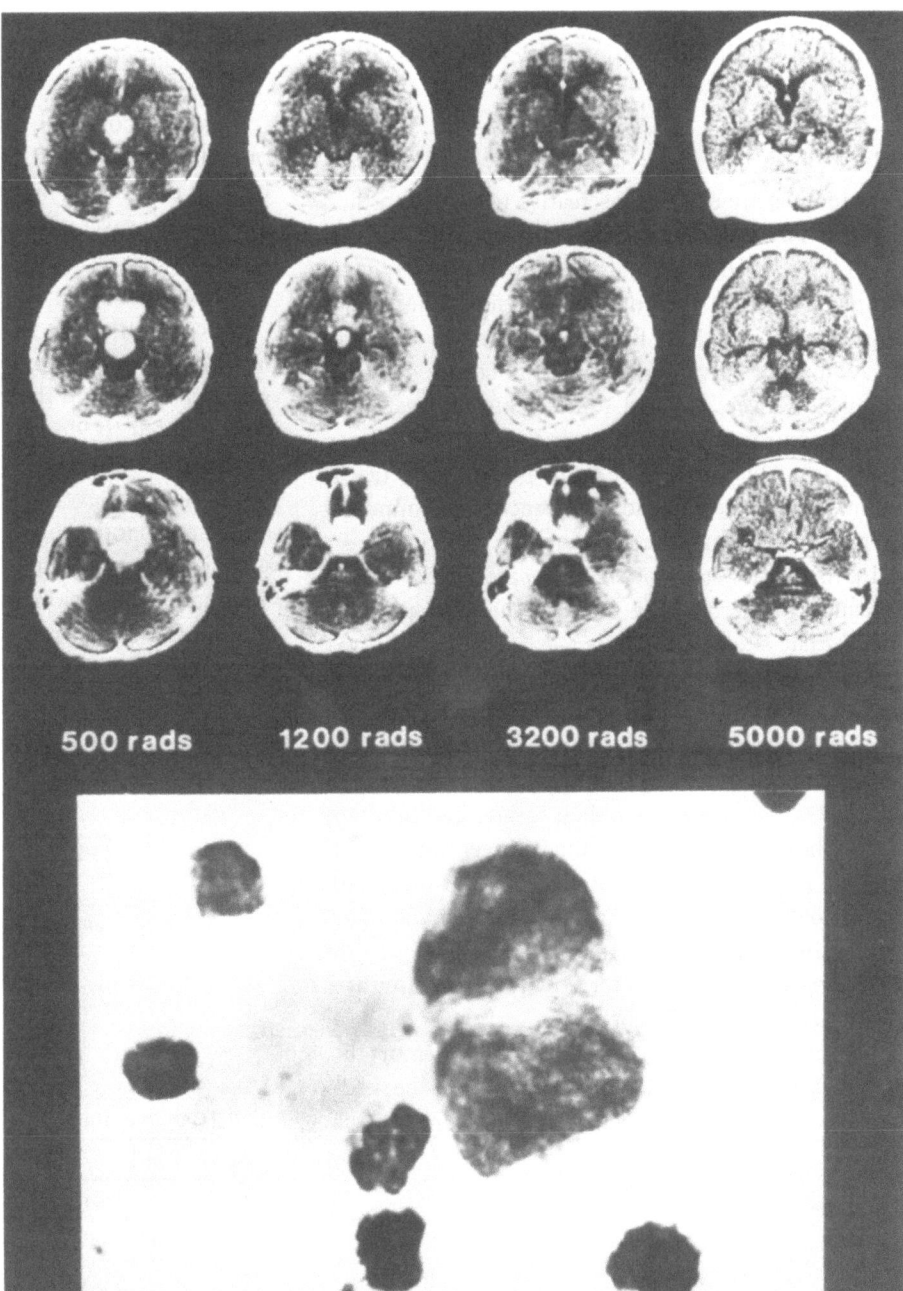

Fig. 1. Case F.Y., germinoma. (Upper) Serial post-contrast CT scans showing complete disappearance of the tumour by radiotherapy without surgery. (Lower) CSF cytology showing the characteristic two-cell-pattern of germinoma. Cytocentrifugation method (Cytospin). May-Grünwald-Giemsa stain, ×100

an elegant way of diagnosis of germ cell tumours. However, we sometimes experienced a discrepancy between the immunohistochemistry and the serum level of these tumour markers in about one third of the cases with immature teratomas. The failure in immunostaining was probably due to sampling error of the surgical specimen, because the cells positively stained were not always distributed uniformly in the tumour, particularly in AFP-producing tumours, whose serum level was only moderately increased. Immunohistochemical

staining was technically difficult in some of the HCG-producing choriocarcinomas, in which the surgical specimens were largely composed of necrotic and haemorrhagic tissues. Accordingly, we feel that the preoperative determination of the serum level of tumour markers is more reliable than the immunohistochemical staining of surgical specimens.

It is known that lymphocytes infiltrating in the germinoma are usually T lymphocytes[9]. We have found that they are more strongly stained by anti-Leu 3 than anti-Leu 2, suggesting that the predominant sub-population of the infiltrating T lymphocytes is the helper/inducer subset[10].

Treatment

Germinomas are extremely sensitive to radiation therapy (Fig. 1). Recent experience with CT has shown that all 10 cases of intracranial germinomas disappeared completely by the conventional external radiotherapy (Fig. 2). Most of them had disappeared by the time 20 Gy had been given during the full course of radiotherapy. We were the first to advocate the usefulness of the diagnostic 20 Gy radiotherapy for suspected germinomas[8], which is now a quite popular practice in Japan. There has been no recurrence of germinomas after complete remission so far in our experience.

In this series, there were five cases with tumour metastasis of germinoma through the CSF pathway (7.2%). The site of metastasis was spinal cord in two, spinal cord and peritoneum in one, frontal in one and peritoneum in one. In four out of the five cases, the metastasis occurred to an area outside of the radiation field. In view of the high incidence of positive CSF cytology, it has been our policy to give radiotherapy to the whole neuraxis of patients with germinoma, because it is generally difficult to treat the metastatic lesions, once they have occurred. Our schedule of radiotherapy for germinomas comprise of 35 Gy to the whole brain, 15 Gy of booster dose to the tumour area and 30 Gy to the spinal cord.

It goes without saying that the operative mortality in the surgery of pineal tumours has been markedly improved since the introduction of microsurgical techniques[12]. We usually use the infratentorial supracerebellar approach for relatively small tumours and the occipital transtentorial approach for larger tumours.

When it comes to the postoperative morbidity in our cases with pineal tumours who are still alive, we have an impression that performance status is generally better in the cases who were irradiated without surgery, although there is a recent reduction in the incidence of postoperative morbidity as well as operative mortality.

The 5-year survival rate of 35 cases of pineal germinomas, treated by radiotherapy with or without surgery, was 73% and the 10-year survival rate was 60%, as compared to 9% and 5%, respectively, in 22 cases of pineal germinomas without radiotherapy. In case of teratomas, however, radiotherapy was not so effective and the general prognosis was poorer than in germinomas. Surgery is still the treatment of choice in teratomas and occasionally, long-term survival was obtained after total removal of well differentiated mature teratomas.

In the recent years, Cisplatin has been used as the so-called PVB therapy in conjunction with Vinblastin

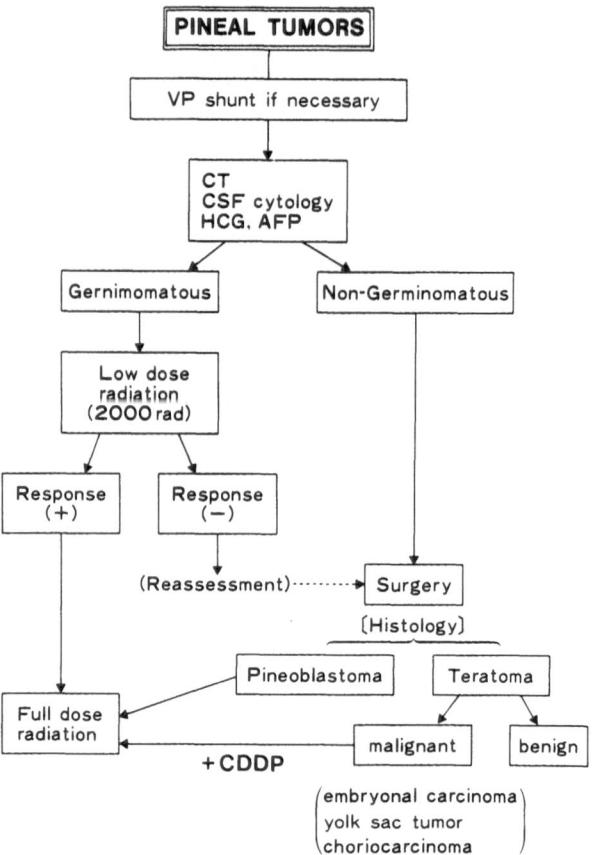

Fig. 3. Our current policy for the treatment of pineal tumours

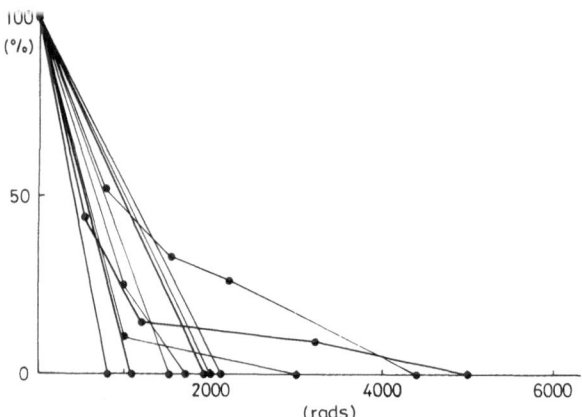

Fig. 2. Effect of radiotherapy for germinomas. The vertical axis represents the size of tumour calculated as the product of two crossing diameters of the tumour on CT

and Bleomycin for malignant teratomas which are producing AFP and/or HCG[6,9]. The increased serum level of AFP and/or HCG is rapidly normalized by the PVB therapy. The long-term prognosis of malignant teratomas are still poor, because the tumour usually recur after a short period of remission.

Conclusion

Our current policy in the treatment of pineal tumours is summarized as follows (Fig. 3);

1. Ventriculo-peritoneal shunting operations should be first performed in the presence of increased intracranial pressure.

2. Attempts at differential diagnosis should be made by CT, angiography, CSF cytology, and measurement of AFP and HCG in the serum and CSF.

3. If a germinoma is strongly suspected, a course of radiotherapy for the whole cerebrospinal axis should be given, provided it is tolerated.

4. If a germinoma is suspected but the possibility of other tumours, particularly other variety of teratomas, cannot be completely excluded, CT is performed again after 20 Gy of diagnostic radiation. If there is no reduction in tumour size at all, it is unlikely that the tumour is a germinoma and surgery should be considered.

5. Although the incidence of CSF metastasis of germinomas in our series was only 7%, we currently apply whole cerebrospinal axis irradiation routinely to patients with germinomas, hoping to prevent CSF dissemination.

6. In conclusion, it should be stressed that surgery is not necessary for germinomas which are extremely radiosensitive, if the diagnosis is confidently made by the modern armamentarium of laboratory investigations.

References

1. Allen JS, Nisselbaum J, Epstein F, Rosen G, Schwartz MK (1979) Alphafetoprotein and human chorionic gonadotropin determination in cerebrospinal fluid: An aid to the diagnosis and management of intracranial germ cell tumours. J Neurosurg 51: 368–374

2. Araki C, Matsumoto S (1969) Statistical reevaluation of pinealoma and related tumours in Japan. J Neurosurg 30: 146–149

3. Brain Tumour Registry in Japan (1984) the Committee of Brain Tumour Registry in Japan, vol 5, pp 911

4. Camins MB, Schlesinger EB (1979) Treatment of tumours of the posterior part of the third ventricle and the pineal region: A long-term follow-up. Acta Neurochir (Wien) 40: 131–143

5. DeGirolami U (1977) Pathology of tumours of the pineal region. In: Schmiedek HH (ed) Pineal tumours. Masson, New York, pp 119

6. Einhorn CH, Donohue J (1977) Cis-diamminedichloroplatinum, vinblastin, and bleomycin combination chemotherapy in disseminated testicular cancer. Ann Int Med 87: 293–298

7. Haase J, Nielsen K (1979) Value of tumour markers in the treatment of endodermal sinus tumours and choriocarcinomas in the pineal region. Neurosurgery 5: 485–488

8. Handa H, Yamashita J (1981) Current treatment of pineal tumours. Neurol Med Chir 21: 147–154 [Japanese]

9. Neuwelt EA, Fenkel EP, Smith RG (1980) Suprasellar germinomas (ectopic pinealomas): Aspects of immunological characterization and successful chemotherapeutic responses in recurrent disease. Neurosurgery 7: 352–358

10. Paine JT, Handa H, Yamasaki T, Yamashita J (1986) Suprasellar germinoma with shunt metastasis: Report of a case with an immunohistochemical characterization of lymphocyte subpopulations. Surg Neurol 25: 55–61

11. Salazar OM, Castro-Vita H, Bakos RS, Feldstein ML, Keller B, Rubin P (1979) Radiation therapy for tumours of the pineal region. Int J Radiol Biol Phys 5: 491–499

12. Stein BM, Fetell MR (1984) Therapeutic modalities for pineal region tumours. Clin Neurosurg 32: 445–458

13. Shinoda J, Miwa Y, Yamada H et al (1985) Immunohistochemical study of placental alkaline phosphatase in primary intracranial germ-cell tumours. J Neurosurg 63: 733–739

14. Takeuchi J, Handa H, Oda Y, Uchida Y (1979) Alphafetoprotein in intracranial malignant teratoma. Surg Neurol 12: 400–404

15. Teilum G (1965) Classification of endodermal sinus tumour (mesoblastoma vitellinum) and so-called "embryonal carcinoma" of the ovary. Acta Pathol Microbiol Scand 64: 407–429

16. Wara WM, Fellows CF, Sheline GE, Wilson CB, Townsend JJ (1977) Radiation therapy for pineal tumours and suprasellar germinomas. Radiology 124: 221–223

17. Yamashita J, Oda Y, Takeuchi J, Nakao S, Iwaki K, Handa H (1979) Cerebrospinal fluid (CSF) cytology using "Cytospin" in patients with brain tumours. Brain Nerve 7: 751–758 [Japanese]

Address for correspondence: Junkoh Yamashita, M.D., Department of Neurosurgery, Kyoto University Medical School, 54 Kawaharacho, Shogoin, Sakyoku, Kyoto 606, Japan.

Acta Neurochirurgica, Suppl. 42, 142–146 (1988)

Infratentorial Epidermoids

A. Bartal, N. Razon, J. Avram, S. Rochkind, and **A. Doron**

Department of Neurological Surgery, Tel Aviv – Elias Sourasky Medical Center, Ichilov Hospital, Tel Aviv, Israel

Summary

Infratentorial epidermoids are rarely seen in the lifetime of a neurosurgeon. Most published series consist of a dozen or so cases. In a period of 20 years we operated on only 6 such patients. Five are doing excellently, one patient died two and a half months after operation of a fulminating infection. The follow-up is of 20 years, 9 years, and 3 years of the first three patients, while the remaining three were operated during the last year. Such an accelerated pace of epidermoid incidence in our department during the last year may be fortuitous, but may also be an indication that, many patients with vague complaints who had an epidermoid, had been missed in the past. Undoubtedly, the CT scan has greatly facilitated the diagnosis of epidermoid cysts, whether infra- or supratentorial. Diagnosis, however, hinges on suspicion or awareness on clinical grounds of the possibility of an infratentorial epidermoid.

The analysis of the clinical presentation in our 6 cases, seems to allow the division of infratentorial epidermoids into those that are posteriorly located, which uniformly manifested at some stage of illness raised intracranial pressure, and the anterior epidermoids in the cerebello-pontine angle characterized by the insidious involvement of cranial nerves.

Computerized tomography, in some cases with the adjunct of Metrizamide cisternography, confirms the diagnosis and delineates the spread of the lesion.

There is no alternative to operation for infratentorial epidermoids, however, the procedure should be carried out judiciously, since a total resection of the capsule is rarely, if ever, possible, and too ambitious surgery may be fatal.

Keywords: Epidermoids; infratentorial cysts; cholesteatoma; CPA tumours.

Introduction

Intracranial epidermoidal cysts are rare[8, 10, 12], and a personal experience with the infratentorial epidermoids is even more limited[2, 8, 14]. Indeed, most recent reports are of single cases[3, 4]. More recently, Berger and Wilson[1], reviewed 14 patients with infratentorial epidermoids seen during a period of 10 years.

We have operated on 6 infratentorial epidermoids since 1966, a period of 20 years. The fact that these lesions are so rare does not detract from the attention they deserve and the challenge they pose.

Indeed, these so slowly growing, benign lesions appear easy to deal with and invite an optimistic, confident approach, aiming at a total excision with the expectation of a cure.

However, in their very slow growth, epidermoids tend to insinuate in every possible recess formed between cranial nerves, vessels, brain stem and openings at the base of the skull and the tentorium. Further, their soft malleable consistency allows for their gradual moulding into the surroundings, rather than displacing and compressing them, a feature probably also accounting for their late manifestation and uncharacteristic signs and symptoms[1, 10, 14, 15].

In the past, these tumours carried a significant morbidity and mortality attributed to the delay in diagnosis[7, 8, 10].

Computerized tomography has greatly facilitated the diagnosis of epidermoid cysts, but the full extent of involvement could not be determined until surgery[1, 2, 5, 6]. Remarkably, all patients were found to have lesions, much more extensive than could be appreciated on preoperative computerized tomography[1, 12].

Further, the hypodense, irregular areas seen in epidermoids on computerized tomography, may not be appreciated, unless suggested clinically. With this in mind we analyzed the clinical presentation of our patients with infratentorial epidermoidal cysts and it seems that it may be practical to consider two clinically distinguishable groups: those posteriorly located, who presented predominantly with the symptoms and signs of raised intracranial pressure and the more difficult to diagnose and dangerous to treat, anteriorly growing epidermoids, that had signs of cranial nerve involvement, invaded the retroclival space and supratentorially, the retro and latero-chiasmatic cisterns and did not feature increased intracranial pressure.

Own Material

If we look at the posteriorly located infratentorial epidermoids, we see that they almost uniformly presented with raised intracranial pressure in addition to vague complaints, at least at the beginning of illness, of general malaise, nausea, headaches, diplopia and transitory "fainting spells", or loss of consciousness.

When laterally situated, the lesions were posterior to the acoustic and trigeminal nerves. The midline lesions penetrated the IVth ventricle in one case and in

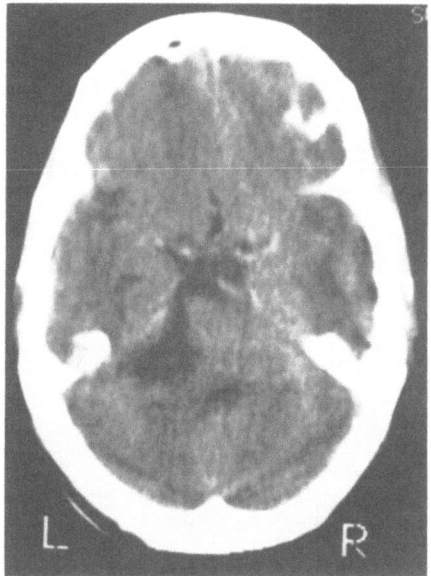

Fig. 3. The same patient as in Fig. 4. The lesion has greatly increased in size, within only three years since previous examination

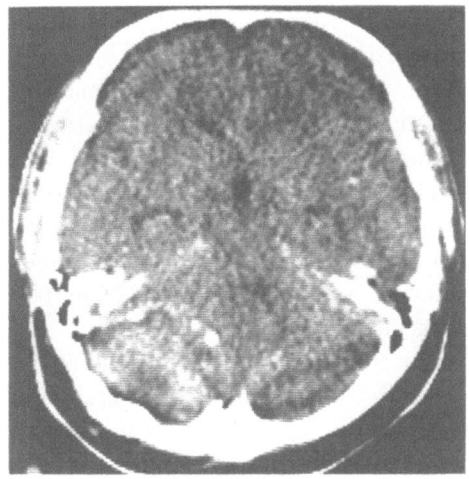

Fig. 1. CT scan showing a hyperdense, posteriorly located infratentorial lesion. The arciform calcification delimits the tumour anteriorly

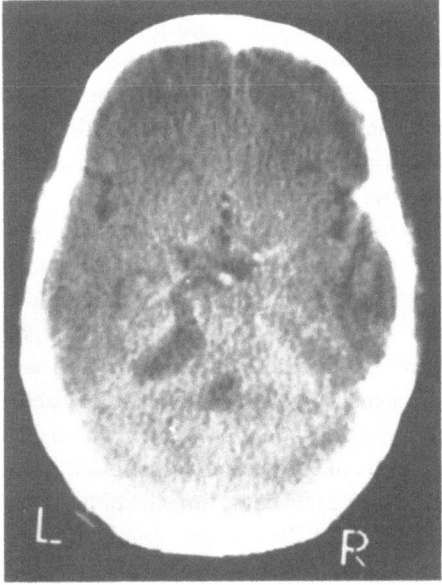

Fig. 2. The CT scan shows a hypodense strip in the left CP angle, which should have led to diagnosis of an anteriorly located infratentorial epidermoid

the other two, invaded the foramen magnum and spread, at the base of the skull, laterally to the posterior pyramid and tentorium on both sides.

The plain X-ray films in posteriorly located epidermoids may show thinning or mottling of the occipital squamous bone, more characteristic of epidural or intradiploic epidermoids[13, 14, 15] rarely an arc-like calcification which, in our case, obviously delimited the borders of the cyst (Fig. 1). Preoperative ventricular drainage of CSF alleviated headaches and by reducing intracranial pressure allowed for more relaxed and safer surgery. The anteriorly located epidermoid cysts have a more subtle, often bizarre presentation. The true nature of the lesion may not be suspected for a long time, although the patients may have quite localizeable complaints. Two of our three patients had been examined innumerable times for unilateral progressive loss of hearing and one had even repeat CT scans done in a major hospital. In retrospect one can now say that the narrow hypodense strip lateral to the pons was already at that time, 3 years ago, CT evidence of the epidermoid (Fig. 2). Interestingly, the CT scans in 1986 were indicative of a noticeable growth of the lesion and now corroborated the diagnosis (Fig. 3). The computed tomography scans of the anteriorly located epidermoids typically show more or less wide hypodense areas along the pyramids in the cerebello-pontine angle, reaching upwards in some cases into the prepontine, the interpeduncular and latero-chiasmatic cisterns.

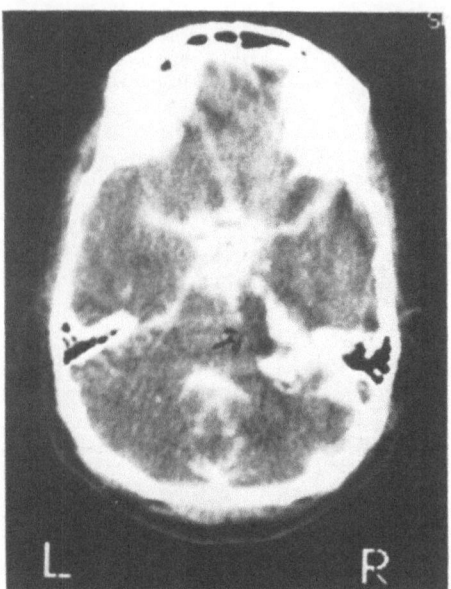

Fig. 4. Metrizamide cisternogram of right anteriorly situated, infratentorial epidermoids

Metrizamide CT cisternography[6] may add valuable information (Fig. 4).

Operation was carried out through the usual straight paramedian suboccipital incision, used for the approach of acoustic tumours but in most posteriorly located tumours a midline incision is recommended.

It is relatively safe and expedient to work with two aspirator tips, with both hands, it is easier to pry loose most finger-like projections out of their recesses, while only occasionally a straight define dura elevator is used to dislodge more resistant parts. The capsule itself may peel off with remarkable ease in places, particularly off larger vessels, the vertebral, the basilar, some cranial nerves, while it may remain unbudgeably resistant in others, particularly the surface of the lateral medulla and pons, the smaller vessels and nerves and very much so in the jugular foramen, an area where it is wiser to leave resistant tumour remnants and capsule alone. In all three anteriorly located epidermoids, the tumour filled up the space around the vertebral and basilar arteries, which it enveloped.

In all anterior cases the tumour infiltrated through the tentorial incisure all along the plane of the clivus in the midline and laterally of the cavernous sinus wall, in two cases beyond the cross-over point of the oculomotor nerve in the incisural cleft.

In one case, the female patient who presented with severe trigeminal neuralgia, the nerve was found to have been grossly splayed by tumour which was also partly embedded in the substance of the pons at the

site of the nerve entry. This tumour rest was pried free and dissected. Remarkably no large or smaller vessels were seen in the area.

At termination of the surgical procedure the remaining cavity is generously rinsed with saline.

Discussion

Intracranial epidermoid cyst have been variously reported to comprise between 1 and 2%[1, 7, 8, 10, 14] of intracranial tumours. These are congenital lesions, resulting from the displaced inclusion of epithelial tissue, during the cleavage of cutaneous ectoderm from the primitive neural crest[1, 8, 13, 14].

Tytus and Pennybaker[14] assumed that the time when epithelial inclusions became displayed, determined their location: those that occured before the 3rd week of embryological development would be midline while inclusions after the 5th gestational week would be laterally located since at that gestational period separation of the neuroectoderm from its cutaneous counterpart is no more along the midline.

The various locations of epidermoid cysts is also explained by some[13, 15] as a result of proliferation of multipotential embryonic epithelial rests in different areas of the base of the skull, from the ethmoids and the orbits anteriorly, the optochiasmatic region, the pyramids and the mastoids to the occiput. The lesion is composed of desquamated epidermal cell debris, originating from an inner epithelial lining of the glistening white tumour capsule[3, 10, 11]. Seepage of cyst content is highly irritative[1, 9] and produces either local reactions with ultimate adhesions or a so-called "aseptic" meningitis[3], typically seen with epidermoid cysts and due to the chemical action of keratin, cholesterin and other proteinateous derivatives on arachnoid membranes and nervous parenchyma[3, 9].

Epidermoid cysts of the posterior fossa have been known to cause trigeminal neuralgia[8], atypical facial pain[2], glossopharyngeal neuralgia and hemifacial spasm[10]. Usually, tumours causing tic douloureux, do so by pushing the trigeminal nerve against blood vessels. In our case, which presented as trigeminal neuralgia, with, neurologically, only a slightly diminished corneal reflex, the trunk of the nerve was infiltrated and splayed by tumour at its entry into the pons. After dissection and removal of the compressing lesions, no small or large arterial or venous vessels were observed in the immediate root entry zone or along the course of the nerve up to the edge of the pyramid.

The surgical treatment for epidermoid cysts in the

Table 1

Patient	Age/Sex Year	Presentation	Duration of symptoms	Site	Result
		Posteriorly located			
M.R.	3 m/m 1966	rec. meningitis high ICP, stupor	3 months rec. meningitis	fourth ventricle	excellent
G.E.	30 y/m 1984	rec. amnesia fits, headache amnesia, raised ICP	20 years suboccipital headaches papilledema obstructive hydrocephalus	intra dural, Lt. cerebellar convexity	excellent
M.Y.	58 y/m 1986	suboccip head. slurred speech diplopia	3 months gait imbalance truncal ataxia partial obstr. hydrocephalus	vermis, both cerebellar hemispheres IVth ventr.	died. sepsis, meningitis acineto bacter
		Anteriorly Located			
Y.T.	39/f 1978	deafness Lt. hearing loss Rt. opistotonic fits instability	10 years recc. paroxismal tonic attacks progressive V, VII, VIII, IX, X palsies, left	CPA Lt. prepontine pontomedull. cisterns for. magnum	good
S.H.	56/m 1986	deafness Lt. tinnitus, gait disturbances Lt. facial and trigeminal n. paresis	10 years tinnitus Lt. progressive hearing loss unstable gait Lt. facial numbness	CPA Lt. extending up above tentorial incisure	good
H.R.	24/f 1986	trigeminal neuralgia, Rt. intractable facial pain	16 months severe pain diminished Rt. corneal reflex	CPA Rt. prepontine supratento- rial exten- sion	excellent

posterior fossa, including those that extend above the incisura, should be radical but one should never strive for a total excision. A complete surgical cure is not feasible, but a radical removal will alleviate the patient's symptoms, allowing to resume work and lead a normal life for many years[1, 7, 8, 10].

The postoperative morbidity from manipulation of cranial nerves was transient in all patients except one with mild facial palsy one year after operation. The patient with severe trigeminal neuralgia is completely free of pain.

We did not use steroids either during the surgical procedure for rinsing or systemically after operation, although we are aware of communications recom-

mending their use[1, 3]. We do give, routinely, pre- and intraoperative antibiotics.

No patient had a pseudomeningocele. One patient had air accumulated, bifrontally, under tension, which required burr holes. One patient who had obstructive hydrocephalus before operation required shunting of CSF. One patient died from a fulminating Acineto-bacter meningitis and septicemia.

In conclusion the surgery of these apparently easy to deal with lesions demands a judicious appraisal before operation and a careful, contained hand during their resection. Indeed, spooning out their content is easy, but it is time consuming and delicate work to follow and extricate their finger-like projections from in between cranial nerves and small and larger vessels, while a total removal of tightly adherent capsule may be fatal. When large enough to penetrate through the tentorial notch, these epidermoidal cysts extend into the interpeduncular and the latero-chiasmatic cisterns. In their downward growth they invade the foramen magnum.

It is therefore important to recognize these lesions early, before they have spread widely, and most importantly, before the lesion became adherently glued to critical vessels, cranial nerves and brain stem. In order to achieve this, a better clinical appraisal is necessary, since very often the CT findings are not conclusive and are easily interpreted as "strips of atrophy", "enlarged cisterns" or "artifacts". If such is the case, then the usually vague and unspecific complaints of most patients, in the early stages of the disease, would hardly lead to pursue more vigorously special investigations to enable an early diagnosis. One can imaging, therefore, that a relatively large number of patients, who have had for many years vague unspecified complaints in association with infratentorial epidermoids, must have been totally missed.

References

1. Berger MS, Wilson CB (1986) Epidermoid cysts of the posterior fossa. J Neurosurg 62: 214–219
2. Bullit E, Tew JM, Boyd J (1986) Intracranial tumours in patients with facial pain. J Neurosurg 64: 865–871
3. Cantu RC, Ojeman RG (1968) Glucocorsteroid treatment of keratin meningitis following removal of a fourth ventricle epidermoid tumour. J Neurol Neurosurg Psychiatry 31: 73–75
4. Clark JB, Six EG (1984) Epidermoid tumour presenting as tension pneumocephalus. J Neurosurg 60: 1312–1314
5. Dee RH, Kishore PRS, Young HF (1980) Radiological evaluation of cerebellopontine angle epidermoid tumours. Surg Neurol 13: 292–296
6. Drayer BP, Rosenblum AE, Maroon JD, Bank WO, Woodford JE (1981) Posterior fossa extraaxial cyst: diagnosis with metrizamide CT cysternography. Am J Roentgenol 128: 431–436
7. Flemming JFR, Botterell EH (1959) Cranial dermoid and epidermoid tumours. Surg Gynec Obstet 109: 403–411
8. Guidetti B, Gagliardi FM (1977) Epidermoid and dermoid cysts. J Neurosurg 47: 12–18
9. Hwang WZ, Hasegawa T, Ito H, Shimoji T, Ikeda K, Yamamoto SH (1985) Intracranial epidermoids: concerning the low absorption value on computerized tomography. Acta Neurochir (Wien) 78: 33–37
10. Obrador S, Lopez-Zafra JJ (1969) Clinical features of the epidermoids of the basal cisterns of the brain. J Neurol Neurosurg Psychiatry 32: 450–454
11. Rosario M, Becker DH, Conley FK (1981) Epidermoid tumours involving the fourth ventricle. Neurosurg 9: 9–13
12. Tan TI (1972) Epidermoids and dermoids of the central nervous system. Acta Neurochir (Wien) 26: 13–24
13. Toglia JV, Netzky MG, Alexander E (1965) Epithelial (epidermoid) tumours of the cranium. J Neurosurg 23: 384–393
14. Tytus JS, Pennybaker J (1956) Pearly tumours in relation to the central nervous system. J Neurol Neurosurg Psychiatry 19: 241–259
15. Ulrich J (1964) Intracranial epidermoids. A study on their distribution and spread. J Neurosurg 21: 1051–1058

Address for correspondence: A. D. Bartal, M.D., Professor and Chairman, Department of Neurological Surgery, Tel Aviv – Elias Sourasky Medical Center, Ichilov Hospital, 6 Weizman St., Tel Aviv 64239, Israel.

Acta Neurochirurgica, Suppl. 42, 147–151 (1988)
© by Springer-Verlag 1988

Cystic Meningiomas

B. Borovich[1], J. N. Guilburd[1], Y. Doron[2], J. F. Soustiel[1], M. Zaaroor[1], J. Braun[3], J. Gruszkiewicz[1], and M. Feinsod[1]

Departments of [1] Neurosurgery, [2] Pathology and [3] Diagnostic Radiology, Rambam Medical Center, Technion Faculty of Medicine, Haifa, Israel

Summary

Four cases of cystic meningiomas were found among 194 meningiomas diagnosed by computed tomography (CT) and operated on during a 7 year period, an incidence of 2%.

The cysts were in all cases peritumoral. The cyst's wall was the brain itself, and the ependymal ventricular wall was part of their medial boundary. They contained xanthochromic fluid with a high protein content.

Three meningiomas were parasagittal and one was adjacent to the pteryon and the external part of the sphenoid ridge. The mural nodules were in 2 cases apparent single nodes although in one it was part of multiple distant and regional growths, in another the tumour was built by the aggregation of 2 nodes, the remaining case was an "en plaque" meningioma. All were definitely attached to the dura.

Histological pattern was different in every case. Those parasagittal were: one pure meningotheliomatous, one mixed meningotheliomatous with pseudo psammomatous and lipoblastic sections and one highly vascular angioblastic; the pteryonal case was psammomatous and microcystic.

CT diagnosis is difficult because glial, metastatic and other tumours may look cystic and resemble cystic meningiomas. Nevertheless in 3 cases the correct diagnosis was suspected preoperatively because the solid portion of the tumour showed intense and homogeneous contrast enhancement with a sharp edge and was located adjacent to the dura. On the other hand in the remaining case, the parasagittal solid tumour was not readily apparent on CT (the "en plaque", case), and the tentative preoperative diagnosis was of an epidermoid tumour.

At operation on the contrary the macroscopic aspect was typical of meningioma and the histological peroperative frozen sections confirmed the diagnosis with ease in all cases.

In the light of the aforesaid findings the authors conclude that every patient with a cystic tumour, no matter the tentative preoperative diagnosis is, should be given the benefit of surgical intervention.

Keywords: Brain tumours; cystic tumours; meningiomas; computed tomography.

Introduction

Though it is well known that every intracranial tumour might be associated with cyst formation, cysts accompanying meningiomas are considered uncommon[3, 8, 10, 15]. Being aware that meningiomas might be cystic is of the utmost importance. Dismissal of a cystic meningioma as a malignant tumour invites disaster because it results in palliative treatment of an otherwise curable lesion[3, 7, 11]. At times, CT differential diagnosis is difficult and errors are still committed[3, 4, 6, 7, 11, 14]. Likewise, peroperative frozen sections of biopsy samples may be inconclusive[6, 7, 11].

This report deals with 5 cases of meningiomas with peritumoural cysts. One of them was not diagnosed preoperatively as a meningioma. A close relationship between cysts and ventricles was noted in all cases. It is suggested that this association may have something to do with cyst enlargement.

Case Reports

Case 1

This 64-year-old left handed woman who was known to suffer from Parkinson's disease, presented with generalized and focal left somatomotor seizures of 5 months duration. Neurological examination disclosed, in addition to her extrapyramidal signs, an expressive dysphasia and left hemiparesis.

A CT brain scan showed a large area of uniformly decreased density with a sharp outline typical of a brain cyst in the right frontal lobe. The cyst was adjacent to the ventricle and was surrounded by another hypodense area of ill defined margins characteristic of brain oedema. Anterior to the cyst and attached to the cranial vault and falx, a contrast-enhanced mass was seen. At higher levels 2 small separate contrast-enhanced falx growths were also evident.

Operation. A right frontal parasagittal craniotomy revealed a firm encapsulated tumour adherent to the convexity dura and falx. The tumour was encircled by a large cyst which contained xanthochromic fluid with a high protein content. The cyst's wall was the brain itself and ventricular ependyma was part of its medial boundary.

The pathological diagnosis was meningotheliomatous meningioma.

Postoperative course: The patient's dysphasia and hemiparesis cleared. Her Parkinson's disease remained unchanged.

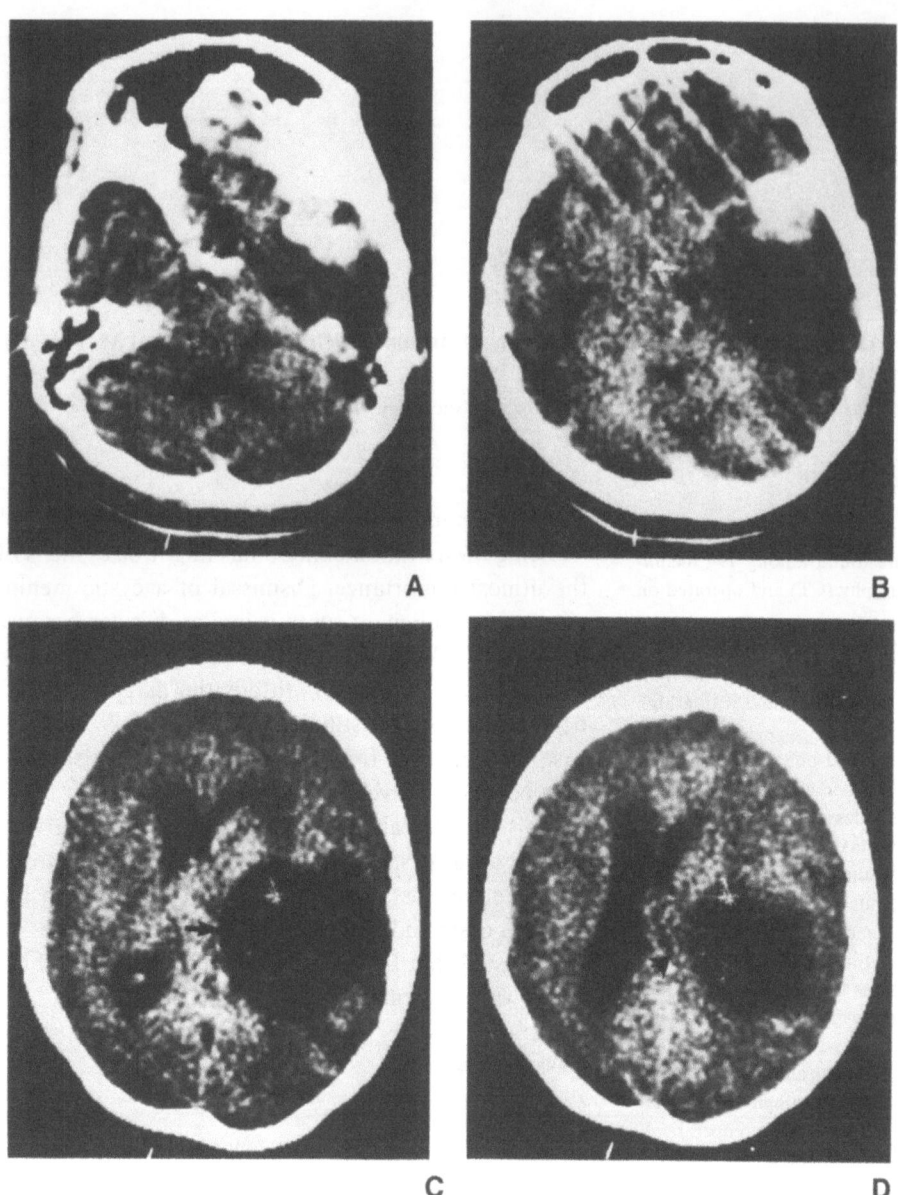

Fig. 1. Case 2. Post-contrast axial scans. Cuts (A and B) demonstrate an enhancing bilobulated nodule adjacent to the ridge. Cuts (C) and (D) show a low density cyst (arrow) surrounded by brain oedema. On cut (D) the cyst is seen contiguous to the body of the ventricle (arrow head)

Case 2

This 54-year-old woman presented with a 4 months history of left hemiparesis and ptosis of the right palpebra. In the last 6 weeks she had become increasingly forgetful, lethargic and urinary incontinence appeared. On examination she exhibited slow cerebration, confusion and disorientation, right side ptosis palpebrae and mydriasis, left hemiparesis, hemihypoesthesia and hemianopsia. Fundoscopy revealed papilloedema.

A CT brain scan revealed a large cystic lesion surrounded by brain oedema in the right fronto-temporal region. Anterior to the cyst there was a bilobulated contrast-enhanced mass attached to the sphenoid ridge. The trigone of the lateral ventricle was contiguous to the internal border of the cyst (Fig. 1).

Operation: A temporo-pterional craniotomy revealed a firm bilobulated encapsulated tumour, adherent to the sphenoid ridge dura, and encircled by a large cyst. The cyst's fluid was xanthochromic with a high protein content. The cyst's wall was the brain itself.

The pathological diagnosis was meningotheliomatous meningioma with numerous pseudo-psammoma bodies – "secretory meningioma"[1] – and some foci of lipoblastic cells.

Postoperative course: Her neurological deficits cleared.

Case 3

This 55-year-old left handed man presented with a 2 year history of progressive decreased attention, forgetfulness and gait disturbances. In the last 2 months he had developed an increasing left hemiparesis and left sided somato-motor and speech arrest seizures. Five years before admission he had an episode of transient left hemiparesis and dysphasia. On examination he was oriented, without clear signs of dementia or dysphasia. A left hemiparesis was evident.

A CT brain scan showed a large right frontal cystic lesion adjacent to the ventricle and surrounded by brain oedema. Anterior to the cyst and attached to the cranial vault and falx, an intensely contrast-enhanced mass was seen (Fig. 2).

Fig. 2. Case 3. Axial computed tomography after injection of contrast material. (A and B) a calcified enhancing mass is seen attached to the falx and the inner table of the calvaria. On the lower cut (A), there also is a sharply defined lucent area (cyst) posterior to the tumour and anterior and contiguous to the ventricle. On the higher cut (B), the ill defined area of hypodensity posterior to the meningioma represents brain oedema (arrow)

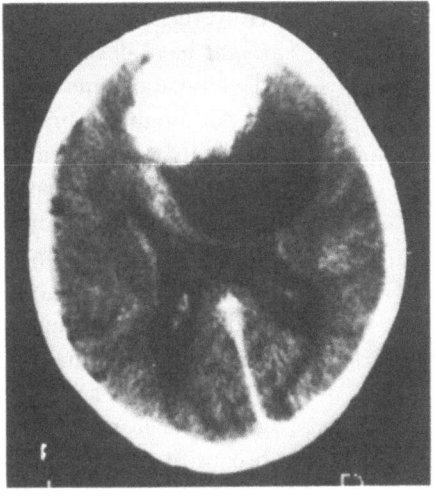 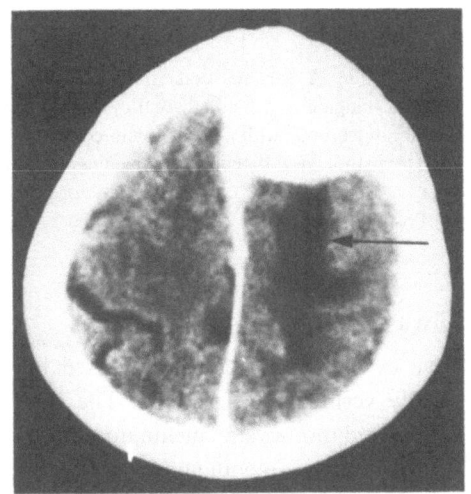

Fig. 3. Case 4. Post-contrast axial computed tomography reveal a sharply demarcated cyst anterior and adjacent to the ventricle. The irregular area of lucency peripheral to the cyst represents brain oedema. On the higher cut (B), there is a thin layer of tissue, dubiously enhanced, adjacent to the inner table of the calvaria – "en plaque" meningioma – (arrow)

Operation: A right frontal parasagittal craniotomy revealed a firm encapsulated tumour adherent to the convexity dura and falx. The tumour was encircled by a large cyst which contained xanthochromic fluid with a high protein content. The cyst's boundaries were the brain itself and the ependymal ventricular wall.

The pathological diagnosis was mixed psammomatous and microcystic meningioma.

Postoperative course: The patient was discharged without residual neurological deficits.

Case 4

This 75 year old man suffering from Parkinson's disease for the last 6 years, presented with a 2 week history of decreased attention, forgetfulness, headaches and progressive left hemiparesis. On examination he was oriented, his recent memory was poor, and he had urinary incontinence. In addition to his long standing extrapyramidal signs a left hemiparesis was evident.

A CT brain scan showed a large right frontal cyst contiguous to the ventricle and surrounded by brain oedema. Anterior to the cyst and immediately beneath the cranial vault, a thin layer of tissue with doubtful contrast enhancement was seen (Fig. 3).

Operation: A right frontal parasagittal craniotomy showed a thin "en plaque" tumour adherent to and invading the convexity dura and bone. The tumour tissue was encircled by a large cyst which contained xanthochromic fluid with a high protein content. The cyst's wall was the brain itself.

The pathological diagnosis was angioblastic meningioma.

Postoperative course: The patient's recent neurological deficits cleared. His Parkinson syndrome remained unchanged.

Case 5

This 58-year-old woman came to our attention because of an isolated generalized seizure. Neurological examination revealed no deficits.

A CT brain scan showed a contrast-enhanced mass attached to the planum sphenoidale. Adjacent to it in the left frontal lobe, a cyst contiguous to the ventricle and encompassed by brain oedema was seen.

Operation: A bifrontal craniotomy revealed a planum sphenoidale meningioma with an adjacent cyst on its left side. The fluid was xanthochromic with a high protein content.

The pathological diagnosis was meningotheliomatous meningioma.

Postoperative course was unremarkable.

Discussion

Five cystic meningiomas diagnosed by CT and surgically verified are described. The clinical history of cystic and non-cystic meningiomas may be similar. However, cystic meningiomas may exhibit a relatively more rapid clinical course all along. In some cases a rapid deterioration constitutes the final stage of a long protracted illness. This clinical pattern makes the correct diagnosis difficult because it mimics that presented sometimes by metastases and gliomas. Three of our cases (No. 1, 2, and 4) became symptomatic 4 and 5 months and 2 weeks respectively prior to admission. Case No. 5 was diagnosed immediately after experiencing a single epileptic attack. Case No. 3 presented a long history (years) with a rapid worsening of symptoms in the last 2 months before hospitalization. In one of Henry's 3 cases[7] and in the case reported by Amano[2], the clinical histories were also short — 5 and 2 months respectively. In Dell's series comprising 8 cases[6], two showed a rapid history of 2 and 4 months; four other patients displayed long histories with quick aggravation of symptoms from 3 months to 2 weeks.

The possible explanations for the rapid progression of symptoms may be:

1. Development of brain oedema around the tumour[3, 6, 11]. Oedema was present in all our cases as evidenced by CT.

2. Rapid cyst expansion. The authors call attention to the close relationship encountered in all our cases between the ventricle and the cyst. This same juxtaposition of ventricle and cyst was noticed by Amano[2]. The penetration of air into the cyst at ventriculography further supports a close connection between cyst and ventricle[5]. This intimate relationship between the two together with the high protein content of the cyst, might result in rapid influx of fluid into the cyst with corresponding swift enlargement.

3. The histological type: (A) secretory meningioma. We wonder if this particular type of meningioma as it was seen in our case 2, might have been partly responsible for the accelerated clinical course[1] and cyst formation. (B) all our cases showed peritumoural cysts.

However, our case No. 3 which was partly microcystic, might have developed a visible intratumoural cyst in the course of time[8, 9].

Although CT brain scan is the most accurate tool presently widely available for the preoperative diagnosis of meningiomas, its overall accuracy ranges between 77 and 92% only[4, 6, 12, 13, 15]. Likewise, peroperative frozen sections of biopsy samples may be inconclusive[6, 7, 11]. This lack of a sure definitive means of differentiating meningiomas from the more common malignant tumours, is of great practical significance. The grave error of dismissing a benign cystic meningioma for a malignant tumour[3, 6 8, 11, 14] is disastrous because it results in palliative treatment of an otherwise curable lesion[3, 6, 7, 11]. Therefore, we strongly advocate definitive surgery in every case of cystic brain tumour, no matter what the tentative preoperative diagnosis might be[3, 7, 11].

References

1. Alguacil-García A, Pettigrew NM, Sima AAF (1986) Secretory meningioma. Am J Surg Pathol 10: 102–111
2. Amano K, Miura N, Tajika Y, Matsumori K, Kubo O, Kobayashi N, Kitamura K (1980) Cystic meningioma in a 10-month-old infant. J Neurosurg 52: 829–833
3. Becker D, Norman D, Wilson ChB (1979) Computerized tomography and pathological correlation in cystic meningiomas. J Neurosurg 50: 103–105
4. Claveria LE, Sutton T, Tress BM (1977) The radiological diagnosis in meningiomas, the impact of EMI scanning. Br J Radiol 50: 15–22
5. Cushing H, Eisenhardt L (1938) Meningiomas: Their classification, regional behaviour, life history and surgical end results. Ch C Thomas, Springfield, Ill, pp 26, 577–588, 785
6. Dell S, Ganti SR, Steinberger A, McMurtry III J (1982) Cystic meningiomas: a clinicoradiological study. J Neurosurg 57: 8–13
7. Henry JM, Schwartz FT, Sartawi MA, Fox JL (1974) Cystic meningiomas simulating astrocytomas. J Neurosurg 40: 647–650
8. Kepes JJ (1982) Meningiomas: Biology, pathology and differential diagnosis. Masson Publishing USA Inc, New York, pp 108–110
9. Kleinman GM, Liszczak T, Tarlov E, Richardson EP (1980) Microcystic variant of meningioma. Am J Surg Pathol 4: 383–389
10. Lee SH, Rao KCGV (1983) Primary tumours in adults. In: Lee SH, Rao KCGV (ed) Cranial computed tomography. McGraw-Hill Book Co, New York, pp 241–293
11. Nauta HJV, Tucker WS, Horsey WJ, Bilbao JM, Gonsalves C (1979) Xanthochromic cysts associated with meningioma. J Neurol Neurosurg Psychiatry 42: 529–535
12. Pullicino P, Kendall BE, Jakubowski J (1960) Difficulties in diagnosis of intracranial meningiomas by computed tomography. J Neurol Neurosurg Psychiatry 43: 1022–1029

13. Quest DO (1978) Meningiomas: an update. Neurosurgery 3: 219–225

14. Rengachary S, Batnitzky S, Kepes JJ, Morantz RA, O'Boynick P, Watanabe I (1979) Cystic lesions associated with intracranial meningiomas. Neurosurgery 4: 107–114

15. Vassilouthis J, Ambrose J (1979) Computed tomography scanning appearances of intracranial meningiomas. J Neurosurg 50: 320–327

Address for correspondence: B. Borovich, M.D., Department of Neurosurgery, Rambam Medical Center, Haifa, Israel.

Acta Neurochirurgica, Suppl. 42, 152–156 (1988)

Validity of Stereotactic Biopsy as a Diagnostic Tool

F. Colombo[1], L. Casentini[1], M. Zanusso[1], D. Danieli[2], and A. Benedetti

[1] Department of Neurosurgery, and [2] Institute of Pathology, City Hospital, Vicenza, Italy

Summary

254 patients affected by intracranial lesions underwent stereotactic biopsy in our department from 1978 to 1986. Target localization was achieved by CT. Multiple biopsy sampling was performed by cup microforceps or sliding cannula. Operative mortality was limited to 2 cases. Definitive tumour diagnosis, including type and approximate grading was obtained in 211 (83%) patients. Diagnostic failures have been investigated from the neuroradiological point of view. Failure rate is low in solid tumours with CT homogeneous appearance and clear-cut borders, gradually increases in non homogeneous tumours, necrotic haemorrhagic or cystic, and is high in non classifiable lesions, generally hypodense at CT, with indefinite borders. In the authors opinion the variability of diagnostic retrieval in different types of lesions must be taken into account when proposing stereotactic biopsy.

Keywords: Brain tumours; stereotactic biopsy; computerized tomography.

Introduction

Since early report of Conway[7], stereotactic biopsy has become a widely accepted fundamental diagnostic step in the work-up of intracranial space-occupying lesions.

After early attempts of intracranial tumours targetting with indirect radiodiagnostic examinations, CT and, more recently, MRI have become the instrumentation of choice for determining stereotactic coordinates. Useful and correct diagnostic yeld has greatly increased with a marked decrease in both mortality and morbidity, nowadays well below 1%[1, 8, 11, 12]. The aim of stereotactic biopsy is to direct subsequent treatment; nature and 3 D configuration confirmation are indispensable in stereotactically related therapeutic procedures such as isotope implantation, stereotactically directed external beam irradiation and radiosurgery[6, 10, 13].

We have employed stereotactic biopsy as a diagnostic tool since 1978[4]. Biopsy sampling has become a preliminary mandatory step in intracranial radiosurgery procedures for tumours since 1982[6]: The indication for and clinical evolution after a largely invasive therapeutic procedure such as radiosurgery are totally dependent on preoperative biopsy determination of the nature and grading of the treated lesion. The aim of this work is to determine the real percentage of correct and valid biopsy diagnosis.

Material and Methods

254 patients harboring intracranial space-occupying lesions underwent stereotactic biopsy in our department since 1978.

Preoperative angiography (either following standard procedure or in stereotactic conditions) has been deemed necessary in 243 patients. Target coordinates have been calculated with the aid of a second generation CT scan in combination with an indirect method until January 1984[3]. More recently we have turned to a completely CT assisted procedure which employs a fourth generation CT machine and a computer program for coordinates determination, similar to that described by Birg[2]. In homogeneous tumours the target was usually selected near the center of the tumour volume. In cases of non homogeneous or multifocal tumours, targets were multiplied to have histological demonstration as representative as possible of different pathological features.

Our original stereotactic apparatus employs base ring fixation in combination with a robotized tool holder, working with a spherical coordinates system[5]. An intracranial lesion can be reached from any selected direction.

Biopsy instrumentation utilized was the cup microforceps (allowing withdrawal of 3 cubic millimeter fragments) and the sliding cannula affording cylindrical 2 × 10 millimeter samples. With the forceps a mean of four-five fragments were obtained each 5 mm of biopsy track. With the sliding cannula, two or three samples were taken each 10–15 mm. The choice between the two instruments was mainly influenced by the physical characteristics of the lesion (dimensions, consistency). Large and soft lesions were biopsied by cannula though and small lesions by microforceps.

Immediately after retrieval part of the specimens was evaluated by smear technique (toluidine blue staining). Biopsy was repeated until a "significant" sample was obtained.

For definitive diagnosis paraffin embedding and the usual staining techniques were employed.

Results

Operative mortality was limited to 2 cases (0.7%). We observed 3 transitory (1.1%) and 2 definitive procedure-related neurological deteriorations.

At the beginning of our experience, while employing the indirect method for calculating target coordinates from CT, in 4 cases we missed the tumour (lesions dimensions ranging from 15 to 22 mm). In 2 of these cases biopsy was repeated and a definitive satisfying diagnosis was finally obtained. Since the indication to biopsy is mainly based on CT examination, before evaluating the validity of stereotactic biopsy we divided the lesions into four classes, according to contrast enhanced CT appearance, irrespective of tumour dimensions:

First class: solid tumours with homogeneous density, clear-cut borders. Technique: one series of samples along a single track passing through the center.

Second class: non homogeneous tumours with indefinite borders. Technique: we tried to obtain samples from different areas, considered representative of the whole tumour mass. Necrotic and haemorrhagic samples were identified by intraoperative examination with the smear technique and discarded.

Third class: cystic tumours. Technique: multiple attempts to obtain samples from the cyst wall (usually with microforceps). Cytological examination of cystic fluid.

Fourth class: hypodense or mixed hypo-hyperdense lesions, usually large with indefinite borders. No clearly localized foci to aim at. Technique: multiple biopsy samples taken "randomly" along multiple tracks, near the center of the lesion.

We have considered as a sound and valid diagnostic response a unequivocal definition of nature and grading, not contradicted by neuroradiological appearance and clinical evolution, confirmed, in cases in which it was possible, by operation, autopsy or repeated (in other departments) biopsy procedure.

Equivocal responses such as "gliosis", "oedematous tissue", "infiltration zone", "tumoural tissue, unknown nature", "probable low-grade glioma" and similar were given in 34 out of 254 patients (Table 1).

34 patients underwent operative bulk removal or extensive tumour volume examination at autopsy (Table 2). We noted 4 discrepancies. One case was an hypodense lesion on which a diagnosis of grade I fibrillary astrocytoma was contradicted by the operative response of anaplastic astrocytoma grade III. Discrepancy of nature (biopsy: glioblastoma-surgical specimen: carcinoma metastatis) and grading (biopsy: glioma grade I – surgery: glioma grade III) was noted in 2 cases on non homogeneous tumours. The fourth case was a cystic tumour (biopsy: colloid cyst – operation: craniopharyngioma).

In 2 patients repeated biopsy in other departments confirmed the diagnosis. Histological diagnosis was macroscopically inconsistent with clinical evolution and/or neuroradiologic appearances in 4 cases out of 220 unequivocal responses (Table 3). One case was a homogeneously contrast enhanced intraventricular mass in which a diagnosis of "plexus papilloma" was given. Angiographic study was negative. In six months this evolved to a multilobulated invasive tumour, with non-homogeneous appearance, an evolution not consistent with biopsy diagnosis. One case was haemorrhagic and another was a largely necrotic lesion. In the first one the diagnosis of "haemorrhagic infiltrated tissue" was contradicted by subsequent appearance of

Table 1. *Stereotactic Biopsy Equivocal Responses—Presumptive Diagnosis.* 34 cases on 254 patients (12.99%)

	Patients no.	Inconclusive responses	%	Reliability (%)
Solid, homogeneous tumors, clear-cut borders	88	3	3.4	96.6
Non-homogeneous tumors (necrotic, haemorrhagic) indefinite borders	115	9	7.8	92.2
Cystic tumors	43	14	32.5	67.5
Non classifiable hypodense or mixed hypo-hyperdense lesions no presumptive diagnosis available	8	8	100.0	0.0
			Mean	86.6

Table 2. *Discrepancy with Operative or Autopsy Verification.* 4 cases on 34 patients (11.76%)

	Patients no.	Nature	Grading	Discrepancy (%)	Reliability (%)
Solid, homogeneous tumors, clear-cut borders	14		1	7.1	92.9
Non homogeneous tumors (necrotic, haemorrhagic) indefinite borders	16	1	1	12.5	87.5
Cystic tumors	4	1		25.0	75.0
				Mean	88.24

Table 3. *Diagnosis Macroscopically Inconsistent with Clinical Evolution and/or Neuroradiologic Aspect.* 4 cases on 220 patients (1.8%)

	Patients no.	Inconsistent responses	%	Reliability (%)
Solid, homogeneous tumors, clear-cut borders	85	1	1.17	98.8
Non homogeneous tumor (necrotic, haemorrhagic) indefinite borders	105	2	1.88	98.1
Cystic tumors	29	1	3.44	96.5
		Mean		98.2

Table 4. *Overall Validity of Stereotactic Biopsy as Diagnostic Tool* 75%

Solid, homogeneous tumors, clear-cut borders	88.6%
Non homogeneous tumors (necrotic, haemorrhagic)	79.1%
Cystic tumors	48.8%
Non classifiable hypodense or mixed hypo-hyperdense lesions no presumptive diagnosis available	0.0%

contrast enhanced ring feature and tumoural evolution. In the second case both neuroradiological appearance and clinical evolution suggested a higher than diagnosed glioma grading.

Overall validity of biopsy was calculated by composition of different percentages. In our series the possibility to obtain a valid response was evaluated as around 75% but significantly differs in different lesion classes (Table 4).

Discussion

The reliability of stereotactic biopsy has been assessed by many authors at very high percentages[1, 9, 11, 12, 13].

On the other hand, for different reasons, operative and autopsy confirmation on the whole bulk of the tumour is very rare. Our experience seems to confirm that in cases of homogeneous, clear-cut tumours the percentage of a valid diagnostic retrieval is high (88.6%) no matter how little the target volume (Fig. 1). Nevertheless we think that, in some cases of other lesion classes (II–IV), a brilliant diagnosis can be obtained from biopsy material only from a Pathologist strongly influenced by the neurosurgical environment. For cystic lesions, sometimes the collection of characteristic crystal into the fluid has been deemed sufficient for diagnosis of certain oncotypes (for example craniopharyngioma). In our 14 cases of probable craniopharyngioma with characteristic fluid, positive tissue demonstration was available in only 3 cases (Fig. 2). In these cases the stereotactic approach always afforded immediate symptomatic relief by cystic fluid aspiration but if one consideres only positive demonstration of characteristic tissue a prerequisite for a sound diagnosis, these biopsies should have been considered failures.

Moreover, sometimes also unequivocal responses must be regarded as suspect (Glioma grading, differ-

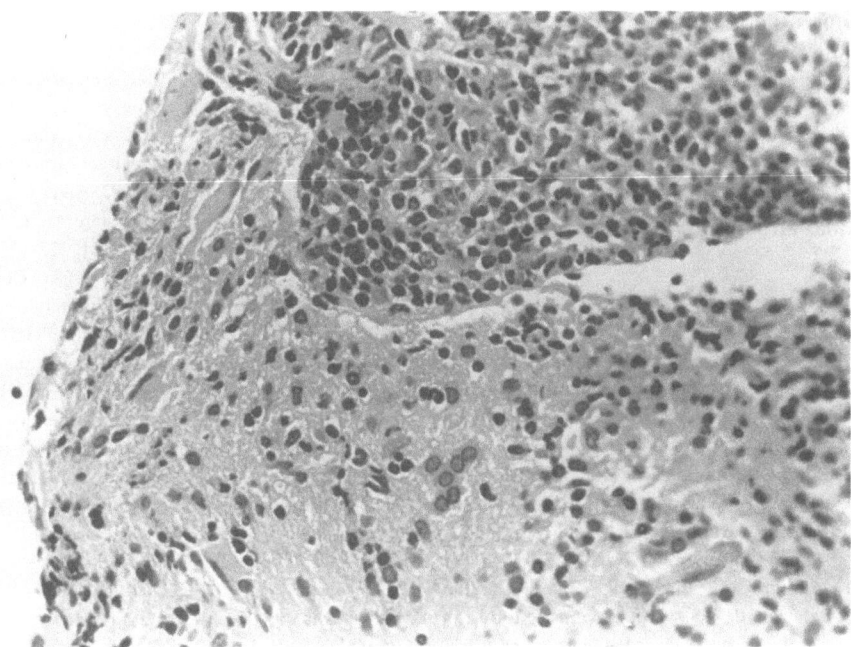

Fig. 1. Bioptic specimen taken by microforceps at the edge of a small (4 × 7 mm) homogeneous, deeply located lesion. Nodule of limphomatous tissue and adjacent cerebral tissue with reactive gliosis (H & E, × 40)

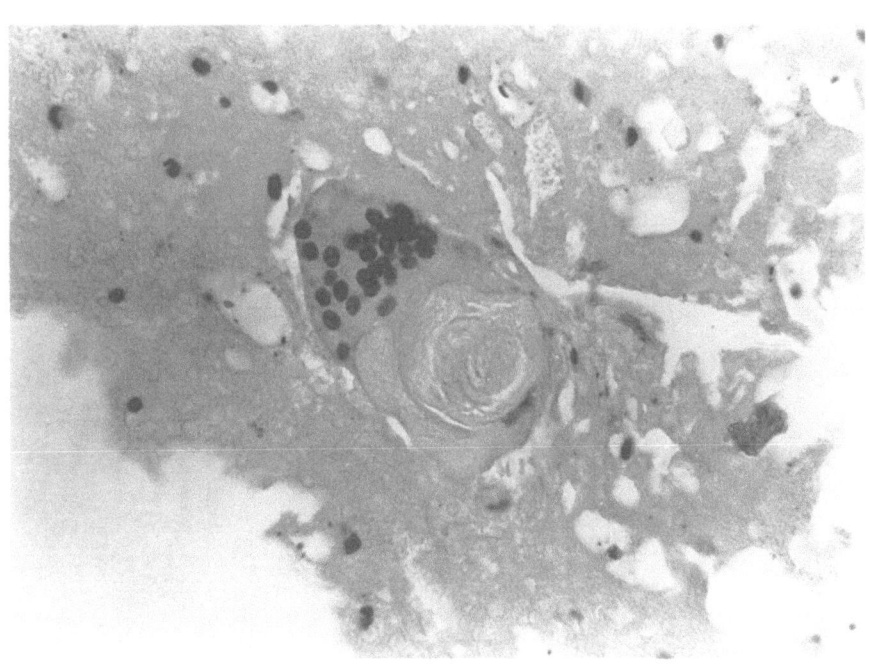

Fig. 2. Bioptic specimen taken by microforceps on the wall of a cystic lesion. Polynucleated cell nest containing a keratin pearl: craniopharyngioma was diagnosed (H & E, × 160)

ence between glioblastoma and metastasis etc.). Since we had operative or autopsy verification only in 34 plus 2 out of 220 valid responses and among these cases we found 2 discrepancies in grading and 2 in nature we wonder how many patients labelled by an authoritative diagnosis have been incorrectly treated or evaluated. In our series the percentage lies around 11%, *i.e.* 24 patients! In some cases of large, indefinite lesions with no foci to aim at and no clinical orientation available

the biopsy is completely useless, and also even a very low risk is unacceptable.

In conclusion, also if we strongly emphasize that stereotactic biopsy is an indispensable step for directing therapy of intracranial tumours, we think also correct diagnostic retrieval to vary largely in different tumour classes identified by CT appearance. Since therapeutic strategies may differ significantly after biopsy and mortality and morbidity are low but not nil, candidates for stereotactic biopsy must be evaluated with regard not only to the feasibility of the procedure but also with respect to the possibility to obtain an unequivocal diagnosis. Moreover, unequivocal diagnosis should be cross-matched with neuroradiolgoical appearance and, possibly, clinical evolution before deciding to undertake largely invasive therapeutic approaches.

References

1. Apuzzo ML, Chandrasoma PT (1987) Image directed stereotactic biopsy: methods, utilization, strategies. In: Tasker (ed) Neurosurgery: state of the art review. Hanley and Belfus, Philadelphia, pp 287–308
2. Birg W, Mundinger F (1982) Direct target point determination for stereotactic operations from CT data and the calculation of setting parameters for polar coordinates stereotactic devices. Appl Neurophysiol 45: 387–395
3. Colombo F, Angrilli F, Zanardo A, Pinna V, Alexandre A, Benedetti A (1981) A new method for utilizing CT data in stereotactic surgery: measurements and transformation technique. Acta Neurochir (Wien) 57: 197–203
4. Colombo F, Casentini L, Visonà A, Benedetti A (1982) Stereotactic biopsy: Diagnostic problems, value of cytological and histological examinations. Zentralbl Neurochir 43: 309–315
5. Colombo F, Zanardo A (1984) Clinical application of an original stereotactic apparatus. Acta Neurochir (Wien) [Suppl] 33: 569–573
6. Colombo F, Benedetti A, Pozza F, Zanardo A, Avanzo RC, Chierego G, Marchetti A (1985) Stereotactic radiosurgery utilizing a linear accelerator. Appl Neurophysiol 48: 183–145
7. Conway LW (1973) Stereotaxic diagnosis and treatment of intracranial tumours, including an initial experience with cryosurgery for pinealomas. J Neurosurg 38: 453–460
8. Kelly PJ, Alker GJ, Kall BA, Goerss JJ (1984) A method of CT based stereotactic biopsy with arteriographic control. Neurosurgery 14: 172–177
9. Kelly PJ (1987) Computerized guidance for stereotactic treatment of brain tumours: from CT guided biopsy to computerized laser resection. In: Tasker (ed) Neurosurgery: state of the art review. Hanley and Belfus, Philadelphia
10. Mundinger F (1979) Rationale and methods of interstitial Ir 192 brachicurietherapy and Ir 192 and I 125 protracted long term irradiations. In: Szikla (ed) Stereotactic cerebral irradiations. Elsevier, Amsterdam, pp 101–117
11. Mundinger F (1985) CT stereotactic biopsy for optimizing therapy of intracranial processes. Acta Neurochir (Wien) [Suppl] 35: 70–74
12. Ostertag C, Mennel HG, Kiessling M (1980) Stereotactic biopsy of brain tumours. Surg Neurol 14: 257–262
13. Szikla G, Blond S, Daumas-Duport C, Missir O, Myahara S, Munari C, Musolino A, Schlinger N (1983) Stereotaxis in the management of brain tumours: three dimensional angiography, sampling biopsies and focal irradiation using the Talairach system. Ital J Neurol Sci [Suppl] 2: 83–96

Address for correspondence: Dr. F. Colombo, Department of Neurosurgery, City Hospital, Vicenza, Italy.

Acta Neurochirurgica, Suppl. 42, 157–160 (1988)

Experiences with CT-guided Stereotaxic Biopsies in 121 Cases

H. Niizuma, T. Otsuki, T. Yonemitsu, M. Kitahara, R. Katakura, and J. Suzuki

Division of Neurosurgery, Institute of Brain Diseases, Tohoku University School of Medicine, Sendai, Japan

Summary

Needle biopsy using a CT-guided stereotaxic technique was performed in 121 cases of suspected brain tumour. Using this technique, it is possible to perform biopsies safely and accurately on even small and deep-seated lesions with a minimum of surgical intervention. However, accurate diagnoses are sometimes not possible when only a small number of viable cells are obtained, such as in lesions containing old haematoma, cystic tumours, previously treated tumours or tumours which are either too hard or too soft. We were able to make accurate diagnoses in 98 of our 121 cases (81%). The accuracy of diagnosis is somewhat hampered by the small volume of sample material obtained using this biopsy technique, and this fact should be kept in mind when evaluating the histological material.

Keywords: Computed tomography; biopsy; brain tumour; stereotaxic operation.

Introduction

Due to the development of CT-guided stereotaxic operations, it has become possible to approach lesions virtually anywhere within the cranium, provided that they can be identified in CT scans. Over the preceding five years, we have employed a CT-guided stereotaxic technique in order to perform needle biopsies in 121 cases of suspected brain tumour. Despite the fact that this technique is extremely accurate and involves minimal surgical intervention, several diagnostic problems have arisen with such biopsies due to the small volume of tissue obtained. In the present study, we discuss the strengths and weaknesses of the CT-guided needle biopsy technique.

Clinical Materials and Methods

Biopsy was considered in a total of 126 cases in which brain tumour was suspected on the basis of clinical symptoms or CT findings. Since notable hypervascularity was identified in the angiograms of five cases, biopsy was performed in a total of 121 cases. Ages ranged from 5 to 77 years; 79 were male and 42 were female. The sites of the lesion are shown in Table 1.

The entire surgical operation is performed in the CT room. In our set-up, Leksell's CT-stereotaxic system[4] is directly attached to the CT bed and the patient's head is immobilized there for the stereotaxic biopsy. After a coordinate indicator has been attached, an enhanced CT is done, the three-dimensional coordinates of the target area are determined and a burr hole is made. Biopsy is then done using a Backlund spiral needle[1]. Depending upon the size and spread of the tumour, between one and three tracks are used. Biopsy specimens are obtained at one centimeter steps starting one centimeter above the region where the tumour is thought to be. Normally, between two and nine samples are obtained to a depth calculated to be two-thirds through the depth of the tumour. For lesions of the posterior fossa, a supine-lateral retromastoid approach is used[8]. Blood pressure is monitored and controlled during the surgery and, in cases with notable intracranial hypertension, prophylactic administration of mannitol and steroid is done for several days following the biopsy.

Results

1. Biopsy

In three cases, the tumour was extremely hard and no biopsy material could be obtained. Two of these cases had a cystic tumour following irradiation therapy, and

Table 1. *Site of the Lesion*

Cerebral cortex & subcortex		66
frontal	24	
temporal	16	
parietal	16	
occipital	6	
corpus callosum	4	
Basal ganglia		17
Thalamus		17
Third ventricle		6
Pineal region		4
Cerebellum		4
Pons		4
Lateral ventricle		3
Total		121

one was a tumour of the pineal region in which calcification of a portion had occurred. In both cases, the spiral needle could not enter the tumour tissue and, therefore, a histological diagnosis could not be made.

There were also cases in which sampling with the spiral needle was difficult due to excessive softness of the tumour. In such cases, the centre of the tumour tissue was often depicted as an area of low CT density without contrast enhancement.

In addition, there were cases in which a cystic tumour with a thin wall was biopsied, but little material was obtained prior to entering the cystic cavity. A total of 18 cases were found to be such cystic tumours with thin walls.

There were six cases in which the flow of an old blood clot was encountered during the biopsy procedure. In one case, a sufficient quantity of biopsy material had already been obtained, but in the remaining five cases, tissue samples were small. However, it was considered dangerous to obtain further material and the operation was completed with mild aspiration of the haematoma.

2. Complications

After biopsy, fresh blood was found to flow from the needle cannula in 11 cases (9%), but haemostasis was obtained by inducing hypotension and waiting for some ten minutes. In CT scans obtained immediately after surgery or on the following day, one case showed a small (1 cm diameter) intracerebral haematoma, and one case, in which biopsy was performed via the lateral ventricle, had a small blood shadow in the posterior horn. In three other cases, high density spots on the biopsy track of a size corresponding to the thickness of the biopsy needle were found. In other words, there were five cases (4%) of minimal bleeding, as seen in CT scans, but none of these cases showed aggravation of symptoms due to the bleeding.

Aggravation of neurological symptoms following biopsy was seen in three cases, one of which showed transient symptoms and two of which had had severe preoperative symptoms, suggesting that the aggravation may have been the natural course of the disease rather than due specifically to the biopsy. There were neither fatal cases nor severely complicated cases in this series. In one case of metastatic brain tumour, there was spread of tumour tissue along the needle track following the biopsy.

3. Histology

Histological diagnosis of the tumour was possible in 87 cases (Table 2). There were 57 cases diagnosed as

Table 2. *Histological Diagnosis*

Tumour		87
high grade astrocytoma	30	
low grade astrocytoma	27	
glioblastoma	14	
oligodendroglioma	4	
malignant lymphoma	3	
unclassified tumour	3	
metastatic tumour	2	
germinoma	2	
embryonal carcinoma	1	
ependymoma	1	
Normal brain tissue		11
Necrosis		6
Blood clot		5
Granuloma, granulation tissue		3
Reactive gliosis or astrocytoma		2
Little or no sampling		7
Total		121

low or high grade astrocytoma and 14 diagnosed as glioblastoma. Although diagnosed as tumour tissue, a detailed classification of the type of tumour was not possible in another three cases.

There were 11 cases in which the biopsy material was diagnosed as normal brain tissue, three of which were ultimately diagnosed from the clinical course and CT scans as cerebral infarction. There were also two cases diagnosed as normal brain tissue in which convulsions were the only symptoms. In CT scans, small low density areas showing no contrast enhancement were seen near the motor area. In both cases, more than two years have elapsed since biopsy, and the patients remain with unchanged CT and clinical findings. Three of these 11 cases had a cystic tumour with thin walls; in two of them, the biopsy was performed subsequent to irradiation therapy. It is thought that in these three cases normal tissue lying above the tumour wall was biopsied. Similarly, the biopsy material was diagnosed as normal brain tissue in two cases which had already undergone steroid therapy for suspected malignant lymphomas and in which the tumour mass was thought to have shrunk considerably. In one of these cases, a second biopsy was performed at a time when the tumour mass had enlarged, and it was then diagnosed as malignant lymphoma. The remaining case involved a lesion extending from the pons to the thalamus. All of the tissue sample was obtained from the thalamus because of the lesser danger of that area and was diagnosed as normal. It is thought that the tissue sample may have been obtained from an inappropriate site.

All six of the cases diagnosed as necrosis showed irregular ring enhancement, while the central region was of low density. Despite the fact that a sufficient volume of tissue was obtained few viable cells were found. Chemo- and radiotherapy was then begun in these cases, but all were dead within one year; we suspect that they were probably malignant tumours.

In five cases, old blood clots were included in the tissue sample. Subsequent craniotomy in two cases revealed arteriovenous malformations. In two cases, there was a localized mass in the thalamus or motor cortex, suggesting the presence of a cavernous angioma, but since the clinical symptoms have been mild, both patients have been kept under close observation without further treatment. In the light of the subsequent course and CT findings in the remaining case, intracerebral haematoma is suspected.

There were two cases in which the histological diagnosis was "reactive gliosis or astrocytoma". In both cases low density areas which did not show enhancement were found to extend over a broad area in CT scans. They were treated as brain tumour. One of the patients, however, was diagnosed as adrenoleukodystrophy at autopsy. In the other case, a second biopsy provided a diagnosis of astrocytoma.

Finally, there were seven cases in which little or no sampling prevented histological diagnosis. These included four with cystic tumours having thin walls and three in which the firmness of the tumour tissue prevented sampling. Among the former cases, three were cystic tumours of the third ventricle or suprasellar region. Histological diagnosis was not possible, but in light of a motor oil-like cystic fluid, they were diagnosed as craniopharyngioma.

In summary, among the cases in which an appropriate diagnosis was not possible, a diagnosis of normal brain tissue was made in six cases, necrosis was diagnosed in six, only old blood clots were found in the tissue sample in four, and inadequate volumes of tissue to make a diagnosis were obtained in seven. In other words, a correct diagnosis could not be made in 23 cases although the CT-guided stereotaxic biopsy technique was effective in diagnosing in the remaining 98 cases (81%).

Study of the influence of the therapy prior to the biopsy on histological diagnosis showed that among the 105 cases in which therapy had not been started prior to biopsy, the rate of accurate diagnosis was 87%, whereas it was only 44% among the 16 cases which had received previous therapy.

Discussion

Methods of performing stereotaxic surgery of brain tumours have been discussed since the time of Spiegel and Wycis *et al.*[10], and the effectiveness of stereotaxic biopsy of tumour tissue is now widely ackowledged. However, prior to the advent of computed tomography, stereotaxic techniques could be used for biopsies only in cases where angiography and ventriculography were sufficient for determining the site of the lesion[2].

Subsequent to the emergence of CT scanning, however, it has been possible to obtain far more information concerning the size, shape and location of intracranial lesions than was possible with the older techniques. Making use of simultaneous CT scans, needle biopsy with a free hand[6, 11] or using a traditional stereotaxic apparatus[3, 9] has also become possible. More recently, CT-guided stereotaxic operations which allow for an accurate approach to and biopsy of even small and deep-seated lesions have been devised[5, 7].

We have performed biopsies using Backlund's spiral needle and have found that, using this needle, even when bleeding occurs after sampling, it remains a small haemorrhage without clinical symptoms. In comparison with other reports, we believe this technique is safer[3, 9].

However safe and accurate this technique is problems remain concerning the correct diagnosis of some tissue samples. From our experience thus far, we believe that the nature of the tumour itself has a large influence on the accuracy of the diagnosis obtainable. The kinds of tumours which are frequently difficult to diagnose are listed in Table 3. With regard to the quality of the histological diagnosis using this method, it must be said that it is inferior to biopsy following craniotomy due simply to the fact that a larger tissue sample can be obtained than by means of needle biopsy. When viable tumour cells are evenly distributed in the tumour tissue, the needle biopsy method is appropriate, but when their

Table 3. *Tumours in Which Histological Diagnoses Are Sometimes Difficult*

1 Hard tumours

2 Extremely soft tumours (necrotic tumours)
   central part of the tumour is
   low density on enhanced CT

3 Previously-treated tumours
   irradiation, chemotherapy and steroid
   therapy (in lymphomas)

4 Cystic tumours

5 Lesions including blood clots

distribution is uneven, the needle biopsy can present problems. This fact was demonstrated by the finding that the number of cases diagnosed as glioblastomas was only one quarter of that diagnosed as astrocytoma. This incidence of glioblastoma is considerably lower than that normally found in the literature[12]. Even in cases where it is apparent that the tissue is tumorous, further classification may be impossible and differentiation between reactive gliosis and low grade astrocytoma may be impossible. Such problems can be resolved by means of multiple sampling at various sites, but, from the perspective of the patient's safety, there are obvious limitations. Even if a diagnosis of high grade astrocytoma were made from the sample obtained, there is the possibility of a further diagnosis of glioblastoma if samples were obtained from a different site. The evaluation of the tumour tissue must be made fully cognizant of the fact that the sample is but one small part of the entire tumour.

References

1. Backlund EO (1971) A new instrument for stereotaxic brain biopsy. Acta Chir Scand 137: 825–827
2. Conway LW (1973) Stereotaxic diagnosis and treatment of intracranial tumours including an initial experience with cryosurgery for pinealomas. J Neurosurg 38: 453–460
3. Edner G (1981) Stereotactic biopsy of intracranial space occupying lesions. Acta Neurochir (Wien) 57: 213–234
4. Leksell L, Jernberg B (1980) Stereotaxis and tomography – A technical note. Acta Neurochir (Wien) 52: 1–7
5. Lunsford LD, Maroon JC (1982) CT localization and biopsy of intracranial lesions. In: Schmidek HH, Sweet WH (eds) Operative neurosurgical techniques. Grune and Stratton, New York, pp 403–418
6. Maroon JC, Bank WO, Drayer BP *et al* (1977) Intracranial biopsy assisted by computerized tomography. J Neurosurg 46: 740–744
7. Niizuma H, Otsuki T, Katakura R, Suzuki J (1986) Aspiration of intracerebral haematomas and biopsy of deep brain tumours using CT-guided stereotaxic technique. In: Samii M (ed) Surgery in and around the brain stem and the third ventricle. Springer, Berlin Heidelberg New York, pp 534–539
8. Niizuma H, Suzuki J (1987) CT-guided stereotaxic aspiration of posterior fossa haematomas – A supine lateral-retromastoid approach. Neurosurg 21: 422–427
9. Ostertag CB, Mennel HD, Kiessling M (1980) Stereotactic biopsy of brain tumours. Surg Neurol 14: 275–283
10. Spiegel EA, Wycis HT, Mark M (1947) Stereotaxic apparatus for operations on the human brain. Science 106: 349–350
11. Yeates A, Enzmann DR, Britt RH *et al* (1982) Simplified and accurate CT-guided needle biopsy of central nervous system lesions. J Neurosurg 57: 390–393
12. Zülch KJ (1965) Brain tumours, their biology and pathology. Springer, New York

Address for correspondence: Hiroshi Niizuma, M.D., Division of Neurosurgery, Institute of Brain Diseases, Tohoku University School of Medicine, Seiryo-cho 1-1, Sendai, Japan 980.

Acta Neurochirurgica, Suppl. 42, 161–165 (1988)
© by Springer-Verlag 1988

Pitfalls in Diagnostic Stereotactic Brain Surgery

G. Blaauw[1] and **R. Braakman**[2]

[1] Department of Neurosurgery, De Wever Hospital, Heerlen, and [2] Department of Neurosurgery, University Hospital Rotterdam, Erasmus University, Rotterdam, The Netherlands

Summary

Some of the problems and dangers inherent to stereotaxis are discussed in a series of 243 procedures. Neurological deterioration, either due to haemorrhage or to traumatic oedema, was rare. One fatal accident occurred. Infections were not present although the patient's head was not shaved. Provided he can manipulate the proper instruments accurately stereotaxis can nowadays be performed by any surgeon who can make a burr hole and perform simple arithmetic calculations. This procedure can be considered to lie within the scope of any neurosurgeon.

Introduction

As a result of the use of body scanners with wide apertures, CT-guided stereotactic brain surgery has become another common neurosurgical tool. Because of its relative simplicity[1], this procedure has now gained wide application.

The procedures in which stereotaxis can play a role are: diagnostic, functional, therapeutic and instrumentational[2]. Various problems can, however, be encountered in the course of these procedures. In this article we will discuss the problems which can be met in stereotactic procedures for diagnosis (acquisition of tissue) and treatment (removal of fluid or installation of radio-active substances for intracavitary radiotherapy[3]).

In particular, attention will be paid to neurological deterioration, mortality, epilepsy and infection. The need for angiography, the occurrence of negative biopsies and proven wrong diagnosis, the choice of the biopsy instrument and the problem of the diagnosis of multiple lesions on CT will also be discussed.

Patients and Methods

Stereotaxis involving use of the Leksell frame was introduced in Rotterdam in 1960. At that time the first generation frame was used, and initially, it was mainly applied in the treatment of patients with

Parkinson's disease. Metallurgic alterations and the use of carbon fibre materials have made the third generation frames suitable for application in present day body CT and NMR scanners, as well as in conventional X-ray machines[4]. During the present study, we have used the Leksell stereotactic system model D for CT (AB Elekta Instrument, Stockholm) and a Philips Tomoscan 350.

The coordinates are calculated from small slices measuring 1.5 mms. Initially this was done visually by comparing the CT image with a transparent diagram. Later, we devised a computer programme, which could be applied via an ordinary personal computer. A third trigonometric method has been tested in the course of this study[5]. All patients were given 16 mg dexamethasone/24 hours for one day or more; this was initiated prior to the procedure.

Results

Between May 1981 and October 1986, seven neurosurgeons from the two hospitals performed 243 stereotactic operations in 211 patients. Table 1 shows the distribution of the various locations. Surgery was carried out mainly because of deep seated lesions on or near the midline of the brain.

Tables 2 and 3 present the histological diagnoses of the material obtained. In the majority of the cases, the procedure was carried out for diagnostic purposes. We

Table 1. *Location of the Lesion*

Left hemisphere	42
Right hemisphere	49
Corpus callosum	11
Suprasellar	58
Third ventricle, pons	22
Basal ganglia	21
Pineal region	11
Pontocerebellar	2
Cerebellar	1
Multiple	26
Total	243

Table 2. *(Distribution of) Diagnosis*

Low grade glioma	23
Middle grade glioma	40
Malignant glioma	34
Glioma uncertain grade	2
Ependymoma	9
Metastasis	23
Meningioma	4
Pituitary tumour	6
Craniopharyngioma	37
Miscellaneous diagnoses	36 (see Table 3)
Uncertain diagnosis	29
Total	243

Table 3. *Miscellaneous Diagnoses*

Colloid cyst	3	one case of:
Various cysts	6	tuberculoma
Neuroblastoma	3	AVM
Germinoma	3	plexus papilloma
Pituicytoma	2	sarcoma
Lymphoma	3	xantogranuloma
Brain abscess	2	fungus (AIDS)
Haematoma	4	Cerves Navarro
Atrophy	2	normal brain tissue

performed 37 procedures in 16 patients with a cystic craniopharyngioma. This requires one injection of Yttrium for intracavitary radiotherapy; sometimes a procedure had to be repeated to evacuate cystic fluid. In three patients a colloid cyst was drained. Three tumour cysts, three arachnoideal cysts, two brain abscesses and four haematomas were also partially evacuated.

Complications

Neurological Deterioration

Neurological deterioration after the procedure was observed in five patients. Haemorrhage occurred in four patients in our series; one of these proved fatal:

A 50-year-old woman with epilepsia tarda, due to a left parietal cerebral tumour, was subjected to a stereotactic biopsy. Postoperatively, she deteriorated rapidly, developed non-reacting pupils and she had extensor spasms. The CT scan revealed an intracerebral haematoma which was removed by craniotomy. Temporarily, she did show some signs of recovery but died within 24 hours. The histological diagnosis was oligo-astrocytoma anaplasticum (grade III/IV).

In another patient, the haematoma originated from a haemangioma, which was not visualized on angiography, performed prior to the stereotactic procedure.

Table 4. *Complications in 243 Stereotactic Procedures*

Intracerebral haematoma		4
(persistent neurological deficit	3	
dead	1	
Neurological deficit, unknown origin		1
< 24 hours' deterioration in conscious level		3
30 seconds' apnoea		1
Epilepsy		6
focal	3	
generalized	3	

The third, male, patient had multiple intracerebral tumours, probably melanoma metastases. A large non-fatal haemorrhage developed after an uneventful biopsy from a superficially situated lesion (Fig. 1).

The fourth haemorrhage occurred in a medical student, who had a low grade ependymoma in the posterior part of the third ventricle.

In the three surviving patients, serious neurological deficit persisted. Such persistant deficit was also seen in a patient who had a deep seated metastasis. She experienced neurological deterioration after the procedure, but at that time the CT scan only showed an increase in the midline shift without signs of haemorrhage. Hence, we concluded that the surgical trauma had caused progression of the peritumoural oedema.

Three patients showed a temporary (< 24 hours) deterioration in the level of consciousness, and one patient had an apnoea for 30 seconds during the procedure.

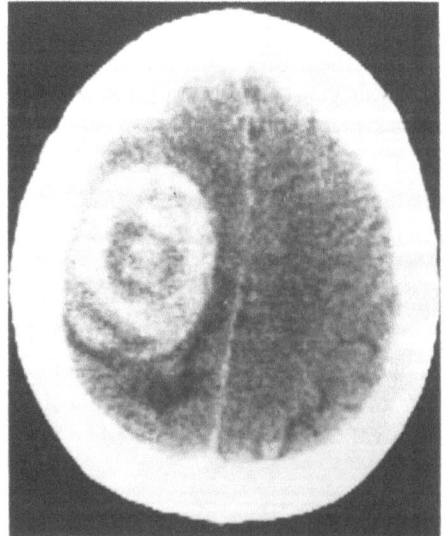

Fig. 1. A large haematoma has developed after stereotactic biopsy of a parietal tumour. The tumour is seen lying within the haematoma

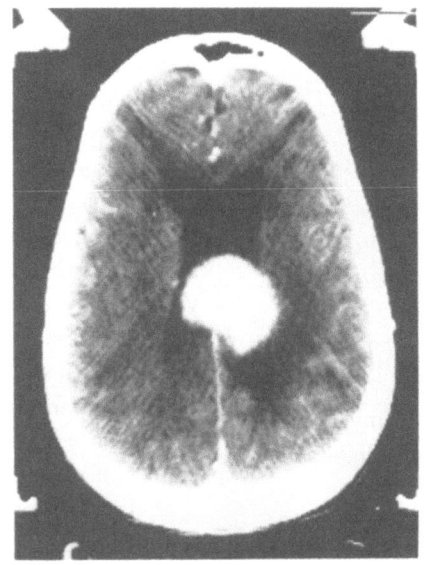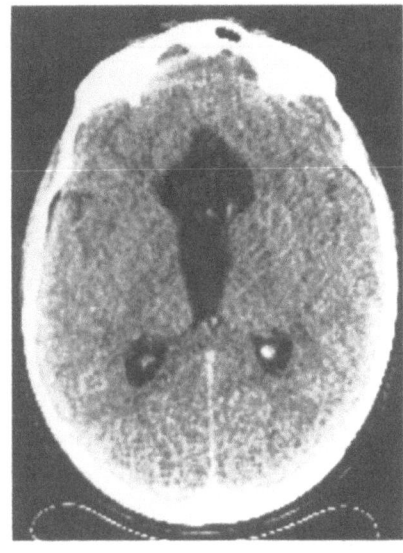

A   B

Figs. 2. (A and B) Due to disarrangement of the console monitor of the CT scanner the head is shown elongated. After correction of h- and v-gain of the monitor a correct image is shown

Epilepsy

In two patients, a generalized seizure was seen during the procedure and in one, immediately afterwards. A focal seizure during the procedure was observed in three patients. In the five patients who experienced a seizure during the operation, the procedure could be completed following injection of anti-epileptic medication. Five of these six patients underwent surgery because of an intracerebral tumour. The sixth patient was operated for a large cystic pituitary tumour, seizures commencing during its aspiration.

Infection

Most patients had their hair washed the night before surgery and the head was preoperatively cleansed with an ordinary disinfectant (70% alcohol). No infections occurred in our series although the heads were not shaved.

Angiography

Cerebral angiography was not undertaken uniformly before each stereotactic procedure. Between 1981 and 1983 the site of a large number of pathological processes near the circle of Willis and the major cerebral vessels was sufficient reason to perform extensive preoperative angiography in about 50% of all patients subjected to a stereotactic procedure. This did not cause a real difference in the calculation of trajectory or biopsy site. Since 1984, angiography has gradually developed as a diagnostic tool when a vascular lesions is suspected or has to be excluded. The indication for angiography is nowadays regarded in this way.

Negative or Uncertain Biopsy

A negative biopsy, *i.e.* a biopsy producing insufficient material for diagnosis or normal brain tissue where pathological brain tissue was expected, occurred in 29 procedures. Initially, the main cause of negative biopsies was disarrangement of the monitor of the CT scanner (Figs. 2 A and B) at the time when we used the visual method for calculating the coordinates. Hence, the picture obtained from the scanner did not fit properly in the ruler. Since we decided to rely on computer software, this error has not recurred.

In 16 of 20 patients in whom the biopsy was negative, the procedure was repeated using a slightly different target, or sometime later. In ten patients, this produced a positive result. The final diagnosis in these ten patients are indicated in Table 5. In the other six patients, the biopsy remained negative. In nine patients the pathologist was uncertain about the diagnosis. In three

Table 5. *Corrected Diagnosis*

Glioblastoma	1
Astrocytoma	3
Oligodendroglioma	1
Chiasma glioma	1
Craniopharyngioma	2
Malignant lymphoma	1
Melanoma	1

Table 6. *Histology of Multiple Lesions*

Metastasis	10
Glioma	9
Lymphoma	3
Meningioma	1
Neuroblastoma	1
AVM	1
Uncertain (lymphoma? metastasis?)	1
Total	26

of these the further course or a craniotomy revealed that the diagnosis was glioblastoma (2) or oligodendroglioma (1).

Multiple Lesions

In 26 patients, the CT scan revealed multiple lesions, sometimes in both hemispheres and occasionally two or three lesions in one hemisphere. The distribution of the histological diagnoses from the material obtained from these patients by the stereotactic procedure is given in Table 6. Note the large percentage in whom the diagnosis – proved to be correct by the subsequent course and CT scan – was glioma.

Biopsy Instrument

The possibility of using different biopsy instruments is clearly in the procedure's favour. Simple aspiration needle techniques are insufficient. Backlund's[6] spiral needle provided a real advantage, but trauma to cerebral vessels was a risk and the acquisition of tissue was often difficult when the pathological tissue was soft, as in gliomas. Further progress was made by the introduction of sampling techniques, *e.g.* using the biopsy forceps or the Sedan or Nashold needle[7]. This is essentially a suction biopsy method with only a slightly harmful effect on vascular structures.

Incorrect Histological Diagnosis

In seven patients, the initial diagnosis of low or middle grade glioma, had to be changed later into malignant glioma on the basis of a craniotomy or the pathological course. Two patients who, after biopsy, were considered to have a glioma, proved to be harbouring a cerebral metastasis.

Discussion

Stereotactic systems were developed for functional brain surgery, *i.e.* the treatment of extrapyramidal movement disorders and psychosurgery. The three-dimensional localization from projections of divergent X-rays on a two-dimensional film caused geometric problems. Therefore, stereotactic methods were not widely used for biopsies.

This changed drastically when CT scanners were introduced, although the initial CT units with their small gantries required special adaptation[8]. The advantages are obvious:

1. Representation of site and size of the disease process without distortion, which occurs when conventional X-ray techniques with divergent X-rays are used. Hence, one achieves greater simplicity, a smaller chance of error and increased safety.

2. Improvement in the quality of the pictures obtained due to the richness of detail of the disease process.

3. Less of a burden to the patient, because of the absence of side effects and hazards of craniotomy. Also, angiography can often be omitted and ventriculography is almost rendered obsolete.

There is no advantage to shaving the patient's head in simple biopsy procedures. Washing the hair and cleansing the head with ordinary disinfectants is sufficient to prevent infections. For these indications, no infections occurred in our series. The insertion of foreign bodies such as catheters, reservoirs, solid isotope sources, etc. requires more stringent aseptic conditions including two-staged stereotactic procedures with adequate sterilization of instruments.

Neurological deterioration after the procedure may be caused by haemorrhage and/or traumatic oedema. Haemorrhage occurred in four patients in our series; in one it proved fatal. In one patient the haematoma originated from a haemangioma which had not been visualized by angiography performed prior to the stereotaxis. Our complication rate is comparable to that reported in earlier publications[9, 10].

References

1. Leksell L, Jernberg B (1980) Stereotaxis and tomography. A technical note. Acta Neurochir (Wien) 52: 1–7
2. Bosch DA (1986) Stereotactic techniques in clinical neurosurgery. Springer, Wien New York
3. Backlund EO (1973) Studies on craniopharyngiomas. III Stereotactic treatment with intracystic Yttrium-90. Acta Chir Scand 139: 237–247
4. Leksell L, Leksell D, Schwebel J (1985) Stereotaxis and nuclear magnetic resonance. J Neurol Neurosurg Psychiatry 48: 14–18
5. Blaauw G, Versteege C, Ammannati F: Calculation of coordinates for stereotaxis. To be published

6. Backlund EO (1971) A new instrument for stereotaxic brain tumour biopsy. Acta Chir Scand 137: 825–827

7. Nashold BS Jr (1982) Brain stem stereotaxic procedures. In: Schaltenbrand G, Walker AE (eds) Stereotaxy of the human brain. Anatomical, physiological and clinical applications, 2nd ed. G Thieme, Stuttgart New York, pp 4475–4483

8. Bergström M, Greitz T (1976) Stereotaxic computed tomography. Am J Roentgenol 127: 167–170

9. Ostertag CB, Mennel HD, Kiessling M (1980) Stereotactic biopsy of brain tumours. Surg Neurol 14: 275–283

10. Apuzzo MLJ, Sabshin JK (1983) Computed tomographic guidance stereotaxis in the management of intracranial mass lesions. Neurosurgery 12: 277–284

Address for correspondence: Dr. G. Blaauw, Department of Neurosurgery, De Wever Hospital, 6418 PC Heerlen, The Netherlands.

Acta Neurochirurgica, Suppl. 42, 166–169 (1988)

The Relevance of Pathological Diagnosis for Therapy and Outcome of Brain Stem Gliomas

J. Artigas[1], **R. Ferszt**[1], **M. Brock**[2], **E. Kazner**[3], and **J. Cervós-Navarro**[1]

Freie Universität Berlin, [1] Insitut für Neuropathologie, [2] Neurochirurgische Klinik im Klinikum Steglitz, [3] Abteilung für Neurochirurgie am Klinikum Charlottenburg, Berlin-West

Summary

41 patients with brain stem gliomas are presented and analyzed. Optimal therapy as well as outcome depend on the histological identity of the tumours. Histological examination cannot be replaced by CT, angiography and NMR. Therefore stereotactic biopsies are recommended in all cases with brain stem gliomas.

Introduction

Gliomas of the brain stem previously regarded as inaccessible to pathological diagnosis and not amenable to efficient therapy are very gradually becoming a more manageable condition.

Often at least partial surgical resection is possible and may be followed by appreciable survival (Lassiter *et al.* 1971) which leads Hoffman *et al.* (1980) among others to advocate agressive therapy.

With the advent of magnetic resonance imaging along with the conventional techniques (CAT, and angiography) the relevance of pathological diagnosis has been questioned (Jacobi *et al.* 1986). In a recent paper Neidhart (1986) on the other hand adequately describes the dilemma of treating brain stem gliomas without morphological tumour definition.

Here we report 41 cases in an attempt to correlate clinical and pathological data and to assess the role pathological diagnosis has to play in the lives of these patients.

Material and Methods

41 patients were seen between 1969 and 1985 diagnosed as brain stem tumours on clinical grounds and were classified by biopsy and/or autopsy. All cases were selected randomly where adequate documentation of the clinical course and of pathology were available. The age of the patients varied from 2 to 78 years, children averaging 8.5 years, adults 38 years and elderly patients 68 years. 30 patients presented with typical symptoms of a posterior fossa tumour *i.e.*

progressive multiple bilateral cranial nerve palsies, pyramidal tract signs and ataxia. The majority of these patients also complained of headaches and vomiting. Atypically various forms of brain failure *i.e.* progressive organic brain syndrome were presenting symptoms in 6 patients similar to the report by Schirmer (1986). 2 individuals showed a subarachnoid haemorrhage primarily. One brain stem glioma was a casual finding at autopsy.

The diagnosis was confirmed by cerebral angiography and pneumencephalography (11 cases) CAT scan (25 cases) and magnetic resonance imaging (3 cases).

Pathological diagnosis was obtained either by stereotactic needle biopsy (3 patients) tumour resection (25 patients) or autopsy (13 patients). The material underwent frozen sectioning and paraffin embedding allowing conventional light microscopic stains (Nissl, H & E, Elastica van Gieson, PAS) immunohistochemistry (fibrillary acidic protein or neurofilaments, pancytokeratine and where necessary leucocytes).

Altogether, precise diagnosis could be performed in 38 of 41 cases while in 3 patients the tumour could only be identified as a low grade glioma.

The morphological tumour classification followed the lines layed down by Zülch (1986). The histological diagnoses are listed in Table 1.

The largest single group of tumours were pilocytic astrocytomas with typically elongated bipolar cytoplasm, spindle shaped nuclei and no mitoses. Typical eosinophilic Rosenthal fibres confirmed diagnosis in 9 cases, 3 of the tumours were cystic, 6 showed exophytic growth. In 9 cases subtotal resection was performed giving a comparatively large tissue volume for pathological diagnosis. Tumours diagnosed as pilocytic astrocytomas were of uniform structure showing little regional variation.

Typical protoplasmic astrocytomas were infrequent, with an eosinophilic multipolar cytoplasm and comparatively pale nuclei. Mitoses were rare, most tumours showing good astroglial differentia-

Table 1. *Precise Histological Diagnosis of the Cases Studied*

12 Pilocytic astrocytomas	3 Protoplasmic astrocytomas
8 Glioblastomas	2 Gangliogliomas
8 Gliomatosis	1 Angioglioma
4 Subependymomas	3 Gliomas

tion. In 3 cases biopsy material was very scant and limited to a number of eosinophilic cells with stunted processes and pale oval nuclei in front of a slightly eosinophilic background, the cells giving a positive GFAP stain. In all 3 cases it is highly probable that we were dealing with protoplasmic astrocytomas though perifocal gliosis could not be ruled out because of the inadequate material. We listed these as "gliomas" not attempting any further differentiation.

We diagnosed gangliogliomas as tumours showing mature gangliom cells and some admixture of glial elements mostly pilocytotic astrocytes with an occasional Rosenthal fibre some microcalcifications and rare microcystic degeneration. Both glial cells and neurons gave a strong GFAP or neurofilament reaction respectively, further indicating the degree of maturity of this tumour.

Gliomatosis was diagnosed according to criteria expounded elsewhere (Artigas *et al.* 1985); essentially there is an infiltration of neoplastic glial cells with little if any anaplasia and intensive GFAP staining. Differentiating gliomatosis from reactive gliosis may be virtually impossible if the clinical course is unknown. Distinguishing gliomatosis from a stage I glioma is equally difficult while the process appears circumscribed and has not yet the great degree of infiltration which is the hallmark of the condition. It is important to stress, however that a few cytological criteria may indicate the nature of the neoplasm quite early in its course (Artigas *et al.* 1985).

Criteria for glioblastomas were a high degree of anaplasia, multiple at times atypcial mitoses, necroses, and pseudopalisading. From 7 glioblastomas seen, 2 morphological groups were formed; 3-tumours had a somewhat atypical appearance with a comparatively high number of pilocytic astrocytes. These tumour invariably and diffusely infiltrated the cerebellar peduncules and the cerebellar white matter which was not the case in the other more typical polymorphous glioblastomas limited to the pons.

Subependymonas were tumours of low cellularity in which aggregates of astrocyte-like cells surrounded by highly fibrillar tissue gave a positive GFAP stain indicating the glial origin of this tissue.

Angiogliomas (Bonnin *et al.* 1983) consist of gemistocytic or fibrillary astrocytes with conspicuous vascular proliferation and a dense reticulin fibre network.

Treatment and Course

25 patients were operated upon, 9 received postoperative irradiation, 9 patients received radiotherapy with doses varying from 40 to 60 gray, 2 patients – those with glioblastoma – received BCNU chemotherapy in addition to radiotherapy. 2 lowgrade astrocytomas were treated by interstitital irradiation following stereotactic biopsy. 11 patients received Holter's drainage when occlusive hydrocephalus developed. 6 patients were only given symptomatic treatment.

6 patients died within four weeks of surgery due to postoperative complications and cannot be further considered here, 5 of them had gliomas of low malignancy. The glioblastoma patients showed an average survival time of five months regardless of whether they received radiotherapy or not while 2 patients receiving both radiotherapy and chemotherapy survived one and two years.

Both patients with gangliogliomas received radiotherapy in addition to partial resection, their survival time averaged seven months. The 7 patients with gliomatosis received various modes of therapy including partial resection, and irradiation or only supportive measures, survival ranging from twenty years to three months. The patient with angioglioma received subtotal resection and survived for fifteen years without any further measured necessary. Of the 3 patients with subependymoma 2 survived surgery, received no additional causal therapy; one survived one year and one is still alive after 12 years. One of the patients with astrocytoma died of postoperative complications the others received either subtotal resection alone or stereotactic biopsy with interstitial irradiation. Survival periods varied from two months to ten years, 2 patients are still alive after seven or two years. The data of all patients are given in Table 2.

Discussion

While locating a brain stem tumour in computerized tomography may cause problems, these growths are readily found by magnetic resonance imaging. However, the exact biological nature of a more or less infiltrating brain stem neoplasm remains difficult to assess. Without the availability of histological data one is forced to evaluate the biological behaviour of the tumour on the basis of the clinical course alone which means that time for treatment is lost. Our material indicates that effective treatment is possible and that survival times vary greatly depending upon the histological nature of the tumour. Most of these tumours were or would have been accessible to stereotactic biopsy. Our material clearly shows that within the comparatively small space of the brain stem and the cerebellar peduncules the regional variability of all the growths is by no means as large as in supratentorial gliomas, a problem hampering stereotactic biopsies in these growths. Furthermore MRI allowes selective localization and biopsy of necrotic, cystic and compact tumour regions which may be evaluated by multiple biopsies. Given the availability of MRI and skillfully applied biopsy techniques, the reliability of this diagnostic approach is not inferior to conventional diagnosis in resected tumour tissue. As a result it should be considered possible to histologically identify the tumour categories described above by a stereotactic method even when the tumour is not amenable to resection.

Histological identification means that the patients can be given a high probability for a long survival

Table 2. *Clinical and Pathological Data*

No.	Age	Sex	Neuropathological diagnosis	Therapy	Postoperative death	Survival period
1	2	f	pilocytic astrocytoma, exophytic	subtotoal resection		10 years
2	78	f	pilocytic astrocytoma, solid (A)	none		2 years
3	54	m	pilocytic astrocytoma, cystic	subtotal resection, shunt		2 years
4	24	m	pilocytic astrocytoma, solid	subtotal resection		2 months
5	42	m	pilocytic astrocytoma, exophytic	subtotal resection		11 months
6	63	m	pilocytic astrocytoma, exophytic	subtotal resection		2 years
7	6	m	pilocytic astrocytoma, solid	subtotal resection		6 months
8	13	m	pilocytic astrocytoma, exophytic	subtotal resection, shunt		2 months
9	8	f	pilocytic astrocytoma, exophytic	subtotal resection	1 month	
10	15	f	pilocytic astrocytoma, solid	subtotal resection		alive after 7 years
11	39	m	pilocytic astrocytoma, solid	stereotactic biopsy, interstit. irrad.		1 year
12	29	m	pilocytic astrocytoma, solid	stereotactic biopsy, interstit. irrad.		alive after 2 years
13	2	f	protoplasmic astrocytoma (A)	none		1 month
14	8	f	protoplasmic astrocytoma	subtotoal resection, shunt	1 day	
15	6	m	protoplasmic astrocytoma	subtotal resection		6 months
16	5	m	ganglioma (A)	irradiation, shunt		8 months
17	13	m	ganglioglioma	subtotal resection, irradiation, shunt		6 months
18	36	m	glioblastoma (A)	irradiation, shunt		7 months
19	5	m	glioblastoma (A)	none		2 months
20	10	f	glioblastoma (A)	none		3 months
21	38	m	glioblastoma	subtotal resection	2 weeks	
22	67	f	glioblastoma (A)	shunt		8 months
23	4	f	glioblastoma (A)	chemotherapy, irradiation, shunt		1 year
24	12	m	glioblastoma (A)	irradiation, chemotherapy		15 months
25	10	m	glioblastoma (A)	shunt		10 months
26	67	f	"glioma" exophytic	subtotal resection	1 month	
27	31	m	"glioma"	subtotal resection		10 months
28	31	m	"glioma"	subtotal resection, irradiation		3 years
29	9	m	gliomatosis (A)	none		
30	72	m	gliomatosis	partial resection	1 month	
31	39	m	gliomatosis (A)	none		3 months
32	30	m	gliomatosis	partial resection		20 years
33	63	f	gliomatosis	irradiation, partial resection		8 months
34	39	f	gliomatosis	irradiation, partial resection		1 year
35	37	m	gliomatosis	irradiation, stereotactic biopsy		4 years
36	39	m	gliomatosis	partial resection		1 year
37	8	m	subependymoma	subtotal resection, shunt		12 years
38	60	f	subependymoma	subtotal resection, shunt	1 week	
39	52	m	subependymoma	subtotal resection, shunt		1 year
40	77	f	subependymoma (A)	none		casual autopsy finding
41	13	m	angioglioma	subtotal resection		alive after 15 years

"(A)" Indicates that the pathological diagnosis was performed at autopsy.

period in subependymonas and angiogliomas while prognosis remains obviously poor in patients with glioblastoma. The average two years survival period for pilocytic astrocytomas makes it relevant to distinguish them from subependymonas on one side and from gliomatosis on the other; the prognosis of the latter remains elusive.

It is also interesting to note that exophytically growing pilocytic astrocytomas seem to have a distinctly more favourable prognosis than those of the cystic type.

We have yet to assess the importance of immuno-histochemical parameters for biological behaviour in a given tumour and since stereotactic biopsies allow ultrastructural examination of the material it will be important to try to identify ultrastructural criteria of long survival times. The block biopsy method (Ferszt and Cruz 1986) has not yet been applied to its full potential. Freeze sectioning needle biopsies instead of smearing them allows full histological diagnosis in an intraoperative setting.

Our findings indicate that distinctly different categories of brain stem gliomas exist and cannot up to now be distinguished by computed tomography, angiography or magnetic resonance imaging. Currently histology helps assess the individual patient's prognosis; with further development of therapeutic approaches such as stereotactic intervention (Pannek *et al.* 1986, Mundinger 1986) and chemotherapy (Neidhart 1986, Haferkamp 1986, Krauseneck 1986): the importance of histological classification seems certain to increase.

References

1. Artigas J, Cervós-Navarro J, Iglesias J, Ebhardt G (1985) Clinico-pathological findings in gliomatosis cerebri. Clin Neuropathol 4: 135–148
2. Bonnin JM, Pena CA, Rubinstein LJ (1983) Mixed capillary hemangioblastoma and glioma. A redefinition of the "angioglioma". J Neuropathol Exp Neurol 42: 504–516
3. Ferszt R, Cruz-Sanchez F (1986) Stereotactic block biopsies; a new tool for intraoperative diagnosis. Symp on Low Grade Gliomas, Budapest 27–30 August
4. Haferkamp G (1986) Chemotherapy of malignant brain stem tumours in adults – new substances, prospect. In: Samii M (ed) Surgery in and around the brain stem and the third ventricle. Springer, Berlin Heidelberg New York, pp 582–587
5. Hoffman HJ, Becker L, Craven MA (1980) A clinical and pathological distinct group of benign brain stem gliomas. Neurosurgery 7: 243–248
6. Jacobi G, Weiermann G, Kornhuber B (1987) Therapy of primary brain tumours. Symposium in Essen, March 1987, Abstract
7. Krauseneck P (1986) Chemotherapy of malignant brain stem tumours in adults. In: Samii M (ed) Surgery in and around the brain stem and the third ventricle. Springer, Berlin Heidelberg New York, pp 575–581
8. Lassiter KRL, Alexander E, Davis CH, Kelly DL (1971) Surgical treatment of brain stem gliomas. J Neurosurg 34: 719–725
9. Mundinger F (1986) CT-guided stereotactic biopsy and interstitial curie-therapy with ^{192}Ir and ^{125}I of non-resectable midline and brain stem gliomas. In: Samii M (ed) Surgery in and around the brain stem and the third ventricle. Springer, Berlin Heidelberg New York, pp 509–517
10. Neidhardt MK (1986) Chemotherapy of childhood brain stem tumours. In: Samii M (ed) Surgery in and around the brain stem and the third ventricle. Springer, Berlin Heidelberg New York, pp 570–574
11. Oppel F, Pannek HW, Voges J, Brock M (1986) Interstitial irradiation of inoperable brain stem tumours. In: Samii M (ed) Surgery in and around the brain stem and the third ventricle. Springer, Berlin Heidelberg New York, pp 526–533
12. Schirmer M (1986) Symptoms of tumours of the brain stem and the third ventricle. In: Samii M (ed) Surgery in and around the brain stem and the third ventricle. Springer, Berlin Heidelberg New York, pp 129–132
13. Zülch KJ (1986) Brain tumours – biology and pathology. 3rd Ed. Springer, Berlin Heidelberg New York

Address for correspondence: PD Dr. R. Ferszt, Freie Universität Berlin, Institut für Neuropathologie, Hindenburgdamm 30, D-1000 Berlin 45.

Acta Neurochirurgica, Suppl. 42, 170–176 (1988)
© by Springer-Verlag 1988

Serial Stereotactic Biopsy of Brain Stem Expanding Lesions. Considerations on 45 Consecutive Cases

A. Franzini, A. Allegranza[1], A. Melcarne, C. Giorgi, S. Ferraresi, and G. Broggi

Departments of Neurosurgery and [1]Neuropathology, Istituto Neurologico "C. Besta", Milano, Italy

Summary

Fourty-five patients affected by brain stem expanding lesions underwent serial stereotactic biopsy between 1978 and 1986. The definitive histological diagnosis allowed the definitive treatment of extrinsic tumours and non-neoplastic lesions. In patients affected from glial tumours the serial stereotactic biopsy allowed the histological grading and the definition of the growth modalities at the superior boundaries of the tumours. These data have been utilized to guide the choice of treatment. The future perspectives of stereotactic biopsies are discussed in view of the therapeutic results obtained in this series and in other series reported in the literature.

Keywords: Brain stem expanding lesions; glial tumours; deep brain abscesses; intraparenchymal spontaneous haematoma; stereotactic serial biopsy.

Introduction

The first extensive review of a singificant number of patients affected by brain stem glial tumours was reported in 1945 by Guillain, Bertrand and Gruner from the University Hospital "Salpetrière" in Paris[14]. The survival of patients ranged between a few months and four years and no surgical or medical treatment could be proposed.

The last extensive review was reported in February 1987 in the U.S.A.[15]; the patients were studied by CT scan and treated by external radiotherapy with or without chemotherapy. Only 17% of cases were still alive at 5 years follow-up.

The choice between surveillance and radiotherapy seems to have very little influence on the prognosis for these patients.

In spite of these considerations a new approach has been suggested to study these lesions starting from the histological diagnosis[2, 7, 10, 19]. The value of this approach is stressed in this report, and 45 consecutive patients affected from brain stem lesions submitted to serial stereotactic biopsy are described with particular regard to the choice of treatment. Future perspectives of interstitial radiotherapy[24], stereotactic radiosurgery and microsurgical stereotactic guided technique[17] are discussed.

Patients and Methods

The Riechert frame and apparatus have been utilized. General anaesthesia has been employed in all cases. Stereotactic ventriculography has been always performed to assess the segmental enlargement of the brain stem and the patency of CSF pathways. The CT and NMR images have been transposed in the stereotactic plain radiographs by mathematical method[13]. All the stereotactic trajectories directed to the brain stem have been performed via precoronal approaches[7]. The electrical impedance has been monitored along the full trajectory from the frontal cortex to the deepest targets in order to reveal the presence of major peritumoural alterations or the presence of intratumoural necrosis or cystic cavities[3, 5]. The Fisher biopsy forceps has been utilized in most cases to obtain small fragments of tissue from each target without damaging functional structures; the Sedan instrument[23] or the similar Nashold instrument has been utilized only to obtain major samples of tissue from the inner part of the tumour when impedance monitoring or intraoperative smear examinations confirmed the absence of non-neoplastic tissue. The number of targets examined ranged between 2 and 5. In 4 patients the intraoperative smear examinations suggested the need for a second trajectory to obtain a more precise spatial definition of the lesion.

Finally a small silver marker (NMR compatible) has been placed at the deepest target to make the postoperative neuroradiological assessment easier. Since 1983 the definitive histological examinations also included the determination of the Labeling Index (LI) by H3-thymidine in vitro procedure[12].

Fourty-five patients, 22 males and 23 females, aged between 3 and 65 years (mean age 41 years), have been operated on between 1978 and 1986. The whole series may be divided into two age groups: 21 patients were under fifteen years of age and 24 patients were adults (Table 2).

The diagnosis of brain stem expanding lesions was obtained by CT examinations usually performed in neurological and neuropediatric services where the patients had been hospitalized following the appearance of the first clinical symptoms. The first recorded symptoms include pyramidal and lemniscal lateralized deficit in 34

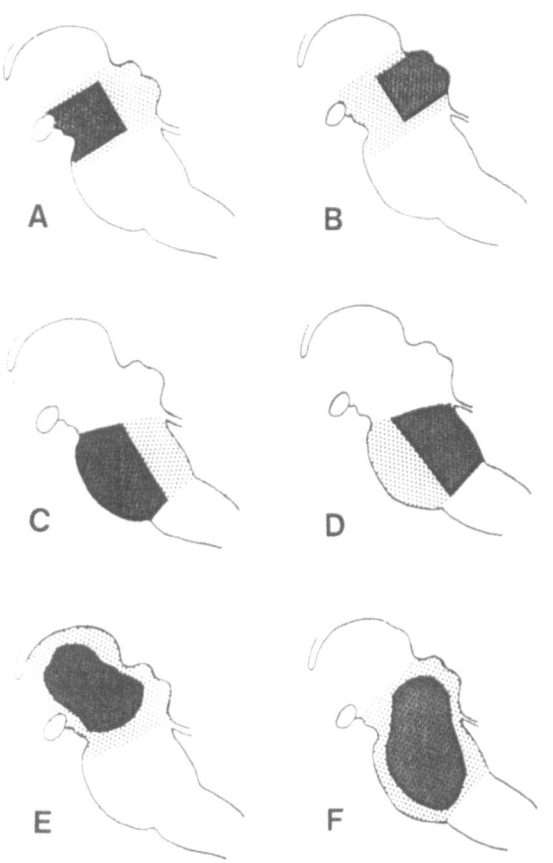

patients, only 9 of these patients presented with deficits of cranial nerves; gait impairment was the first symptom in 11 patients. The time lapse between the onset of symptoms and the neuroradiological diagnosis ranged between 10 years and a few days (mean 37 weeks).

Since 1985 the preliminary neuroradiological investigations included also NMR (9 patients). Angiography was performed in 6 patients to rule out the clinical suggestion of arteriovenous malformation.

The location of the lesions is summarized in Fig. 1 and Table 1.

The CT patterns without and with intravenous contrast medium are summarized in Tables 3 and 4.

Results

The relationships between the histological diagnosis and the site of the lesions are listed in Table 1. The

Fig. 1. Schematic drawing of the localizations of 45 brain stem lesions submitted to serial stereotactic biopsy. These data have been derived from CT slices reconstruction and from NMR sagittal sections (in 9 cases). (A) 3 patients (1 bilateral and 2 lateralized lesions); (B) 4 patients (3 bilateral and 1 lateralized lesion); (A and B) 11 patients (9 bilateral and 2 lateralized lesions); (C) 4 patients (bilateral lesions); (D) 2 patients (bilateral lesions); (C and D) 5 patients (bilateral lesions); (E) 6 patients (4 lesions involved the right thalamopeduncular region and 2 lesions the left thalamopeduncular region); (F) 10 patients (bilateral lesion). The alphabetical sequence utilized to illustrate this figure is the same maintained in the text and Table 1 to show the relationships between localization and histological diagnosis of these series of patients

Table 1. *Histological Diagnosis Obtained in 45 Consecutive Patients Submitted to Serial Stereotactic Biopsy*. The site of the lesions within the brainstem is indicated by alphabetical sequence referred to Fig. 1. (A) anterior mesencephalon; (B) quadrigeminal plate, posterior mesencephalon; (A and B) whole mesencephalon; (C) anterior pontine region; (D) posterior pontine region; (C and D) whole pons; (E) mesencephalon plus thalamopeduncolar region on one side; (F) Whole pontine and mesencephalic regions: Note that pilocytic astrocytoma has been considered separately from mature astrocytoma because of the characteristic histological pattern and behaviour. The classification of "mature" astrocytoma includes differentiated astrocytoma without anaplastic features

	A	B	A + B	C	D	C + D	E	F	Total
Mature astrocytoma	2	1	2	1	1	1	2	4	14
Pylocytic astrocytoma		1	1	1	1	1			5
Ependymoma						1			1
Anaplastic astrocytoma	1		4				3	4	12
Glioblastoma			1					1	2
Primitive neuroepithelial tumour						1			1
Primitive lymphoma							1		1
Metastatic tumour						1			1
Neurinoma (III° cranial nerve)			1						1
Cholesteatoma		1							1
Primitive abscess			1	1					2
Spontaneous haematoma			1	1				1	3
Arachnoid cyst		1							1
Total expanding lesions	3	4	11	4	2	5	6	10	45

Table 2. *Histological Diagnosis Related to the Age of Patients.* Note the prevalence of pilocytic and mature astrocytoma in young patients and conversely the prevalence of extrinsic tumours and non-neoplastic lesions in adult series

	Under 15 years old	Adults
Mature astrocytoma	10	4
Pylocytic astrocytoma	4	1
Ependymoma	—	1
Anaplastic astrocytoma	5	7
Glioblastoma	1	1
Primitive neuroepithelial tumour	1	—
Primitive lymphoma	—	1
Metastatic tumour	—	1
Neurinoma (III° cranial nerve)	—	1
Cholesteatoma	—	1
Primitive abscess	—	2
Spontaneous haematoma	—	3
Arachnoid cyst	—	1
Total	21	24

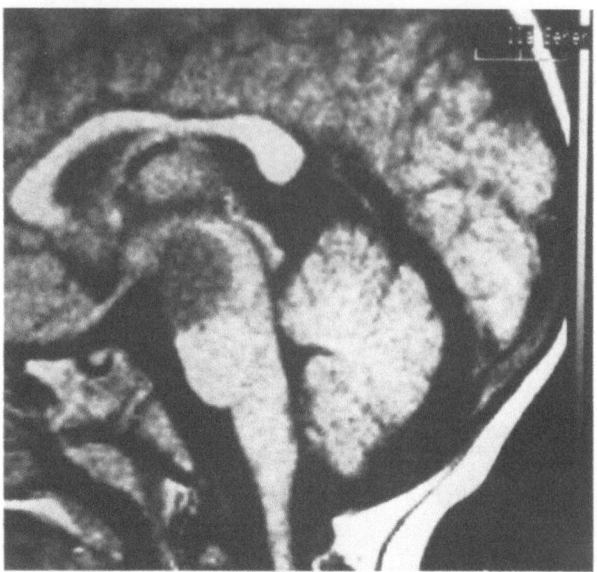

Fig. 2. NMR 0.5 T, SE 350/30, T 1 weighted. Midline sagittal section showing pontomesencephalic expanding lesion which was a pilocytic astrocytoma at definitive histological diagnosis

relationships between the histological nature of the lesions and the age of patients are shown in Table 2. The relationships between the CT patterns and the definitive histological diagnosis are set out in Tables 3 and 4. The NMR pattern of two patients harbouring pilocytic astrocytoma are shown in Figs. 2 and 3.

The findings of stereotactic ventriculography are summarized in Table 5.

Patients with non-neoplastic lesions underwent treatment immediately after the biopsy. In two patients with primitive brain abscesses an intracavitary catheter has been implanted for repetitive antibiotic administrations and multiple aspirations of purulent material[8]; in both

Table 3. *C.T. Patterns Detected Preoperatively and Without Intravenous Contrast Medium.* The lesions considered hypodense showed Hounsfield values lower than normal tissue and higher than C.S.F. The lesions considered hyperdense were non-homogeneous and showed irregular density areas suggesting small calcifications in 2 cases. The lesions considered cystic showed rounded areas with density values similar to C.S.F.

	Hypodensity	Hyperdensity (dishomogeneous)	Cysts
Mature astrocytoma	10	1	3
Pylocytic astrocytoma	3		2
Ependymoma	1		
Anaplastic astrocytoma	11		1
Glioblastoma	1	1	
Primitive neuroepithelial tumours			
Primitive lymphoma		1	
Metastatic tumour		1	
Neurinoma (III° cranial nerve)			1
Cholesteatoma			1
Primitive abscess	1		1
Spontaneous haematoma		3	
Arachnoid cyst	1		
Total	19	8	9

Table 4. *C.T. Patterns Detected Preoperatively and After Intravenous Contrast Medium*. Note the small number of hypodense lesion in comparison with those reported in Table 3; in fact most of lesions which were hypodense without i.v. contrast medium showed consistent enhancement or appeared as "ring lesions" characterized by an hyperdense rim surrounding an hypodense zone. Intralesional cysts have also been revealed in lesions which had appeared as diffuse hypodensity in the basal C.T. and lesions which had appeared cystic showed irregular enhancement. No unequivocal or clearcut relationship between C.T. findings and definitive histological diagnosis could be found in these series

	Hypodensity	Enhancement	Cysts	Ring
Mature astrocytoma	2	8	3	1
Pylocytic astrocytoma	1	2	2	
Ependymoma		1		
Anaplastic astrocytoma	1	9	1	1
Glioblastoma		1		1
Primitive neuroepithelial tumour		1		
Primitive lymphoma		1		
Metastatic tumour		1		
Neurinoma (III° cranial nerve)				1
Cholesteatoma	1			
Primitive abscess				2
Spontaneous haematoma		1		2
Arachnoid cyst			1	
Total	5	25	7	8

Table 5. *Patency of C.S.F. Pathways as Demonstrated by Stereotactic Ventriculography and C.T. Preoperative Examinations*. The number of patients who required C.S.F. shunts is indicated in brackets. Asterisk shows that the patient has been relieved of hydrocephalus following the surgical ablation of the lesion or following stereotactic procedures which reduced the volume of the lesion

C.S.F. Pathways	Stereotactic ventriculography		C.T. Hydrocephalus
	Stenosis-Dislocation	Obstruction	
Mature astrocytoma (14 cases)	5	3	6 (6)
Pylocytic astrocytoma (5 cases)	2	—	1 (2)
Ependymoma (1 case)		1	1 (1)
Anaplastic astrocytoma (12 cases)	4	4	6 (6)
Glioblastoma (2 cases)	1	1	1
Primitive neuroepithelial tumours (1 case)	—	1	1*
Primitive lymphoma (1 case)	—	—	—
Metastatic tumour (1 case)	—	—	—
Neurinoma (III° cranial nerve) (1 case)	—	1	1*
Cholesteatoma (1 case)	1	—	1*
Primitive abscess (2 cases)	1	—	1*
Spontaneous haematoma (3 cases)	3	—	—
Arachnoid cyst (1 case)	—	1	1*
Total	17	12	19 (15)

patients the lesion disappeared completely after 1–4 months of treatment; they are well at 5 years follow-up. Three patients with spontaneous intraparenchymal haematomas and negative digital and conventional angiography underwent single aspiration in 1 case and repeated aspirations in 2 cases. In the first patient a complete clinical and neuroradiological recovery has been obtained at 4 years follow up, while in two cases rebleeding occurred within the same site and these patients had major neurological deficits in spite of the

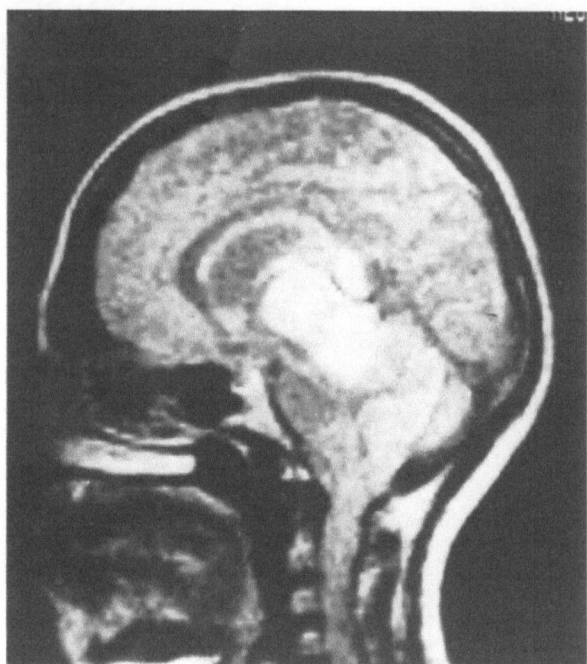

Fig. 3. NMR 0.15 T, SE 1000/100, T 2 weighted. Midline sagittal section showing mesencephalic thalamopeduncolar lesion which was a pilocytic astrocytoma at definitive histological diagnosis

repetition of the stereotactic procedures. Nevertheless in the succeeding months a slow recovery of neurological deficits took place in both patients. In these patients the aspiration of blood was performed through a fenestrated cannula or through an implanted silicone catheter[18]; the timing of treatment allowed us to aspirate liquified clots. Therefore a screw device[1, 9, 16] has never been applied within the brain stem. The patient with an arachnoid cyst was relieved of hydrocephalus and intracranial hypertension by a single evacuation of the cavity; he is still free from symptoms at 6 years follow-up without signs of the lesion at CT scan examination. The histological diagnosis in this case was obtained by examination of a thin sample from the wall of the cyst.

Aspiration procedures via intracavitary implanted devices have been performed also in 6 patients with cystic glial tumours; this treatment was merely palliative in anaplastic astrocytomas, but it has been utilized for periodic fluid aspiration in mature astrocytomas. In one patient with a pilocytic astrocytoma the intracavitary device has been utilized for the introduction of radiocolloids within the cyst (this procedure was performed in another neurosurgical department where the patient was referred to complete the treatment)[24].

At the beginning of our experience, external radio-

therapy seemed the only treatment available also for low grade gliomas, but the results of external radiotherapy in mature astrocytoma were poor: out of 5 treated patients, just one survived at 3–8 years follow-up. In anaplastic astrocytoma, glioblastoma and primitive neuroectodermal tumour (PNET) the external radiotherapy caused only a transitory reduction of the mass which lasted about 6 to 32 months before the recurrence.

Since 1983 the patients with mature astrocytomas have been only clinically and neuroadiologically followed-up and external radiotherapy has been avoided in these patients. Four patients out of 5 included in these series are still living. One of the survivors, who had clinical symptoms suggesting malignant evolution of the tumour, underwent a second stereotactic biopsy two years after the first procedure and the presence of anaplastic features and neovascularization was detected. The follow-up in these patients ranged between 1 and 3 years. Finally 4 patients with low grade gliomas and 4 patients with pilocytic astrocytomas have been submitted to interstitial irradiation in other neurosurgical centres, (particularly the Neurosurgical Service of St. Anne Hospital in Paris for radiocolloids treatment, and the University Hospital in Homburg/Saar for Iodine 125 seeds application). The follow-up is short but the results in these groups of slow-growing tumours are very promising.

Finally the microsurgical radical resection of the tumour was the chosen treatment of third nerve neurinoma, cholesteatoma and metastatic tumour. The patient with primitive lymphoma has been successfully treated by corticosteroid, chemotherapy and external irradiation.

Fifteen patients among the whole reported series underwent ventriculo-peritoneal shunt because of symptomatic hydrocephalus.

No mortality related to serial stereotactic biopsy occurred in these patients. Two patients developed permanent worsening of cranial nerve deficit (one oculomotor nerve, the second oculomotor and trigeminal nerve) following the procedure. The side effects were less, and the accuracy of stereotactic biopsy of brain stem lesion was better than in other hemispheric or basal ganglia tumours investigated by our group with the same methodology[6]. Moreover, the risks of stereotactic serial biopsy are considerably less than those of open surgical exploration[20]. The feasibility of cell kinetics investigation by 3H-thymidine and LI determination has been demonstrated also in brain stem tumours.

Discussion and Conclusion

The great variety of histological diagnosis obtained by serial stereotactic biopsies in these series of patients allows important considerations and comments:

1. A considerable number of mature astrocytomas (47%) and pilocytic astrocytomas (24%) have been found in patients under fifteen years of age (Table 2);

2. A certain number of non-neoplastic lesions (25%) have been found in adult patients and the stereotactic procedures always allowed the definitive treatment of these lesions (Table 2);

3. In the whole series of patients, the clinical symptomatology and the neuroradiological examinations were unable to predict the histological diagnosis; the CT patterns were not unequivocal regarding the nature of the lesions (Tables 3 and 4).

In our opinion, these data provide firm grounds to advise against blind external radiotherapy both in young and in adult patients harbouring brain stem expanding lesions[15, 26], because the results of radiotherapy of low grade gliomas are poor both in the literature reports[25, 26] and in our own material[7].

In conclusion, the histological diagnosis and the assessment of the volume obtained by serial stereotactic biopsy may greatly contribute to the search for the "new effective methods needed for treatment of brain stem tumours" recently hoped for and forcast by the U.S.A. Children Cancer Study Group after an extensive review of the literature and after a careful analysis of a large casuistic of patients treated by conventional methods[15].

Acknowledgements

This study has been supported in part by the Consiglio Nazionale delle Ricerche, Rome, Italy: Grant Number: 86.00633.44 and by the Associazione "Paolo Zorzi" for Neurosciences.

References

1. Backlund EO, Holst H von (1978) Controlled subtotal evacuation of intracerebral haematomas by stereotactic technique. Surg Neurol 9: 99–101

2. Beatty RM, Zervas NT (1983) Stereotactic aspiration of a brain stem haematoma. Neurosurgery 13: 204–207

3. Benabid AL, Persat JC, Chirossel JP, Rougemont J, Barge N (1978) Délimitation des tumeurs cérébrales par stéréo-impedo-encéphalographie (SIEG). Neurochirurgie 24: 3–14

4. Betti O, Derechinski VE (1984) Hyperselective encephalic irradiations with linear accelerator. Acta Neurochir (Wien) [Suppl] 33: 385–390

5. Broggi G, Franzini A (1981) Value of serial stereotactic biopsies and impedance monitoring in the treatment of deep brain tumours. J Neurol Neurosurg Psychiatry 44: 397–401

6. Broggi G, Franzini A, Giorgi C (1987) Méthodologie et fiabilité prognostique de la biopsie stéréotaxique dans les tumeurs d'origine gliale. Rev EEG Neurophysiol Clin 17: 55–58

7. Broggi G, Franzini A, Giorgi C (1983) The value of stereotactic biopsy in management of brain stem lesion. Ital J Neurol Sci [Suppl] 2: 51–56

8. Broggi G, Franzini A, Peluchetti D, Servello D (1985) Treatment of deep brain abscesses by stereotactic implantation of an intracavitary device for evcuation and local application of antibiotics. Acta Neurochir (Wien) 76: 94–98

9. Broseta J, Gonzales-Darder J, Barcia-Salorio JL (1982) Stereotactic evacuation of intracerebral haematomas. Appl Neurophysiol 45: 443–448

10. Coffey RJ, Lundsford ID (1985) Stereotactic surgery for mass lesions of the midbrain and pons. J Neurosurgery 17: 12–18

11. Daumas-Duport O, Monsaingeon V, Szenthe L, Szikla G (1982) Serial stereotactic biopsies: a double histological code of gliomas according to malignancy and 3D configuration, as an aid to therapeutic decision and assessment of results. Appl Neurophysiol 45: 431–437

12. Franzini A, Broggi G, Allegranza A, Melcarne A, Ventura L, Costa A (1986) Cell kinetics of gliomas by serial stereotactic biopsy. Bas Appl Histochem 30: 203–207

13. Gleason CA, Wise BL, Feinstein B (1978) Stereotactic localization (with computerized tomographic scanning), biopsy and radiofrequency treatment of deep brain lesions. Neurosurgery 1: 217–222

14. Guillain G, Bertrand I, Gruner J (1945) Les gliomes infiltrés du tronc cérébral. Masson, Paris

15. Jenkin RDT, Boesel C, Ertel I, Evans A, Hittle R, Ortega J, Sposto R, Wara W, Wilson C, Anderson J, Leikin S, Hammond D (1987) Brain stem tumours in childhood: a prospective randomized trial of irradiation with and without adjuvant CCNU, VRC and prednisone. J Neurosurg 66: 227–233

16. Kandel IE, Peresedov VV (1985) Stereotactic evacuation of spontaneous intracerebral haematomas. J Neurosurg 62: 206–213

17. Kelly PJ, Bruce AK, Goerss BS, Earnest F (1986) Computer assisted stereotactic laser resection of intra-axial brain neoplasms. J Neurosurg 64: 427–439

18. Matsumoto K, Hondo H (1984) CT-guided stereotactic evacuation of hypertensive intracerebral haematomas. J Neurosurg 61: 440–448

19. Ostertag CB, Mennel HD, Kiessling M (1980) Stereotactic biopsy of brain tumours. Surg Neurol 14: 275–283

20. Riegel DH, Scarff TB, Woodford JE (1979) Biopsy of pediatric brain stem tumours. Childs Brain 5: 329–340

21. Rorke LB, Gilles FH, Davis RL, Beker LE (1985) Revision of the World Health Organization classification of brain tumours for children brain tumours. Cancer 56: 1869–1876

22. Russel DS, Rubinstein LJ (1977) Pathology of tumours of the nervous system, 4th ed. Williams and Wilkins, Baltimore, pp 146–282

23. Sedan R, Peragut JC, Vallincioni P (1975) Présentation d'un appareillage original pour biopsie cérébrale et tumorale en condition stéréotactiques. Communication à la Société de Neurochirurgie de Langue Française. Paris, Décembre

24. Szikla G, Betti O, Szenthe L, Schlienger M (1981) L'expérience actuelle des irradiations stéréotaxiques dans le traitement des gliomas hémisphériques. Neurochirurgie 27: 296–298

25. Villani R, Gaini SM, Tomei G (1975) Follow-up study of brain stem tumours in children. Childs Brain 1: 126–135

26. Whyte TR, Colby MYJ, Layton DD (1969) Radiation therapy of brain stem tumours. Radiology 693: 413–416

Address for correspondence: Prof. G. Broggi, M.D., Department of Neurosurgery, Istituto Neurologico "C. Besta", Via Celoria, 11, I-20133 Milano, Italy.

Acta Neurochirurgica, Suppl. 42, 177–181 (1988)
© by Springer-Verlag 1988

Stereotactic Biopsy and Treatment of Brain Stem Lesions: Combined Study of 33 Cases (Bologna – Marseille)

F. Frank[1], **A. P. Fabrizi**[1], **R. Frank-Ricci**[1], **G. Gaist**[1], **R. Sédan**[2], and **J. C. Peragut**[2]

[1] Division of Neurosurgery, Bellaria Hospital, Bologna, Italy. [2] Functional Neurosurgical Service, C.H.U. Hôpital de la Timone, Marseille, France

Summary

Intraaxial brain stem tumous are rarely treated by open surgical technique. In Bologna and Marseille, 23 and 10 stereotactic biopsies respectively were performed in patients with brain stem mass lesions. The mortality due to biopsy was 3% (1 pat.); while the morbidity was temporary in 6 cases (18%) and permanent in one patient (3%). The approach to the brain stem was via a frontal burr hole. 7 times, after biopsy and histological diagnosis, radioisotope implant of the neoplasm was performed with [125]I (iodine).

From the histological diagnoses of the lesions the following was found: only 40% of the young patients had highly malignant tumours; 83% of the adults had neoplasms (not all of the malignant type), while 17% of the verified lesions were non-neoplastic. Since a diagnosis of the lesion nature is impossible with current neuroradiological means, the authors noting a variety of masses found in their experience, emphasize the importance of stereotactic biopsy, as a scarcely invasive method to give a precise diagnosis and a possible treatment.

Keywords: Brain stem biopsies; anatomical definition of brain stem; histopathological diagnoses; radioisotope implantations.

Introduction

Stereotactic biopsies of intrinsic brain stem mass lesions are at high risk for the patient, as shown by Benabid et al.[1]. Besides few contrary opinions (Munari et al.)[9], brain stem biopsies should be performed when there exists an uncertain neuroradiological diagnosis and when the clinical conditions permit.

The aim of this study is to demonstrate: 1. The necessity of stereotactic brain stem biopsies to obtain a precise histopathological diagnosis of mass occupying lesions for a correct medical and/or surgical treatment. 2. The possibility, after biopsy, of direct stereotactic treatment (abscess or haematoma evcuation, radioisotope implantation). 3. The possibility to classify type of brain stem lesions according to age groups.

Materials and Method

Contrary to many authors[9, 10] that include the diencephalon in their studies of the brain stem, our group considers only the infratentorial components (mesencephalon, pons and medulla) as brain stem.

Thirty-three stereotactic biopsies were performed with two different procedures: the Reichert method in Bologna, and the Talairach method in Marseille. All the operations were performed under local anaesthesia and mild sedation, except for children who underwent general anaesthesia for the procedure. A frontal approach was used for all cases with the burr hole placed one centimeter in front of the coronal suture. Average tissue specimens obtained in Bologna and Marseille were 3 and 2 respectively. The Bologna group uses fine

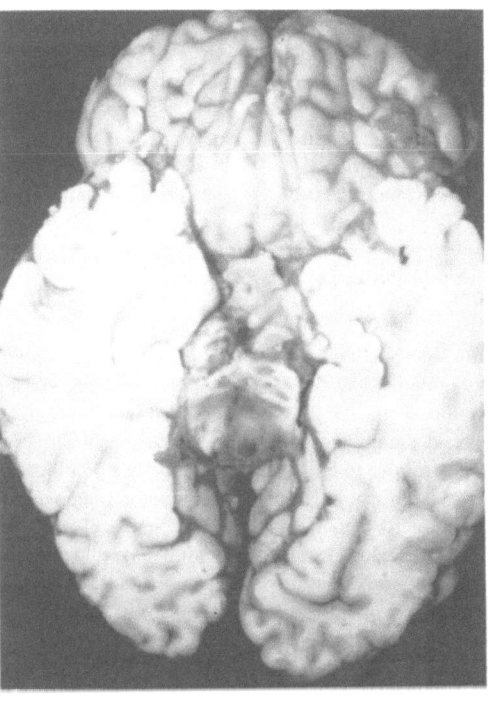

Fig. 1. 58-year-old male. Pontine malformation "telangectasia pontis" seen at autopsy

Table 1. *Histological Diagnoses of Brain Stem Masses in Young Adults and Adults (Over 15 years old): 23 Patients*

Malignant astrocytoma	11	Lymphoma	1
Low grade astrocytoma	3	Haemorrhagic infarction	1
Ependymoma	1	Arachnoidal cyst	1
Metastasis	2	"Telangectasia pontis"	1
Meningioma	1	Cavernous angioma	1
Total			23 patients

Table 2. *Histological Diagnoses of Brain Stem Masses in Children (less than 15 years old): 10 Patients*

Malignant astrocytoma	4
Low grade astrocytoma	4
"Ectopia vermis"	1
Ependymoma	1
Total	10 patients

Table 3. *Follow-Up (7 Patients) After ^{125}I Implantations (12 Months–6 Years)*

Case no.	Sex	Age	Histological diagnoses	Length of follow-up	Results
1	f	26	ependymoma	6 years	good
2	m	15	gr. II astro-cytoma	4 years	good
3	m	8	pilocytic as-trocytoma	2 years	poor
4	f	5	pilocytic as-trocytoma	12 months	poor*
5	f	29	gr. II astro-cytoma	3 years	good
6	m	12	gr. II astro-cytoma	3 years	fair
7	m	18	gr. II astro-cytoma	2 years	fair

* Patient died probably due to vasogenic oedema.

micro-alligator forceps (Fischer Inc., Freiburg im Breisgau, Federal Republic of Germany), and the Marseille group uses special aspi-rating/cutting forceps (Sédan forceps, Marseille, France). All tissue samples were taken along the major axis of the lesions.

One patient died due to biopsy (3%), harbouring a pontine mal-formation (telangectasia pontis) not visible on vertebral angiography (Fig. 1). The morbidity of our series presents: 5 cases of acute tri-ventricular hydrocephalus (15%), treated with ventriculo-systemic shunts; one patient with transitory hemiparesis; and one case of permanent ipsilateral cranial nerve palsies, where the cannula pen-etrated the brain stem.

Tables 1 and 2 present the patients treated in Bologna and in Marseille. The two tables are divided into age groups (over and under 15 years old) and histopathological diagnoses obtained after serial biopsies.

Table 3 presents a 1 to 6 year follow-up of 7 patients that under-went ^{125}I radioisotope implantations.

The biopsies performed with the Reichert stereotactic system receive direct data from the computerized tomography (CT) scanner [(and in the last 3 patients CT and RMI (magnetic resonance im-aging)]; while the Talairach method requires adjunct opaque ven-triculography. All the patients underwent vertebral angiography either prior or during intervention for correct vascular evaluation of the mass occupying lesions, and to rule out vascular malformations (artero-venous malformations (AVM) and aneurysms).

Discussion

The usefulness of stereotactic cerebral biopsies has been well documented in literature[1, 2, 3, 9, 10]. However, little has been published employing stereotaxy in brain stem masses[9, 10]. "True" brain stem biopsies are not free of operative risk, and the outcome of the method should be a diagnosis, with the maximum accuracy and the minimum risk for the patients' well being. However, brain stem biopsies are mandatory because our series shows 15% of the patients did not harbour neoplasms, but other lesions (haematoma, abscess, granuloma, cavernous angioma, etc.) (Fig. 2). In these cases "blind" radiotherapy would have been inappropriate and del-eterious. Furthermore, current neuroradiological methods (including MRI) can give neither neoplastic type nor grade of malignancy; therefore, without bi-opsy the therapeutic protocol of the neoplasm may be altered. Fig. 3 shows a CT image of a brain stem mass lesion with negative angiography, the biopsy proved the lesion to be a meningioma originating from the choroid plexus of the fourth ventricle, and successfully excised.

The grading of the neoplasms is an important index for the radiotherapeutic protocol. Low malignancy neoplasms are treated by fractionated radiation doses given in various cycles; while high malignancy tumours receive a smaller dose in only one cycle of treatment with very short time lapse between sittings. Our series also shows 2 cases of secondary metastatic brain stem tumours in absence of the primary lesions. After the biopsies a complete screening located the primary tu-mour (2 cases of lung cancer).

We agree with many authors[5, 7, 8, 12, 13] that extraaxial brain stem lesions should be approached by open sur-gical techniques; however, we do not agree with others[4, 6, 11] on the surgical exploration of intrinsic brain stem masses, presenting elevated mortality and morbidity. The morbidity in stereotactic biopsies should not be overlooked. However, major surgical procedures may be avoided, if the lesion is non-tumoural (haematoma or abscess). Also, low dose radioisotopes may be placed in certain small low grade malignancy neoplasms.

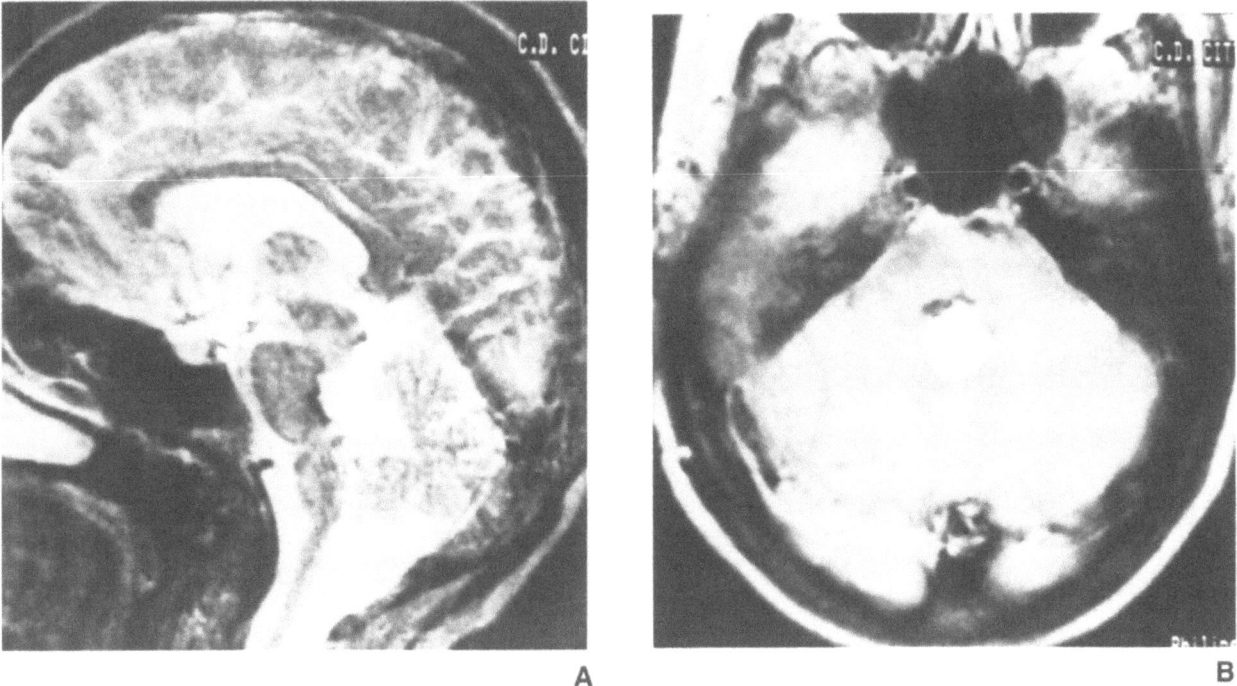

Fig. 2. 54-year-old female. Spontaneous haematoma of the pons seen on RMI. (A) Lateral view. (B) Sagittal view

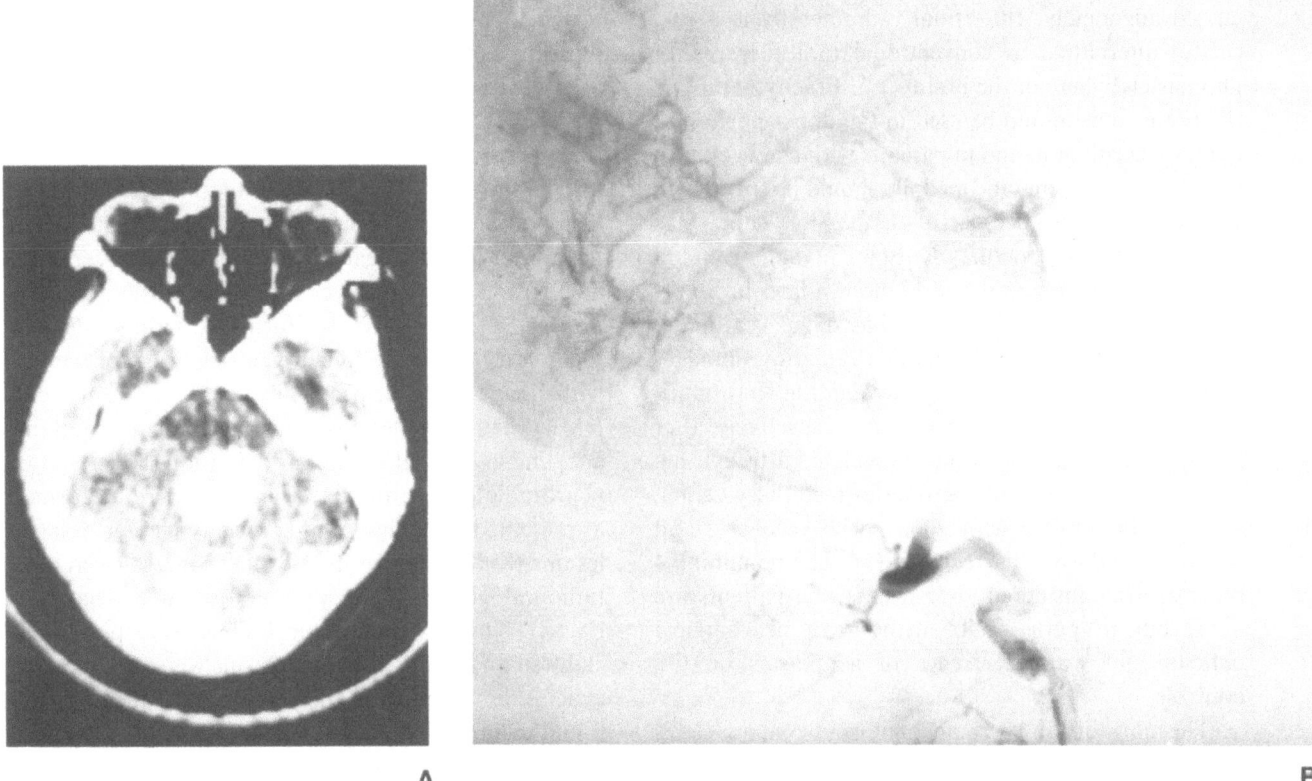

Fig. 3. 42-year-old female. Meningioma of the pons after biopsy. (A) CT image. (B) Lateral view of vertebral angiography

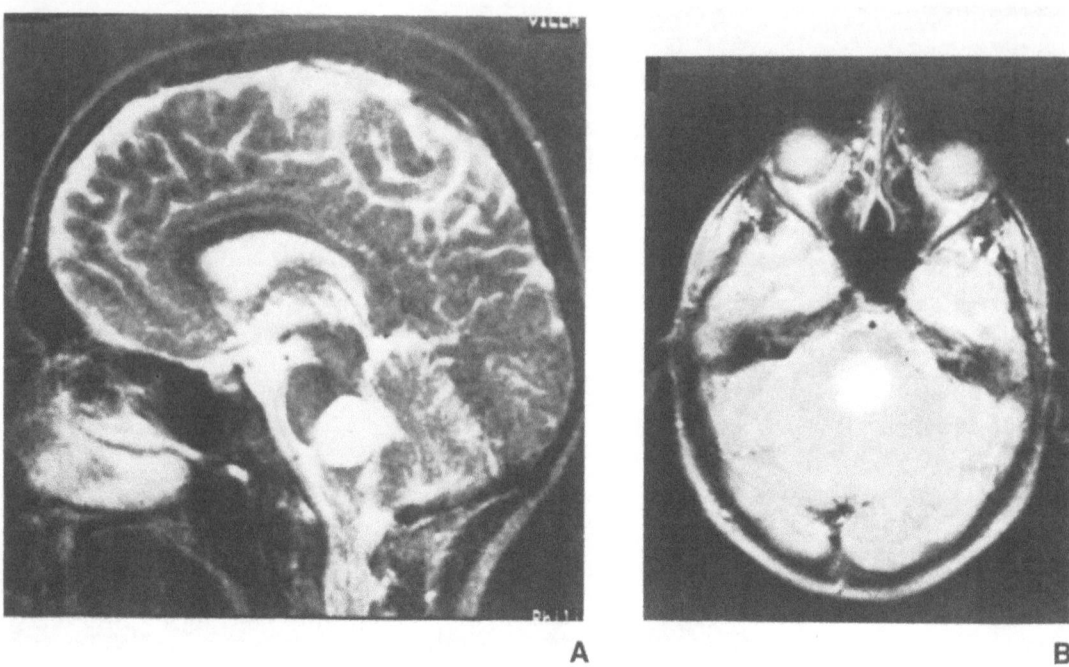

Fig. 4. 18-year-old male. RMI of a fibrillary astrocytoma of the pons. (A) Lateral view. (B) Sagittal view

Our experience with brachytherapy in the treatment of low grade brain stem gliomas is limited (7 patients were treated, Table 3), and therefore can not be discussed adequetely. Interstitial radiotherapy is a successful alternative to conventional radiotherapy for gliomas elsewhere in the brain[2, 3, 10]. Brachytherapy in the brain stem should be used in the upper portion of the mesencephalon, and in patients with stable clinical conditions. The pons and medulla should not be treated with radioisotopes because of the possible damage to highly functional structures. One patient underwent ^{125}I implant in the pons, and died 12 months later probably due to vasogenic oedema. [125]I has a 2.5 centimeter radius of activity, and the gamma radiation damages and destroys large amount of surrounding vital tissue. Two other patients, 4 and 6 years respectively after implantation, harbouring a pedunculated astrocytoma and an ependymoma of the anterior medullary velum, are in good health, with good social recovery, and without neoplasm recurrence on CT. The remaining 4 patients are considered poor results: the patients are alive, but in poor health with grave neurological deficits, and with recurrence of neoplasms on CT controls.

The mortality and permanent morbidity in our series of brain stem biopsies is relatively low (6%). The transitory morbidity of acute obstructive hydrocephalus (5 patients — 15%) was not due to haemorrhagic infarction of the brain stem as seen on CT images. The acute syndrome, appearing a few hours after biopsy, is probably due to the impaction of small blood clots in the foramen of Magendie. The clinical history of these patients show rapidly evolving coma, totally reversible after ventriculo-systemic shunts. Permanent morbidity persisted in one patient (3%) with ipsilateral third and fourth cranial nerve palsies.

The operative risks accepted by the patients led to histopathological classification of lesions that could have been hypothetical with CT and RMI. Vertebral angiography did not give further diagnostic information in our series. CT also furnished indirect data of the lesions, when the fourth ventricle was not seen or with images of nonspecifically enlarged brain stem. RMI, when possible to obtain, gave precise lesion images and axact spatial tumour definition (Fig. 4). After neuroradiological studies 33 biopsies were performed as presented in Tables 1 and 2. Twenty-eight neoplastic lesions (85%) were verified; while 5 masses were non tumoural (inflammatory, malformative, haemorrhagic). The non neoplastic lesions were present in children and adults. The neoplasms were of the astrocytic series except for 2 metastases and 1 meningioma found in adults. The young patients harbored low malignancy glial neoplasms in 50% of the cases; while few of these neoplasms were found in adults, who had a prevalence of highly malignant lesions.

Our results are important for the following reasons: 1. It is not logical to label and condemn a lesion as tumour without a biopsy. 2. In case of neoplasm, the biopsy permits a more specific treatment. 3. Low malignancy neoplasms, especially in young patients (50% of our series) may be treated by radiosurgery.

References

1. Benabid AL (1985) Stereotaxic brain biopsy. In: the meeting of the French Study Group on stereotaxic brain biopsy, held in Marseille, France in February 1985. (Personal communication)
2. Frank F, Fabrizi AP, Frank-Ricci R, Sturiale C, Nuzzo G (1985) Treatment of low malignancy brain neoplasms by means of stereotactic interstitial radiotherapy. Appl Neurophysiol 48: 121–126
3. Frank F, Fabrizi AP, Frank-Ricci R, Gaist G (1985) The treatment of low-grade malignant brain tumours with stereotactic brachytherapy. In: Gerosa MA (ed) Brain tumours biopathology and therapy. Pergamon Press, Oxford New York, pp 185–193
4. Hendrick EB, Hoffman HJ, Humphreys RF (1975) Treatment of infratentorial gliomas in childhood. Recent results. Cancer Res 51: 102–106
5. Hoffman HJ, Becker L, Craven MA (1980) A clinically and pathologically distinct group of benign brain stem gliomas. Neurosurgery 7: 243–248
6. Lassiter KRL, Alexander E Jr, Davis CH Jr (1971) Surgical treatment of brain stem gliomas. J Neurosurg 34: 719–725
7. Littman P, Jarret P, Bilamink LT (1980) Paediatric brain stem gliomas. Cancer 45: 2787–2792
8. Milhorat TH (1975) Pontine glioma. JAMA 232: 595–596
9. Munari C (1987) Stereotactic approach to occupying space lesions in the posterior fossa. In: The meeting of the American society for stereotactic and functional neurosurgery held in Montreal, Canada, in June 3–6. (Personal communication)
10. Ostertag CB (1983) Biopsy and interstitial radiation therapy of cerebral gliomas. Ital J Neurol Sci [Suppl] 2: 121–128
11. Reigel DH, Scarff TB, Woodford JE (1979) Biopsy of pediatric brain stem tumours. Childs Brain 5: 329–340
12. Strange P, Wohlert L (1982) Primary brain stem tumours. Acta Neurochir (Wien) 62: 219–232
13. Tomita T, McLone DG, Naidich TP (1984) Brain stem gliomas in childhood – Rational approach and treatment. J Neurooncol 2: 117–122

Address for correspondence: Dr. F. Frank, Department of Neurosurgery, Bellaria Hospital, Via Altura 3, Bologna, I-40139, Italy.

Acta Neurochirurgica, Suppl. 42, 182–186 (1988)
© by Springer-Verlag 1988

Brain Stem Expansive Lesions:
Stereotactic Biopsy for a Better Therapeutic Approach

F. Giunta, G. Marini, G. Grasso, and **F. Zorzi**[1]

Department of Neurosurgery, University of Brescia, and [1] Service of Pathology, Spedali Civili, Brescia, Italia

Summary

CT diagnosed brain stem malignant lesions were in the past almost always treated with radiation therapy (RT). Eventually this turned out to be a grave mistake. With stereotactic serial biopsies of all brain stem expanding lesions we have been able to verify the histological nature in all cases but two and to prevent a blind therapeutic approach. In 24 patients bearing CT diagnosed expansive lesions into the brain stem 68 samples were taken during 24 stereotactic procedures. In 8 patients surgical removal of the expanding lesion was attempted after stereotactic biopsy.

Introduction

Brain stem expanding lesions were treated in the past in almost every cases with Radiation Therapy (RT) after neuroradiological studies had made the diagnosis of malignant neoplasm[21, 23, 33]; this was because of the technical impossibility of a surgical approach[40]. Eventually this turned out to be a grave mistake. Ventriculography and vertebro-basilar angiography showed only the exophytic intra fourth ventricle mass together with an "enlarged" brain stem, but today Computer Tomography (CT) and Magnetic Resonance (MR) imaging reveal the precise localization of an even a small brain stem lesion and can differentiate the normal brain from the suspected pathological tissue. This remarkable feature has allowed the development of the surgical approach either to remove pathological tissue or simply to obtain surgical samples for histological study[1, 16, 24, 31, 32, 38].

With stereotactic serial biopsies into all brain stem expanding lesions every neurosurgeon could be able to verify the histological nature of the lesion and to prevent a blind therapeutic approach. We believe that the stereotactic procedure should be familiar to every neurosurgeon because 1. of its exceptional ability to give an histological diagnosis without major risk, 2. it stim-

ulates further surgery in benign lesions and 3. it is eventually curative by itself.

Materials and Methods

24 patients bearing brain stem expansive lesions demonstrated by neuroradiological studies (CT, MR, Cerebral Angiography) underwent a stereotactic serial biopsy procedure at the Neurosurgical Clinic of University of Brescia, from May 1983 to June 1987, to verify the exact nature of the disease in order to choose the best therapeutic approach. During the same period a total of 217 stereotactic procedures in 213 patients were performed.

68 biopsy samples of 24 brain stem expansive lesions were taken during 24 stereotactic procedures on a number of 32 calculated targets on CT scans. 10 patients were male and 14 female. The patients' ages varied from 2 to 64 years, 12 of them were children.

Surgical stereotactic procedures were carried out under local anaesthesia in all adults but not in children. A mild sedation with systemic analgesia was sometimes necessary.

A trans-cerebellar approach was used in 17 lower brain stem, cerebellar, and pineal region lesions. All patients were treated in the sitting position[17, 28].

A trans-frontal approach, in the supine position, was performed in 7 patients with mesencephalic lesions.

A Backlund Biopsy Kit with the Leksell and G Stereotactic Systems[18] was used.

Results

History, CT findings, stereotactic approach, pathological data, post-biopsy treatment and follow-up of all patients are shown in Table 1.

Pain was experienced by two of the patients with pineal region expansive lesion (pineocytoma and germinoma) during sampling with the spiral. The pain was referred to as a deep cephalalgia every time the spiral was inserted into the neoplastic mass and when it was withdrawn. One patient (case 15) suffered intense vertigo for some minutes and vomited during the withdrawal of the spiral with the biopsy sample.

Table 1. *Stereotactic Approach to Brain Stem Expanding Lesions*

Case Sex/Age	Main Symptom Karnofski	CT Findings Ventriculi	Date STE = samples Approach	Pathol. Findings	Post-biopsy Treatm.	Survival Follow-up Status
1. M.L. F/7	May '79 Parinaud K = 80	hypodense mesencephalon hydrocephalus	7/20/83 biopsy = 3 frontal	low grade astrocytoma	shunt 48 Gy	43 mo recurr. alive
2. M.G. M/2	Gen. '84 vomit K = 60	hypodense L cerebellum hydrocephalus	6/20/84 biopsy = 4 cerebellar	non (?) neoplastic tissue	shunt	15 mo recurr. died
3. T.R. M/35	April '79 tetraparesis K = 40	hypodense R pons & cerebellum	6/21/84 biopsys = 5 cerebellar	anaplastic astrocytoma	—	11 mo progress. died
4. A.L. F/60	May '84 Parinaud K = 60	hyperdense pineal reg. hydrocephalus	7/19/84 biopsy = 3 frontal	papillary ependymoma	shunt removal RT 54 Gy	35 mo well alive
5. G.D. F/12	July '84 tetraparesis K = 40	hyperdense R pons & oblongata	8/28/84 aspirat. cerebellar	haemorraghic clot	—	34 mo well alive
6. B.C. M/4	Sept. '82 strabysmus K = 60	irregular pons-bulbar hydrocephalus	9/11/84 biopsy = 4 cerebellar	anaplastic ependymoma	shunt removal	1 mo progress. died
7 S.L. F/3	Aug. '84 R hemiparesis K = 50	hypodense L pons & oblongata	11/20/84 biopsy = 2 cerebellar	pilocytic astrocytoma	—	5 mo progress. died
8. P.A. F/42	Sept. '83 papilledema K = 60	hyperdense pineal reg. hydrocephalus	12/19/84 biopsy = 2 cerebellar	pineocytoma	shunt removal	30 mo well alive
9 P.A. F/10	Oct. '84 L hemiparesis K = 70	hyperdense R mesencephalon	4/5/85 biopsy = 6 frontal	pilocytic astrocytoma	stereot. removal	26 mo well alive
10. C.F. M/22	Apr. '85 papilledema K = 50	hyperdense pineal reg. hydrocephalus	5/21/85 biopsy = 2 cerebellar	germinoma	shunt RT 58 Gy	25 mo well alive
11. M.D. F/7	May '85 L VI & VII K = 80	enhanced pons hydrocephalus	6/06/85 biopsy = 3 cerebellar	anaplastic astrocytoma		8 mo well alive
12. L.A. M/19	Nov. '84 papilledema K = 80	hyperdense pineal reg. hydrocephalus	8/26/85 biopsy = 2 frontal	anaplastic ependymoma	shunt RT 60 Gy	22 mo well alive
13. R.E. F/64	July '85 gait K = 80	hypodense pons-bulbar hydrocephalus	10/16/85 biopsy = 3 cerebellar	non (?) neoplast. tissue	shunt	20 mo progress. alive
14. V.P. F/59	Aug. '85 L hemiparesis K = 60	enhanced R mesencephalon	10/29/85 biopsy = 3 frontal	protoplasm. astrocytoma	stereot. removal	9 mo pneumonia died
15. F.M. F/25	June '85 R hemiparesis K = 70	hyperdense R pons	12/03/85 biopsy = 2 cerebellar	pilocytic astrocytoma	shunt RT 48 Gy Ir 192	18 mo progress. alive
16. G.E. F/10	Gen. '86 papilledema K = 90	hyperdense R mesenceph. hydrocephalus	7/15/86 biopsy = 2 frontal	pilocytic astrocytoma		11 mo well alive

Table 1 (continued)

Case Sex/Age	Main Symptom Karnofski	CT Findings Ventriculi	Date STE = samples Approach	Pathol. Findings	Post-biopsy Treatm.	Survival Follow-up Status
17. V.S. M/7	May '86 R dystonia K = 40	cystic pons & mesencephalon	9/16/86 aspirat. cerebellar	pilocytic astrocytoma	suboccip. removal mural n.	9 mo unchanged alive
18. C.A. M/5	Sett. '86 L hemiparesis K = 60	enhanced L pons & oblongata	10/14/86 biopsy = 4 cerebellar	fibrillary astrocytoma	RT 60 Gy	8 mo recurr. alive
19. B.I. M/36	Oct. '81 R hemiparesis K = 50	hypodense L pons & cerebell. ped.	12/16/86 biopsy = 3 cerebellar	low grade astrocytoma	—	6 mo unchanged alive
20. B.E. F/52	Nov. '85 tetraparesis K = 40	hyperdense clival hydrocephalus	4/16/87 biopsy = 1 cerebellar	psammomatous meningioma	suboccip. removal	1 mo unchanged alive
21. D.S. M/3	Gen. '87 gait K = 90	hypodense pons & cerebellum	4/28/87 biopsy = 3 cerebellar	fibrillary astrocytoma	—	1 mo well alive
22. M.L. F/17	Feb. '87 gait K = 70	hypodense R pons & mesencephalon	5/26/87 biopsy = 2 cerebellar	ependymal cyst	suboccip. removal	1 mo well alive
23. P.Q. M/50	Gen. '87 L hemiparesis K = 50	hyperdense R mesenceph.	6/10/87 biopsy = 3 frontal	anaplastic astrocytoma	RT	unchanged alive
24. G.L. F/12	Feb. '87 gait K = 60	hypodense L pons & oblongata	6/16/87 biopsy = 5 cerebellar	pilocytic astrocytoma		unchanged alive

RT = radiotherapy, L = left, R = right, mo = months, STE = stereotaxia, progress. = disease progression, well = normal life.

We have had no mortality; 3 patients (case 5, 17, and 22) showed an improved neurological status because of the reduction of the expanding mass; 21 remained unchanged. No morbidity was observed.

Histological diagnosis was reached in all patients but two (8%); in these the tumour could not be correctly identified (case 2 and 13). Fifteen months later the first of these developed a rapid growth of tumour and a decompressive suboccipital craniectomy revealed an anaplastic astrocytoma. The second is still alive but with progression of the disease; the clinical evidence suggests a slow growing astrocytoma.

Benign expansive lesions were recognised in 16 patients: 1 haemorrhagic clot, 1 papillary ependymona, 1 pineocytoma, 1 psammomatous meningioma, 1 ependymal cyst and 11 slow growing astrocytoma. Malignant tumours were diagnosed in 6 patients: 1 germinoma, 2 anaplastic ependymoma and 3 anaplastic astrocytoma. All of them were referred for radiotherapy. Radiotherapy in slow growing astrocytoma is still con-

troversial, but it was performed in 3 out of our 11 cases. Malignant evolution was observed in two patients (case 1 and 18) and progression of the illness was noted in one patient (case 15). The follow-up of the remaining patients is too short to reach any conclusion.

Obstructive hydrocephalus was observed in 11 patients and a ventriculo-peritoneal shunt was implanted.

Stereotactic findings stimulated a major surgical approach in 8 patients where a craniotomy or craniectomy with microsurgical removal was attempted. In 6 cases a suboccipital craniectomy was performed: three patients (case 4, 8, and 22) improved, two (case 17 and 20) remained unchanged and one (case 6) died one month after major surgery. In two cases (9 and 14) a transventricular approach was attempted and with the aid of the stereotactic needle the pilocytic astrocytoma into the thalamo-mesencephalon was reached and removed. But only one of these two patients (case 9) recovered.

Discussion

Midbrain, pons, oblongata and cerebellar peduncles are generally considered "off limits" for conventional surgical approach. However, many cases have now been reported of intra-axial expanding lesions successfully operated: haemorrhagic clot[4, 7, 15, 16, 25, 26; 27, 30, 32, 36], inflammatory lesions[14, 39], lipoma[22] and low grade astrocytomas cystic or not[1, 2, 3]. A small biopsy was the only result in many cases, but it was very useful in guiding further treatment.

It was stated that: "theoretically only the brain stem is surgically inaccessible"[10] and "le plan du foramen ovale de Pacchioni semble actuellement la frontière à ne pas depasser"[8]. The question now is whether all stereotactic systems are suitable for a posterior fossa approach in view of their design oriented only to thalamic surgery.

Stereotactic biopsy samples are considered when an histological diagnosis is necessary to plan the best therapeutic strategy. Sampling pathological tissues is without risk if cerebral vessels are not damaged and normal brain structures are respected.

A trans-frontal approach[12] is widely used by the stereotactic neurosurgeon in mesencephalic expansive lesions and in these cases we also followed this approach.

A trans-tentorial approach with entry point in the parieto-occipital region was originally used for cerebellar stimulation. This route appears suitable for biopsy[3].

A trans-cerebellar direct approach[12, 17, 28] seems the unique way to approach expansive lesions of the pons and medulla. It is also particularly suitable for cerebellar peduncles or deep nuclei.

The stereotactic serial biopsy procedure is the safe way to discover the nature of an expansive lesion, and sometimes it stimulates open microsurgical approach to remove benign lesions. Albright *et al.*[2] reported that malignant brain stem tumours can show a significantly longer survival after total or partial removal in comparison to biopsy only.

We recommend that benign expansive lesions be removed if they are well delimited. Stereotactic approach along a needle positioned into the centre of the tumour can help the procedure.

References

1. Albright AL, Guthkelch AN, Packer RJ, Price RA, Rourke LB (1986) Prognostic factor in pediatric brain stem gliomas. J Neurosurg 75: 751–755
2. Albright AL, Sclabassi RJ (1985) Use of the Cavitron ultrasonic surgical aspirator and evoked potentials for the treatment of thalamic and brain stem tumours in children. Neurosurgery 17: 564–568
3. Apuzzo MJ, Sabshin JK (1983) Computed tomographic guidance stereotaxis in the management of intracranial mass lesions. Neurosurgery 12: 277–285
4. Arseni C, Stanciu M (1973) Primary haematomas of the brain stem. Acta Neurochir (Wien) 28: 323–330
5. Baghai P, Vries JK, Bechtel PC (1982) Retromastoid approach for biopsy of brain stem tumours. Neurosurgery 10: 574–579
6. Beatty RM, Zervas NT (1983) Stereotactic aspiration of a brain stem haematoma. Neurosurgery 13: 204–207
7. Becker DH, Silverberg GD (1978) Successful evacuation of an acute pontine haematoma. Surg Neurol 10: 263–265
8. Benabid A, Blond S, Chazal J *et al* (1985) Les biopsies stéréotaxiques (BS) des néoformations intra-crâniennes. Réflexions à propos de 3052 cas. Neurochirurgie 31: 295–301
9. Berger MS, Edwards MSB, LaMasters D *et al* (1983) Pediatric brain stem tumours: radiographic, pathological and clinical correlations. Neurosurgery 12: 298–302
10. Bosch DA (1980) Indications for stereotactic biopsy in brain tumours. Acta Neurochir (Wien) 54: 167–179
11. Chou SN, Erickson DL, Ortiz-Suarez HJ (1975) Surgical treatment of vascular lesion in the brain stem. J Neurosurg 42: 23–31
12. Coffey RJ, Lunsford LD (1985) Stereotactic surgery for mass lesions of the midbrain and pons. Neurosurgery 17: 12–18
13. Danzinger J, Allen KL, Bloch S (1974) Brain stem abscess in childhood. Case report. J Neurosurg 40: 391–393
14. Dierssen G, Trigueros F, Sanz F, Coca JM, Orozco M (1978) Surgical treatment of a mesencephalic tuberculoma: Case report. J Neurosurg 49: 753–755
15. Doczi T, Thomas DGT (1979) Successful removal of an intra-pontine haematoma. J Neurol Neurosurg Psychiatry 42: 1058–1061
16. Durward QJ, Barnett HJM, Barr HWK (1982) Presentation and management of mesencephalic haematoma: report of two cases. J Neurosurg 56: 123–127
17. Fulton DS, Levin VA, Wara WM *et al* (1981) Chemotherapy of pediatric brain stem tumours. J Neurosurg 54: 721–725
18. Giunta F, Marini G, Cani A (1986) Stereotactic brain stem biopsies by trans-cerebellar approach. 7th Meeting European Society for Stereotactic and Functional Neurosurgery, Birmingham
19. Giunta F, Marini G, Benussi G (1986) Stereotactic head-holder for close and open surgery. 7th Meeting European Society for Steretotactic and Function Neurosurgery, Birmingham
20. Golden GS, Ghatak NR, Hirano A *et al* (1972) Malignant gliomas of the brain stem. A clinicopathological analysis of 13 cases. J Neurol Neurosurg Psychiatry 35: 243–248
21. Hacker RJ, Fox JL (1980) Surgical treatment of brain stem carcinoma: case report. Neurosurgery 6: 430–432
22. Halperin EC (1985) Pediatric brain stem tumours: Patterns of treatment failure and their implications for radiotherapy. Int J Radiat Oncol Biol Phys 11: 1293–1298
23. Halmagyi GM, Evans WA (1978) Lipoma of the quadrigeminal plate causing progressive obstructive hydrocephalus. J Neurosurg 49: 453–456
24. Hitchon PW, Abu-Yousef MM, Graf CJ, Turner DM, Van Gilder JC (1983) Management and outcome of pineal region tumours. Neurosurgery 13: 248–253

25. Hoffman HJ, Becker L, Craven MA (1980) A clinically and pathologically distinct group of benign brain stem gliomas. Neurosurgery 7: 243–248

26. Humphreys RP (1978) Computerized tomographic definition of mesencephalic haematoma with evacuation through pedunculotomy. Case report. J Neurosurg 49: 749–752

27. Koos WT, Sunder-Plassman M, Salah S (1969) Successful removal of a large intrapontine haematoma: case report. J Neurosurg 31: 690–694

28. LaTorre E, Delitala A, Sorano V (1978) Haematoma of the quadrigeminal plate. Case report. J Neurosurg 49: 610–613

29. Mathisen JR, Giunta F, Marin G, Backlund E-O (1987) Transcerebellar biopsy in the posterior fossa: 12 years experience. Surg Neurol 27: 297–299

30. Murphy MG (1972) Succesful evacuation of acute pontine haematoma: case report. J Neurosurg 37: 224–225

31. Obrador S, Dierssen G, Odoriz BJ (1970) Surgical evacuation of a pontine-medullary haematoma: case report. J Neurosurg 33: 82–84

32. O'Laoire SA, Crockard HA, Thomas DGT, Gordon DS (1982) Brain stem haematoma. J Neurosurg 56: 222–227

33. Pak H, Patel SC, Malik GM *et al* (1981) Successful evacuation of a pontine haematoma secondary to rupture of a venous angioma. Surg Neurol 15: 164–167

34. Ravetto F, DiCagno L, Bosco M, Madon E, Gajno TM (1979) I tumori infiltranti del tronco encefalico in eta'evolutiva. Min Ped 31: 95–110

35. Russell B, Rengachary SS, McGregor D (1986) Primary pontine haematoma presenting as a cerebellopontine angle mass. Neurosurgery 19: 129–133

36. Robert CM Jr, Stern WE, Brown WJ, Greenfield MA, Bentson JR (1975) Brain stem abscess treated surgically. Surg Neurol 3: 153–160

37. Scott BB, Seeger JF, Schneider RC (1973) Successful evacuation of a pontine haematoma secondary to rupture of a pathologically diagnosed "cryptic" vascular malformation: case report. J Neurosurg 39: 104–108

38. Soffer D, Sahar A (1982) Cystic glioma of the brain stem with prolonged survival. Neurosurgery 10: 499–502

39. Stroink AR, Hoffman HJ, Hendrick EB, Humphreys RB (1986) Diagnosis and management of pediatric brain stem gliomas. J Neurosurg 65: 745–750

40. Van Gilder JC, Allen WE, Lesser RA (1974) A pontine abscess: survival following surgical drainage: case report. J Neurosurg 40: 386–390

41. Villani R, Gaini SM, Tomei G (1975) Follow-up of brain stem tumours in children. Childs Brain 1: 126–135

Address for correspondence: Filippo Giunta, M.D., Neurochirurgia, Spedali Civili, I-25100 Brescia, Italy.

Acta Neurochirurgica, Suppl. 42, 187–192 (1988)
© by Springer-Verlag 1988

Experience of Boron-neutron Capture Therapy for Malignant Brain Tumours – with Special Reference to the Problems of Postoperative CT Follow-ups

H. Hatanaka

Teikyo University Hospital Tokyo, Japan

Summary

Boron-neutron capture therapy (BNCT) is theoretically a highly selective treatment of infiltrating tumours, in that the tumoricidal heavy particle radiation is limited to a sphere of 10 microns around a tumour cell which is loaded with non-radioactive boron-10 atoms.

There were 73 gliomas among the 83 cases treated by boron-neutron capture therapy. For grade III–IV cerebral gliomas, 5 and 10 year survival rates were an unimpressive 19 and 10% respectively. This was the result of technical problems such as unsatisfactory reactors and inadequate craniotomies for the majority of the patients. If the analysis was limited to those whose tumours had been irradiated with more than 2.5×10^{12} neutrons/cm^2 (yielding more than 3,000 rem or more), the 5 and 10 year survival were almost 100 and 50%. The longest surviving glioblastoma (grade IV) patient has lived in a satisfactory manner for the past 15 years. For the cases who had been treated with borderline doses (lethal or sublethal), interpretation of the postoperative CTs was frequently intriguing. Several cases had to undergo re-opening and occasionally even another BNCT, only to find no viable tumour tissue. Death occurred in some, either due to discontinuation of supportive treatments by local physicians, or due to excessive therapies by the author directly involved in the patient's care, both of whom had erroneously believed in recurrence. At autopsy, residual tumour cells were recognized only in the areas where the above-mentioned neutron fluence had not been delivered at the time of the treatment. Delayed cerebral necrosis was not recognized except for the cases which had purposely been given a larger neutron dose than the current regimen.

If a larger reactor like those at MIT or Brookhaven of 1951–61 series is available, and if epithermal neutron facility is constructed, any deep-seated tumours will be good targets of BNCT. Application of larger diameter craniotomy and heavy water to replace the water in the cranial space will improve the thermal neutron penetration into deeper part of the brain, and will guarantee a better survival rate.

Keywords: Glioma; cerebral tumour; Boron-neutron capture therapy; malignant brain tumour.

Introduction

Boron-neutron capture therapy (BNCT) is theoretically a highly selective and effective treatment for in-filtrating tumours, in that its tumoricidal heavy particle radiation is limited to a sphere of 10 microns around a tumour cell which is loaded with non-radioactive boron-10 atoms[2, 7, 8, 9, 12]. In the nineteen years since 1968, eighty-three patients with malignant brain tumours have been treated in Tokyo by the author[3, 4, 5, 6, 7, 8]. There were 73 gliomas among the 83. As detailed in *Boron-Neutron Capture Therapy for Tumours* (edited by Hatanaka, 74 contributors, 463 pages, Nishimura & Co., Niigata City, 1986), 5 and 10 year survival rates of patients with grade III–IV cerebral gliomas were an unimpressive 19% and 10%, respectively (38 patients)[5]. This was the result of two technical problems[5, 10]: 1. unsatisfactory reactors which had not been designed for medical purposes and could not yield more deeply penetrating faster neutrons (epithermal neutrons), and 2. inadequate craniotomies which were too small to permit the delivery of neutrons over a broad enough area to cover the entire original tumour bed in the brain. Craniotomies can be improved if the initial surgeon is aware of the need. Reactors can also be improved, if they are at least ten times larger than the 100 kw reactors which the authors have used in Tokyo[13] and if they are in the neighborhood of hospitals. Physicists are already aware of the design necessary for such a reactor[1, 11, 14, 15]. What is needed is only a decision on the part of the governments constructing such reactors[2, 14].

If the analysis of the clinical results is limited to those whose tumours were irradiated with neutron fluence of more than 2.5×10^{12} neutrons/cm^2 [10] (yielding more than 3,000 rem equivalent dose by a single radiation inside the tumour cells but far less in the normal brain matter), the 5 and 10 year survival rates are almost 100 and 50%[5]. In fact all 3 patients who have

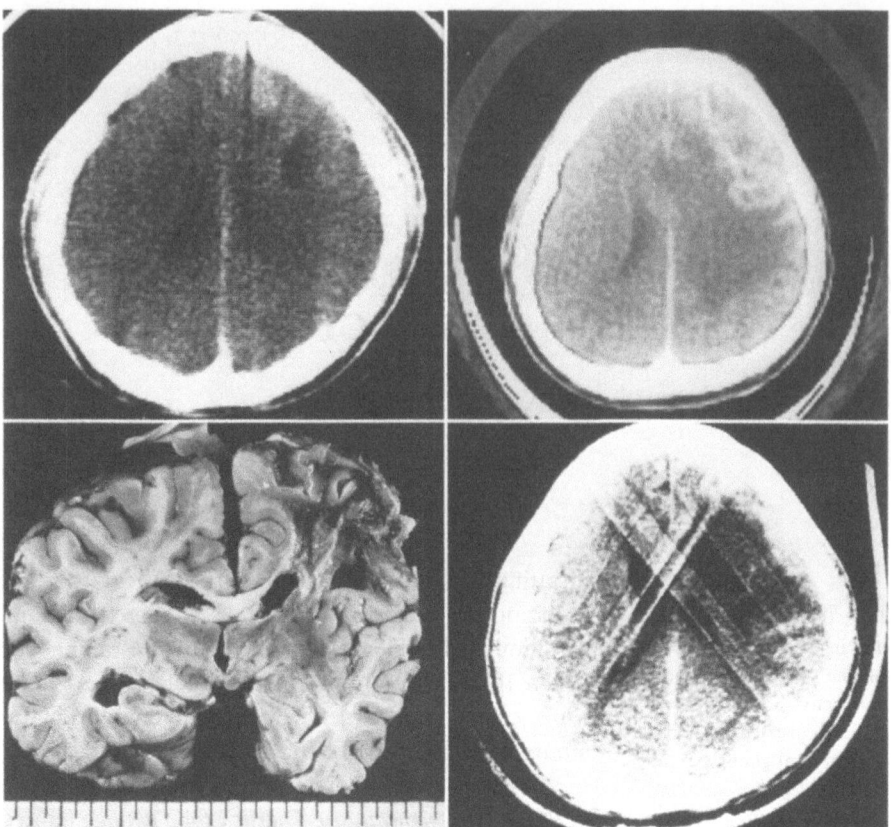

Fig. 1. A grade III astrocytoma of a 37-year-old male. Clockwise from upper left. (i) Soon after the first craniotomy and partial excision of the grade III astrocytoma. (ii) Eighteen months after BNCT. Epileptic seizures were rather frequent. Both contrast-enhancement and the volume of the right frontal lobe increased, and a low density area expanded. These changes were limited to the tumour side hemisphere, although both frontal lobes had been irradiated. Hence it can be suggested that these changes do not indicate any radiation damage to the normal brain matter. Since the radiation dose had been considered tumouricidal, no therapeutic measure was taken. (iii) Forty-two months after BNCT. A low density area was recognized in place of the previous contrast-enhancing lesion. The swelling of the right frontal lobe disappeared. Ventricles are symmetric. Seizures became infrequent. (iv) After 5 years the patient started deteriorating and died. Autopsy revealed microscopic subarachnoid dissemination of tumour cells. In the primary tumour site no tumour cell was recognized, and multiple porencephalic cavities were noted

survived more than 10 years had tumours (grade IV glioblastoma, grade III astrocytoma, and chondrosarcoma) within 6 cm from the surface of the brain, and in locations convenient for direct exposure to incoming neutrons[5] (frontal or parietal in a narrow irradiation chamber). Also important was the size of their craniotomies which were all considerably larger than the diameter of their tumours[5].

The present paper is intended to emphasize the need for cautious evaluation of clinical and radiological follow-up findings when tumours have been treated with borderline doses of neutrons from inadequately small reactors.

Example I

A 37-year-old male with a grade III astrocytoma in the right frontal lobe. In September, 1979, a partial excision of the tumour was per-

formed, followed by BNCT. At the bottom of the cavity formed by the surgery, a neutron fluence of 1.6×10^{13} n/cm^2 was delivered*. This neutron fluence was considered sufficient to destroy the tumour cells within a sphere 5 cm larger than the cavity. 18 months after the treatment, contrast-enhancement of the lesion became more marked than before BNCT. Because the neutron intensity was sufficiently tumouricidal, the patient was kept under observation. Another year later, this contrast-enhancement was replaced with a low density area. After surviving 5 years the patient deteriorated without any CT evidence of recurrence. At his death, autopsy disclosed subarachnoid dissemination of the tumour cells with the largest colony in a basal cistern. The primary tumour in the right frontal lobe had been completely dissolved and replaced with multiple pores. No tumour cell was recognized in the primary tumour site (Fig. 1).

Example II

A 41-year-old man with a grade III astrocytoma. The patient was treated in June, 1979 (estimated neutron fluence: 2.5×10^{12} n/cm^2

---

* At 6 cm below the surface 2.5×10^{12} n/cm^2 nvt.

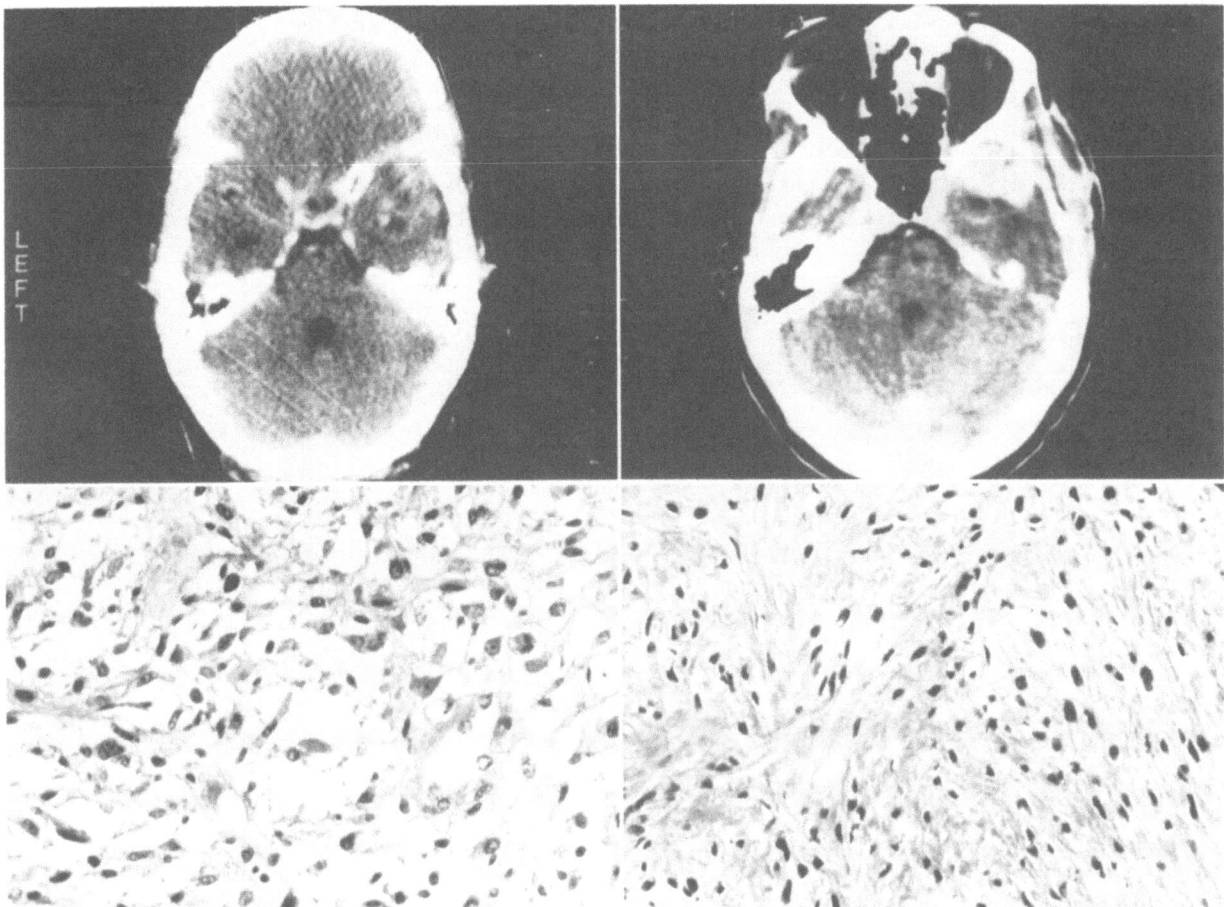

Fig. 2. Glioblastoma of a 54-year-old male. Top left: CT before craniotomy. The tumour is faintly contrast-enhanced and recognized in the right middle fossa. Bottom left: Histology obtained by the first tumour excision. Top right: CT 9 months after BNCT. A well circumscribed, strongly contrast-enhanced mass is seen attached to the sphenoid wing and the dural floor of the middle fossa. It was a reddish firm tumour and was totally removed. Bottom right: Histology of the "tumour" removed by the second-look operation. The photograph here is a most fibrous part. Proliferation of fibres probably contributed to the solid nature of this highly degenerated tumour. Pyknosis of the cell nuclei is remarkable. They are small and dark stained. In most other parts the number of cells was significantly small

at the depth 7 cm from the cortical surface). By now the author knows that this dose should be more than sufficient for this type of tumour, but in 1979, soon after the advent of CT imaging, when a new contrast-enhanced ring was recognized in the original tumour site in the left frontal lobe, the author immediately suspected a recurrence, and decided to reopen the head. Even BNCT was repeated. The "tumour tissue" obtained at this time turned out, upon later histological investigation, to be simply radionecrotic tumour tissue. The skin flap which had been shielded against neutrons but reflected for a considerable time developed an ulcer due to circulatory disturbance. The ulcer was infected and led to phlegmone of the scalp. The infection eventually spread to the brain matter which harbored necrotic tumour tissues, and the patient died. The patient had had absolute lymphocytopenia since before the first craniotomy. This immune insufficiency must have contributed to the infection. At autopsy no viable tumour was confirmed.

Example III

A 32-year-old Caucasian with grade IV glioblastoma in the left temporal lobe underwent craniotomy and subtotal removal of the

tumour. After surgery, the family claimed he had some difficulty finding words, and he was referred to the author. Speech did not improve even after BNCT conducted in June. (The year is concealed to hide the identify of the patient.) In September, he was entrapped by self-destructive thoughts, and frequently asked his physicians for some means of suicide in case of a relapse. In October, CT scanning was done on the instructions of the author, but a local radiologist carelessly told the patient that a regrowth of his tumour was evident. The CT films were sent to the author only a month later. The author called him and told him there was no evidence of a recurrent tumour, but it was too late. He was wandering about looking for some place to commit suicide. In one unsuccessful attempt, he was admitted to a hospital emergency department. Finally in the second attempt he succeeded in killing himself at 6 months after treatment. Throughout his clinical course, he did not present neurological deterioration except for the slight speech difficulty. The author tried to convince him by calling, or by cabling to tell him that his neurological problems could be related to the oedema around the necrobiotic process and could possibly be only transient. The CT films obtained in October showed an increased contrast-enhanced lesion in the left temporal lobe, but this was not sufficient evidence

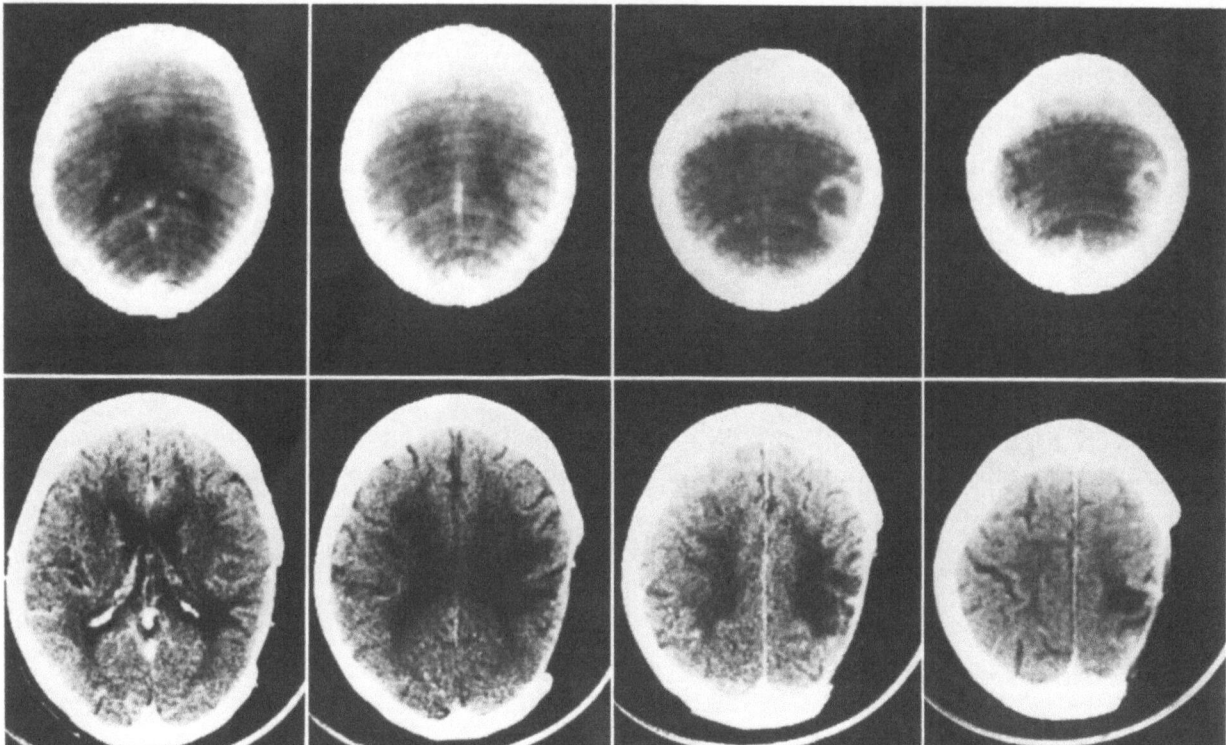

Fig. 3. Grade III astrocytoma of a then 60-year-old female who has been alive without any neurological deficit over 10 years. Her clinical course was one of the most uneventful. The upper row CT was made in June, 1977. The lower row CT was obtained in February, 1987. At the age of 70 she is still working as a cleaning woman and occasionally helps one of the authors as a messenger. This uneventfulness is probably related to the small size of the glioma and to its superficial location. Even a small 100 kw reactor could afford sufficiently tumouricidal neutron fluence. (Cf. Another patient with glioblastoma who has survived more than 15 years is described elsewhere[5])

of recurrence, particularly because the neutron intensity for the tumour bed as measured by gold wire had been a fully tumouricidal dose of 4×10^{12} n/cm^2 at the deepest part of the tumour extension (5 cm from the surface). Although this case lacks pathological evidence for the author's assertion, judging from the radiation dose delivered to the tumour bed, the period of 4 months after BNCT was too short for the adequately treated glioblastoma to reveal itself as recurrence on a CT image, even if the tumour were to come back.

Example IV

A 54-year-old Caucasian with a glioblastoma in the right temporal lobe. In October, 1983, craniotomy and tumour excision were performed, after which the patient was referred to the author for BNCT which was conducted in November, 1983. Ten months later, he complained of headache, and the CT revealed a suspicious contrast-enhanced mass in the original tumour site. This time, however, the mass appeared more irregular in shape and quite strongly contrast-enhanced. The author reopened the head of the patient and removed a red, well circumscribed walnut-size mass with a consistency of a fibrous meningioma which was firmly adherent to the dural floor of the middle fossa. Macroscopically, a sphenoid ridge meningioma was suspected, but the histology turned out to be that of a highly degenerated astrocytoma which was quite different from the histology of the original tumour. This second tumour in most parts presented low cellularity, and these cells were highly pyknotic (remarkably smaller nuclei than the cell nuclei before BNCT, and hy-

perchromatic − staining much darker). Loss of tumour texture and proliferation of fibers were noted. The fiber proliferation was responsible for the resemblance of the mass to a fibrous meningioma. Only a small area, probably located most remotely from the neutron collimator, lacked pyknotic cells, although the number of cells was small. This portion of the mass resembled a low-grade astrocytoma (grade I of Kernohan's). After the macroscopical total extirpation, the author at first wished to keep the patient under observation. After discussion with the patient who demanded radical treatment, another BNCT was given. Since the first BNCT regimen had been a borderline dose*, an increased neutron fluence was delivered to the tumour bed (5×10^{12} n/cm^2). The last CT obtained in Tokyo lacked any tumour contrast-enhancement, but revealed symmetrically enlarged lateral ventricles and an accumulation of cerebrospinal fluid under the skin flap of the craniotomized area. The fluid had also filled the subcutaneous space of the check and neck. The patient was sent home to his native country with a document of request to the local surgeons to perform a shunt operation. Because the patient was a bachelor and lived alone only with servants, no further detailed medical information was obtained, due to the lack of cooperation from the local neurologist who reportedly insisted that "The tumour is spreading down to the neck under the skin". According to the friend who had accompanied him to Tokyo, the ventriculoperitoneal shunt requested by the author was not installed. This so-called "tu-

* Estimated to be 2.5×10^{12} n/cm^2.

mour" claimed by the local neurologist must have been the subcutaneously accumulated cerebrospinal fluid. The patient died 16 months after the first craniotomy. This is a typical unhappy example of the outcome for patients treated by borderline doses of neutrons from small reactors (Fig. 2).

Discussion

Postoperative CT follow-up of BNCT patients has special problems. The first is that many neurosurgeons or physicians do not believe in "a cure" for glioblastoma. They tend to expect a recurrence, and immediately assume one if they happen to see a suspicious CT film. They fail to recognize a bionecrotic process which may eventually turn out to be only necrosis of the tumour – in cases where the neutron dose and the patient's immune response are both adequate (Fig. 3).

The second problem is that CT images are often ambiguous in cases receiving borderline (lethal or sublethal) doses of radiation by BNCT. In these cases, where suspicious contrast-enhancement or an increase of volume develops (which after conventional therapies mean a recurrence), the interpretation of postoperative CTs – even by an experienced eye – is extremely tricky – particularly if there is some neurological deterioration. Several cases had to undergo re-opening, only to find no viable tumour cells. Death occurred in some cases, either because of discontinuation of supportive treatments by local physicians, or because of excessive therapies by the author, all of whom had erroneously believed in recurrence. At autopsy, residual tumour cells were recognized only in areas where the minimum requirement of neutron fluence (= intensity) of 2.5×10^{12} n/cm^2 had not been delivered at the time of the treatment. Delayed cerebral necrosis was not recognized except in the cases which had purposely been given a larger neutron dose than the current regimen. (The pathological studies are detailed in the above-mentioned *Boron-Neutron Capture Therapy for Tumours*).

If a larger reactor like those used a MIT or Brookhaven during the 1953–1961 American series is available, and if an epithermal neutron facility is constructed, every tumour, deep-seated as well as shallow ones, will be good targets for BNCT.

Larger craniotomies[5] and heavy water[2] to replace the water in the cranial space will improve the thermal neutron penetration into deeper parts of the brain, and will guarantee a better survival rate.

References

1. Brugger RM, Less TK (1986) Intermediate energy neutron beams for neutron capture therapy – A progress report. In: Hatanaka H (ed) Neutron Capture Therapy (the Proceedings of the 2nd International Symposium on Neutron Capture Therapy, 1985, Tokyo). Nishimura & Co, Niigata, pp 110–116

2. Hatanaka H (1986) Introduction. In: Hatanaka H (ed) Boron-neutron capture therapy for tumours. Nishimura & Co, Niigata, pp 1–28

3. Hatanaka H, Amano K, Kamano S, Fankhauser H, Hanamura T, Sano K (1978) Boron-neutron capture therapy in relation to immunotherapy. Acta Neurochir (Wien) 42: 57–72

4. Hatanaka H, Amano K, Kamano S, Tovaryš F, Machiyama N, Matsui T, Fankhauser H, Hanamura T, Nukuda T, Ito N, Sano K (1982) Boron-neutron capture therapy vs. photon beam for malignant brain tumours – 12 years experience. In: Brock M (ed) Modern neurosurgery, I. Springer, Berlin Heidelberg New York, pp 122–132

5. Hatanaka H, Kamano S, Amano K, Hojo S, Sano K, Egawa S, Yasukochi H (1986) Clinical experience of boron-neutron capture therapy for gliomas – A comparison with conventional chemo-immuno-radiotherapy. In: Hatanaka H (ed) Boron-neutron capture therapy for tumours. Nishimura & Co, Niigata, pp 349–379

6. Hatanaka H, Sano K (1972) A revised boron-neutron capture therapy for malignant brain tumours. In: Fusek I, Kunc Z (eds) Present limits in neurosurgery. Avicenum, Czechoslovak Medical Press, Prague, pp 83–85

7. Hatanaka H, Sano K (1973) A revised boron neutron capture therapy for malignant brain tumours. I. Experience on terminally ill patients after Cobalt-60 radiotherapy. Z Neurol 204: 309–332

8. Hatanaka H, Sweet WH (1975) Slow neutron capture therapy for malignant tumours. Its history and recent developments. Biomedical Dosimetry (IAEA-SM-193/79). International Atomic Energy Agency, Vienna, pp 147–198

9. Hatanaka H, Watanabe N (1970) Brain tumour treatment with slow neutrons (in Japanese). Geka-shinryo (= Surgical Practice) 12: 1041–1055

10. Hatanaka H, Urano Y (1986) Eighteen autopsied cases of malignant brain tumours treated by boron-neutron capture therapy between 1968 and 1985. In: Hatanaka H (ed) Boron-neutron capture therapy for tumours. Nishimura & Co, Niigata, pp 381–416

11. Kanda K, Kobayashi T (1986) Physics studies on neutron field and dosimetry for neutron capture therapy in Japan. In: Hatanaka H (ed) Boron-neutron capture therapy for tumours. Nishimura & Co, Niigata, pp 255–270

12. Locher GL (1936) Biological effects and therapeutic possibilities of neutrons. Am J Roentgenol Radium Therapy 36: 1–13

13. Matsumoto T, Aizawa O, Suetomi F, Kadotani H (1986) Dosimetry and analyses for boron-neutron capture therapy in the treatment of brain tumours. In: Hatanaka H (ed) Neutron capture therapy (the Proceedings of the 2nd International Symposium on Neutron Capture Therapy, 1985, Tokyo). Nishimura & Co, Niigata, pp 182–187

14. Oka Y, An S (1986) Design of facilities for boron-neutron capture therapy – epithermal neutron sources. In: Hatanaka H (ed) Boron neutron capture therapy for tumours. Nishimura & Co, Niigata, pp 243–254

15. Wakabayashi H, Yoshii K, Sasuga N, Yanagi H (1986) Mixed

dose distribution of fast neutrons and boron-neutron capture therapy for the fast neutron beam from YAYOI reactor. In: Hatanaka H (ed), Boron-neutron capture therapy for tumours. Nishimura & Co, Niigata, pp 271–282

16. Sweet WH, Javid M (1952) The possible use of neutron capturing isotopes such as boron-10 in the treatment of neoplasms. I. Intracranial tumour. J Neurosurg 9: 200–209

Address for correspondence: Prof. Hiroshi Hatanaka, M.D., D.Sc., Teikyo University Hospital, 2-11-1 Kaga, Itabashi-Ku Tokyo 173, Japan.

Acta Neurochirurgica, Suppl. 42, 193–197 (1988)

Radiosurgery in Cerebral Tumours and AVM

V. Valentino

Institute of Neurosurgery of the University, Rome, Italy

Summary

In collaboration with the Institute of Neurosurgery, La Sapienza Rome University, we have treated 214 patients with stereotactic irradiation. The series began in March 1984 and includes 198 cerebral tumours of different histology and 16 AVM. 73% of the patients had been operated on before irradiation. From this first experience the following considerations can be drawn: (a) radiosurgery is not an alternative to neurosurgery except for particular cases; neurosurgery is therefore essential because the smaller the target area the higher the efficiency of stereotactic irradiation; (b) compared to conventional radiotherapy, damage to the brain is minimized as shown by NMR. The follow-up time is too short to allow any definite conclusion. However, positive effects have been observed in malignant gliomas and single metastases. In craniopharyngiomas and pituitary adenomas, tumour growth was arrested or decreased with the disappearance of the tumour in 3 adenomas and 1 craniopharyngioma. With regard to the response of meningiomas to irradiation we have shown that radiosurgery is able to cause a decrease in tumour size as well as reduced contrast enhancement, probably due to vascular changes and fibrosis. In AVM the efficiency of radiosurgery has been further confirmed.

Keywords: Radiosurgery; cerebral tumour; arteriovenous malformation; indication; result.

Introduction

"Radiosurgery" is a term introduced by Leksell[3] to signify external irradiation to be performed as a single high dose on a defined target, using a stereotactic procedure.

From March 1984–March 1987 we used radiosurgery and/or stereotactic hypofractionated irradiation in 214 patients, comprising 198 cerebral tumours and 16 AVM.

Material and Methods

Besides 16 AVM, there were 90 gliomas, 31 metastases, 29 meningiomas, 26 pituitary adenomas, 8 pinealomas, 6 craniopharyngiomas, 6 dysgerminomas and two glomus jugulare tumours. Seventy-three per cent of these patients had been operated on previous to stereotactic irradiation, thus the diagnosis was supported by his-

tology in 149 cases. In two patients stereotactic biopsy was performed. In the remaining 47 cases there was no histological support, and, apart from clinical evidence, diagnosis was based only on plain films of the skull, CT scan, and/or angiography and NMR.

One single high dose was used when the lesion did not exceed 25 mm in diameter. In lesions greater than 25 mm we used two different techniques related to the type and site of the lesion:

– Hypofractionated stereotactic irradiation plus a single high dose.

– Single high doses to multiple target fields using 10–25 mm diameter circular collimator channels.

All three techniques are in harmony with what has already been proven in radiobiology, *i.e.* the smaller the irradiated area the higher the efficiency of radiation.

A 6 MeV Linear Accelerator was the radiation source and irradiation was performed through several arches of different degree arranged on the basis of the site and size of the lesion.

The complete irradiation dosage varied according to the type of the lesion as well as the selected technique. In all patients we used the Greitz-Bergström's fixation head system[2] which has shown to be accurate. Furthermore this system is non-traumatic, easily removable and reproducible, thus being the most adaptable to the different mentioned methods of irradiation.

Results and Remarks

In such a miscellaneous collection of cerebral tumours, the results cannot be analysed altogether.

It is apparent that in malignant gliomas and solitary metastases the advantages of radiosurgery are all concentrated on minor discomfort to the patient as well as the possible acute positive effect on the tumour and oedema, a favourable though transient event which has been observed (Fig. 1).

Dysgerminoma is known to be a very radiosensitive tumour. In our 6 cases a 10–15 Gy single dose was enough to allow in a few days or weeks complete clinical remission and disappearance of the tumour at CT scan. This neoplasm is also known to disseminate readily, and therefore precautionary irradiation of the neuraxis has been recommended. However, this may be a pre-

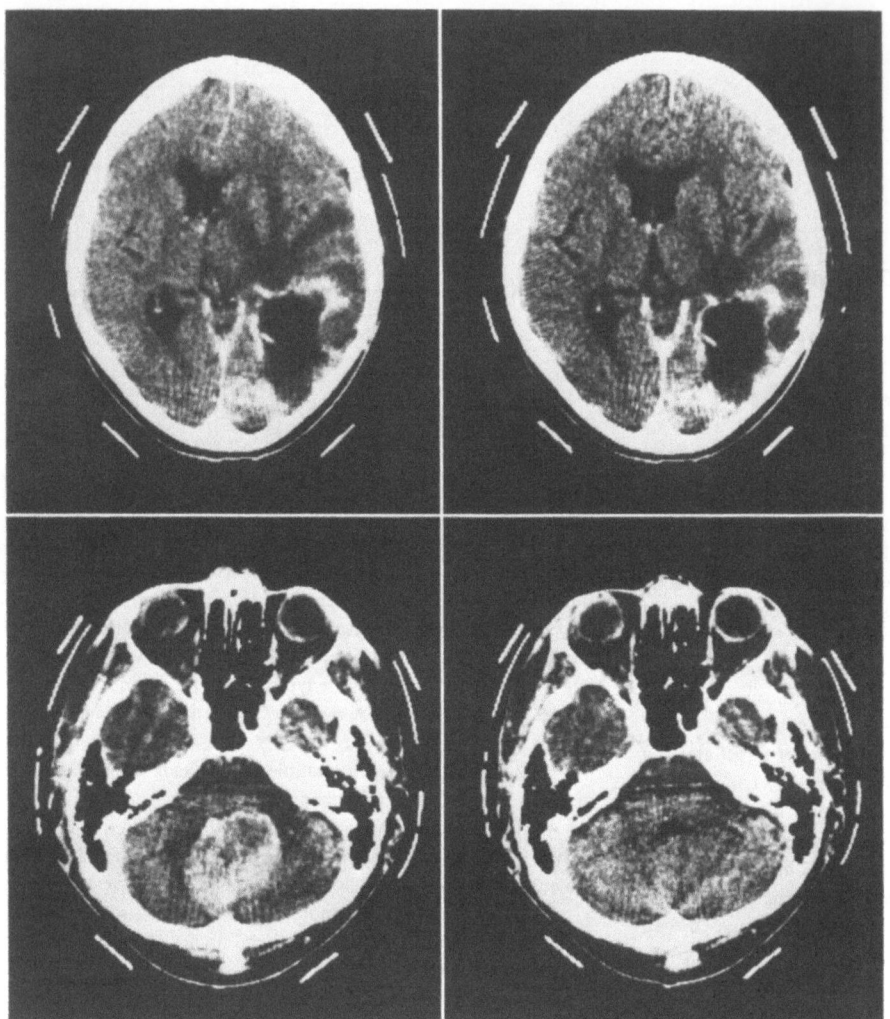

Fig. 1. Upper: Recurrent glioblastoma multiforme before (left) and one week after stereotactic radiotherapy (right). Lower: Solitary metastases from pulmonary carcinoma before (left) and one month after radiosurgery (right)

judice. Conventional irradiation of the neuraxis was performed in only two of our 6 cases. In spite of this, CFS dissemination of tumour occurred shortly after in both these patients, showing that, as in metastases, conventional irradiation is unable to prevent the occurrence of other possible localizations.

Radiosurgery was successful in the 8 cases of expanding processes in the pineal region as well as in the two cases of glomus jugular tumours. Tumours in the region of the pineal gland and quadrigeminal plate are clinically classified as pinealomas, but their histological characteristics can be very variable to such an extent that even the biopsy specimen is sometimes not conclusive (Backlund[1]). Of our 8 cases of pinealomas, only one underwent stereotactic biopsy and the tumour was said to be a dysgerminoma. In the other seven patients the diagnosis of pineocytoma in 3 and pinealoblastoma

in the remaining 4 was reached on the basis of the patient's age, neuroradiological appearances and laboratory tests.

In pituitary adenomas, both cromophobe and hypersecreting, as well as in craniopharyngiomas — a total of 32 patients — tumour growth decreased (Fig. 2) in all except two cases; these were concerned with a recurrent invasive adenoma and a cystic craniopharyngioma. In 4 patients (three pituitary adenomas and one craniopharyngioma) there was neuroradiological disappearance of the tumour.

The response of meningiomas to conventional irradiation has long been disputed and is controversial. In our series there are 29 cases of meningiomas. In 6 cases of recurrent tumour, stereotactic irradiation was unable to prevent re-growth of the neoplasm in the previous or a different site. However, in 11 patients

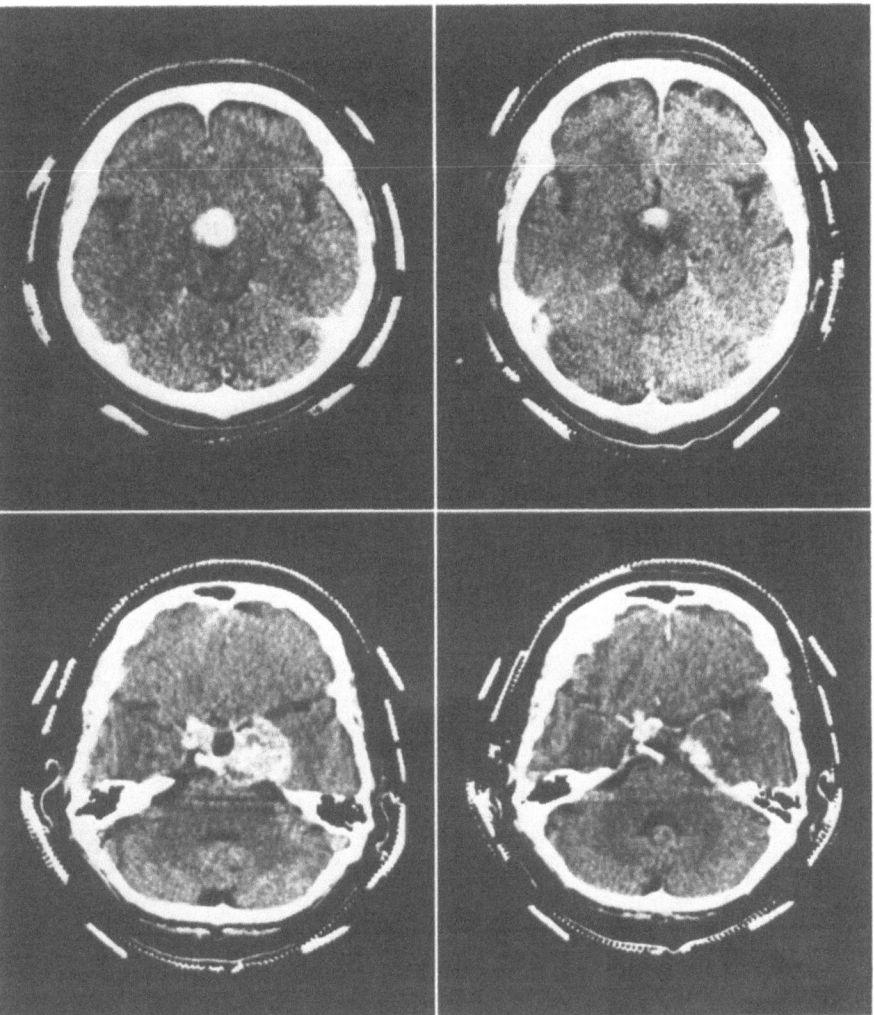

Fig. 2. Upper: Cromophobe pituitary adenoma before (left) and 6 months after radiosurgery (right). Lower: Recurrent craniopharyngioma before (left) and 3 months after stereotactic radiotherapy (right)

tumour growth was arrested. In the remaining 14 there was a varying degree in decrease of tumour size (Fig. 3 upper).

Radiosurgery in AVM can be considered as a routine technique when it seems appropriate versus more conventional methods, such as neurosurgery or embolization. In 1979 Backlund[1] reported a series of 32 cases controlled by angiography from 3–60 months after radiosurgery, and grouped the patients according to the type of irradiation which had been performed, as follows:

A. The whole cluster of pathological vessels was covered by a sufficient dose.

B. Only a part of the cluster could be covered.

C. Only the feeders were irradiated.

D. Only some of several feeders could be included within the radiation field.

In this reported series, control angiography showed no changes of the AVM in 12 patients: 6 of them were registered in group B and 4 in group D. Total obliteration of the AVM had taken place in 10 of 11 patients belonging to group A. The single dose varied between 50 and 125 Gy. The author also said that in AVM "an optimal therapeutic situation is present when all shunting compartments can be covered by radiation". Encouraged by this report, we developed two different techniques of irradiation related to the size of AVM. In 6 cases of small malformation, 25 Gy were delivered in a single high dose to a 15–25 mm field including the feeding arteries and a greater part of the entire cluster. In the other 10 cases of large AVM we used a technique of combined radiosurgery and hypofractionated stereotactic irradiation, as follows: a single high dose of 25 Gy was given on the main feeding arteries where

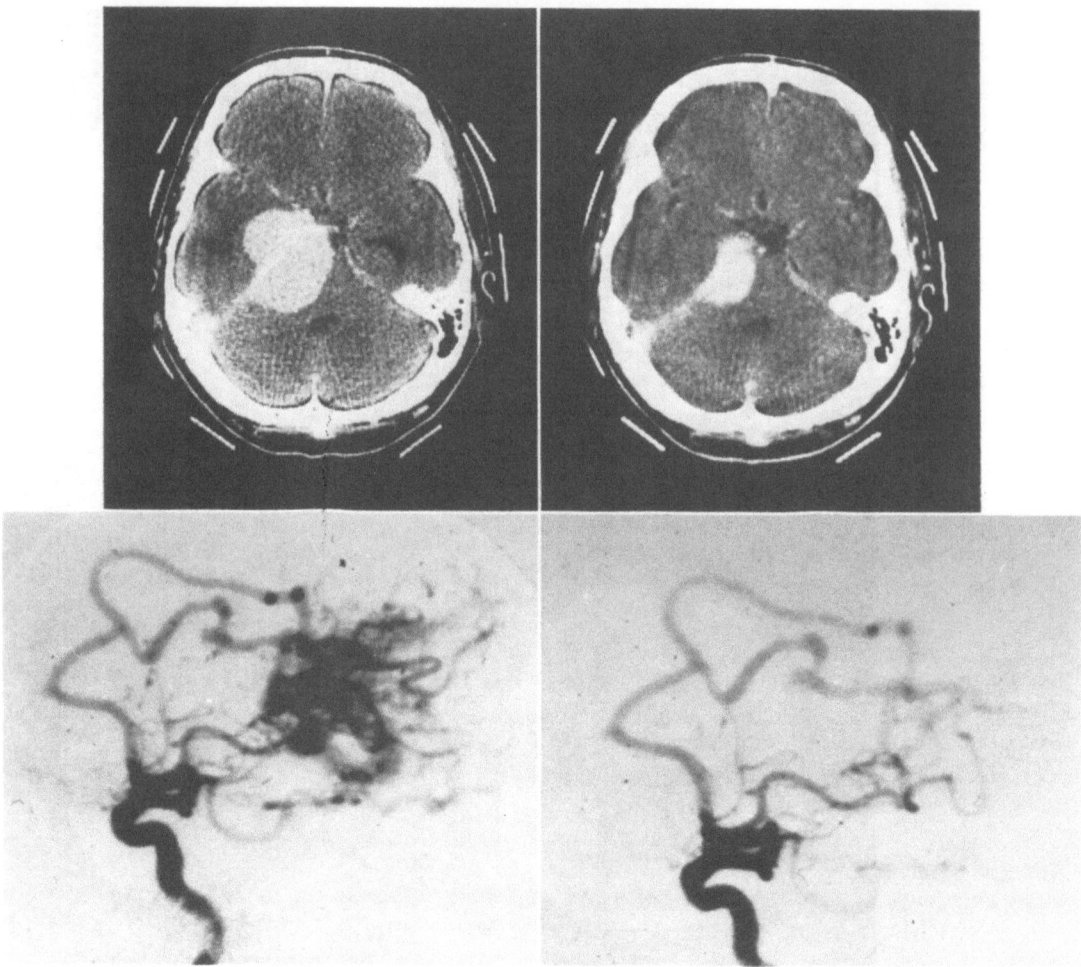

Fig. 3. Upper: Tentorial meningioma before (left) and 6 months after radiosurgery. Lower: AVM before (left) and 18 months after combined radiosurgery and stereotactic radiotherapy

they entered the angioma; then, two weeks later, 4 doses of 5 Gy were delivered twice a week to an area covering both the angioma and feeding arteries. Both methods were shown to be successful in 12 patients in whom control angiography was performed (Fig. 3 lower). In the remaining 6 cases the follow-up has not been long enough for angiographic control.

Conclusions

Radiosurgery is not a standard but an individually planned technique. It can often lead to gradual shrinkage of the tumour and AVM usually without side effects, but it is not an alternative to neurosurgery. Neurosurgery should be suggested whenever possible, but the smaller the pathological area the easier the radiosurgical approach with an enhanced biological efficiency of the radiation. These aspects are both concerned with the problem of necrosis, a problem which

does not appear fully resolved even if the rules for a step dose-gradient versus the tumour's surrounding structures have been strictly respected. In a review and reanalysis of the literature on adverse effects of radiation on the brain, Sheline et al.[5] groupes reactions according to time of appearance, as follows:

— Acute reactions: during the course of irradiation.

— Early delayed reactions: from a few weeks to a few months later.

— Late delayed reactions: from several months to years later.

In 1979 Martins et al.[4] reported 6 cases of necrosis of the brain following radiotherapy in patients who were operated on. In 5 of them the diagnosis was verified histologically. According to these AA. in necrosis of the brain due to irradiation "symptoms most often begin 9 months to 2 years after radiotherapy".

In our entire series there were two cases of acute transient increase of intracranial pressure which could be ascribed to an acute reaction to radiosurgery. As to delayed necrosis, among 41 patients still alive with at least a 2 year follow-up, there was one case of necrosis proven histologically. It should therefore be underlined that whenever possible the pathological area be reduced in size, thus allowing a lower radiation dose and reducing the risk of necrosis.

References

1. Backlund EO (1979) Stereotactic radiosurgery in intracranial tumours and vascular malformations. In: Advances and technical standards in neurosurgery, vol 6. Springer, Wien New York

2. Greitz T, Bergström M, Boëthius, Kingsley D, Ribbe T (1980) Head fixation system for integration of radiodiagnostic and therapeutic procedures. Neuroradiology 19: 1

3. Leksell L (1951) The stereotaxic method and radiosurgery of the brain. Acta Chir Scand 316: 102

4. Martins AN, Johnston JS, Henry JM, Stoffel TJ, Di Chiro G (1979) Delayed radiation necrosis of the brain. J Neurosurg 47: 336

5. Sheline GE, Wara WM, Smith V (1980) Therapeutic irradiation and brain injury. Int J Radiation Oncology Biol Phys 6: 1215

Address for correspondence: Prof. Dr. Vincenzo Valentino, Via Cadlolo 90, D-00136 Roma, Italy.

IV. Benign Intracranial Cystic Lesions, Hydrocephalus, CSF-Volumes

Acta Neurochirurgica, Suppl. 42, 201–204 (1988)

Arachnoid Cysts of the Middle Cranial Fossa: Experience with 77 Cases Treated Surgically

E. Galassi, G. Gaist, G. Giuliani, and **Eu. Pozzati**

Division of Neurosurgery, Bellaria Hospital, Bologna, Italy

Summary

Arachnoid cysts of the middle cranial fossa (Sylvian cysts), represent the most common type of intracranial leptomeningeal malformation. Among the 102 intracranial arachnoid cysts operated on at the authors' institution from January 1970 to August 1986, the 77 cases (75%) located in the middle cranial fossa are reviewed. The higher incidence in the first two decades of life (51 cases) as well as the marked predilection for the male sex (60 cases) and the left hemisphere (55 cases) are confirmed in the authors' experience. As for clinical presentation cranial deformities, symptoms of raised intracranial pressure and epilepsy constituted the most frequent features. In 13 patients a complicating lesion was associated: subdural or intracystic haematomas in 7 cases, subdural hygromas in 4 cases and, extradural haematomas in 2 cases. Based on the apperance at CT scan and the results at CT cisternography the authors proposed a classification into three basic types of increasing severity and different pathophysiologic conditions. All the patients underwent craniotomy, excision of the cyst walls and perforation into the basal cisterns. There was one postoperative death (mortality rate of 1.3%) due to meningitis. The remaining clinical results were gratifying in all three types of lesion; on follow-up CT scans the cysts of type I° and II° exhibited a steady tendency to reduction or obliteration while cerebral reexpansion seemed less evident in the third, most severe, type. The authors compare and discuss the options of radical open surgery versus shunting procedures.

Introduction

According to Robinson's classical review[18] arachnoid malformations would constitute approximately 1% of the whole intracranial space occupying material. More recently the routine use of Computerized Tomography (CT) has led to an increased recognition of these lesions which have been found to be more frequent than thought previously[4, 5, 21]. Although arachnoid cysts of the middle fossa (Sylvian cysts) represent the most commonly reported type of intracranial leptomeningeal malformation[4, 5, 6, 15, 16, 18] their natural history and lifelong prognosis are still incompletely known. Indeed Sylvian cysts may become symptomatic at any time in life or remain occult being diagnosed fortuitously, or at autopsy[17, 22]. Consequently surgery is not always mandatory, and there are diverse opinions regarding the best therapeutic method; surgical options have included burr hole evacuation[17, 18], craniotomy for cyst excision and fenestrating procedures[1, 3, 4, 5, 8, 14, 19] and extrathecal CSF shunting[8, 11, 12, 13, 21, 23].

This controversy prompted us to present our experience with 77 cases of middle fossa cysts, all radically and uniformly treated by craniotomy.

Summary of Cases

From January 1970 to August 1986, 102 patients with intracranial arachnoid cysts were observed and operated on at our Neurosurgical Division. Cases not verified by surgery and histology were not considered. The middle fossa cysts constituted the vast majority of our material (77 cases corresponding to 75%) followed by the infratentorial locations (15 cases). Ten of our sylvian cysts were diagnosed during the last eight months; this fact provides evidence of the increased occurrence of these lesions in our practice. The salient clinical data of the 77 patients are summarized in Table 1. Noteworthy findings have been the higher incidence in the first two decades of life (51 cases = 66%) as well as the marked predilection for the male sex (60 cases) and for the left side (55 cases). We found a history of meningitis in 2 cases, difficult delivery in 16 cases and of a previous head injury in 18 cases. In most of the cases these traumatic events occurred only a few days or weeks before the initial symptoms thus conceivably decompensating a previously asymptomatic condition. In three instances the clinical picture presented and acutely deteriorated soon after the precipitating cranial trauma; in 1 case this was due to massive intracystic

Table 1. *Summary of Clinical Features in 77 Patients with Middle Fossa Arachnoid Cysts*

Age (Years)	Pat. (No)	Sex	Pat. (No)	Symptoms and signs	Pat. (No)
0–10	24	male	60	cranial deformity	35
11–20	27	female	17	raised intracran. press.	29
21–30	7			epilepsy	24
31–40	10	Side	Pat. (No)	focal signs	15
41–50	2			psycho-motor retardation	6
51–60	6	left	55	optic atrophy	5
61–70	1	right	22	post-traumatic coma	3
				ictal onset	1
		Complicating lesion	Pat. (No)	precocious puberty	1
				facial pain	1
		subdural haematoma	5	asymptomatic	4
		subdural hygroma	4		
		intracystic haematoma	2		
		extradural haematoma	2		

bleeding, and in 2 cases to an overlying extradural haematoma.

Complicating lesions were noted in 13 patients (19%): subdural haematomas in five cases, subdural hygromas in four, intracystic haematomas in two and extradural haematomas in two. In one other case there was an associated temporal low grade glioma. Twenty-nine patients had symptoms or signs of intracranial hypertension, twenty four had epileptic seizures, fifteen had focal neurological deficits, six had developmental delay, five had optic atrophy. Three patients with post-traumatic haemorrhagic complications appeared comatose upon admission. An unusual ictal onset, due to a spontaneous intracystic haemorrhage, occurred in one case. Cranial deformities (35 cases) included macrocrania, gross calvarial asymmetry with enlargement of a whole hemicranium and, most frequently, localized bulging of the temporal squama. Four asymptomatic cases were fortuitously discovered either after mild traumas or because of nonspecific psychological complaints. Computerized Tomography (CT) scans were obtained in 70 cases. The cysts appeared as extraparenchymal, sharply defined, non-enhanced masses of the CSF density which constantly occupied the anterior portion of the temporal fossa frequently extending along the Sylvian fissure and even widely abutting over the cerebral convexity. Metrizamide CT cisternography was added in 8 patients in order to obtain a dynamic assesment of the CSF circulation around and inside the pathologic cavity. According to the criteria that we have proposed in a previous paper[5] we could classify our cases into three types of increasing severity. In the mildest form, type I, (18 cases) the lesion is small, lenticular and confined to the most anterior part of the

middle fossa; mass effects are lacking and a free communication of the cyst with the basal cisterns is demonstrated by early simultaneous metrizamide filling. Type II malformations (28 cases) are medium sized, triangular or trapezoidal in shape and completely open the Sylvian fissure. Mass effects are more common, although usually moderate. Metrizamide cisternography reveals a less adequate communication with a delayed entry and washout of the contrast. The most severe type III cysts (24 cases) appear as large areas with rounded or oval contours and relevant displacements of the midline structures. The temporal fossa and the Sylvian fissure are entirely involved by the malformation which extends variably over the cerebral convexity. Metrizamide cisternography shows an absent or markedly delayed staining of the cyst. These findings suggest an inverse relationship between the size and expanding qualities of middle fossa arachnoid cysts and the existence and ease of their communication with CSF pathways.

All our cases underwent the following surgical procedure: craniotomy, evacuation of any complicating haematomas, excision of the removable arachnoid linings and deep perforation of the inner membrane into the basal cisterns in order to enlarge or create a free communication with the CSF pathways. In two patients this open operation was undertaken after unsuccessful cysto-peritoneal shunting; a burr hole aspiration had been followed by rapid recurrence in one other instance. Surgical inspection often revealed tenuously supported bridging veins traversing the dome of the cyst. These thin vessels provide a possible source of the intracystic and subdural haematomas which frequently complicate Sylvian cysts. There was a death in one of the earliest cases in the series caused by severe purulent meningitis (mortality rate of 1.3%). Two other patients had major postoperative complications: one extradural and one intracystic haematoma; these clots were promptly evacuated with complete recovery. The remaining postoperative courses were satisfactory; transient neurological problems (III nerve palsy, aphasia, seizures) have been seen occasionally but have rapidly and completely resolved. No permanent neurological worsening developed after surgery.

The follow-up period ranged from 8 months to 17 years. Fifty-one patients exhibited a full recovery with disappearance of their preoperative disturbances while fifteen showed significant improvement. Eight cases, including the four, who were asymptomatic preoperatively, remained unchanged after surgery. Two cases have been lost to follow-up. At postoperative CT scan study (51 cases) a marked decrease in the cyst size, often leading to a complete obliteration, occurred in 37 instances. In fourteen patients the cyst reduction was only moderate.

While type I and II malformations showed a steady tendency to disappear, brain reexpansion was less definite and often incomplete in the largest type III lesions.

Discussion

Arachnoid cysts of the middle cranial fossa represent a primary congenital anomaly of the developing leptomeninges[16, 18, 22] with a striking unexplained predilection for the male sex and the left hemisphere[4, 5, 15, 17, 18, 21].

The vast majority of these lesions (almost 2/3 in our experience) are detected in the first two decades of life[4, 5, 15, 18, 21]. Although some Sylvian cysts may remain well tolerated for a long time throughout life[2, 3, 9] and even appear as incidental findings in necropsies[17, 22], their potentially threatening qualities have been repeatedly stressed since alarming and possibly fatal deteriorations are known to follow either a rapid enlargement of the lesion[2, 8, 9, 19, 20, 22] or else a complicating intracranial haemorrhage[1, 4, 7, 8, 14, 20]. Our CT cisternographic observations support the hypothesis that the expansion of the pathological cavity is related to its progressive segregation from the subarachnoid spaces with a secondary CSF accumulation due to either secreting, or unidirectional trapping, mechanisms[5, 10].

Surgical treatment seems obviously warranted in symptomatic and complicated cases, but it constitutes a debated issue with regard to patients who are diag-

nosed fortuitously or who show only cosmetic defects[5, 8, 11, 18, 20, 21].

We concur with other authors that all Sylvian cysts exerting detectable mass effects call for a surgical treatment even if clinically silent[4, 8, 10, 11].

In the remaining asymptomatic cases the operative risks and the hazards of a conservative management in a potentially harmful and unstable condition should be carefully weighed[4, 7, 8, 10].

The choice of the optimal surgical method still remains another controversial topic. Recent papers[11, 12, 13, 21, 23] have advocated the superiority of primary cysto-peritoneal shunting alone or eventually combined with ventriculo-peritoneal shunting when hydrocephalus is associated. Although these series seem mostly heterogeneous, including different locations of intracranial arachnoid cysts, the CSF diversions proved safe and frequently effective. Nevertheless some failures caused by shunt malfunction or by infection and defective cyst collapses have also been recorded[8, 11, 20, 21, 23].

The awareness of the possibility of shunt dependency, also of the well-known complications related to shunting devices in hydrocephalic children led us to favour a more radical, and hopefully curative, approach.

Noticeably, unlike midline and infratentorial locations[6], hydrocephalus is usually absent in Sylvian cysts since changes in the CSF kinetics are mainly limited to the cyst cavity and its periphery[5]; therefore we feel that the extensive resection of the secreting membranes and the establishment of a wide communication with the subarachnoid pathways can adequately normalize the fluid circulation[3, 4, 5, 10], and avoid the need for an extrathecal CSF shunting. Furthermore only craniotomy allows the appropriate management of complicating haematomas[1, 4, 7, 14].

It would seen preferable to employ shunts secondarily in some type III lesions when follow-up CT scans show only a partial cyst regression: in these cases the shunt can further assist the reexpansion of a severely compressed parenchyma.

Our experience with a series of 77 Sylvian cysts homogeneously treated by craniotomy indicates the possibility of a high rate of long standing benefits and of cyst regressions at radiological follow-up with a seemingly low incidence of surgical complications.

Apart from the case of a child who died for an aspecific infective complication – which is possible with other forms of surgical therapy – we did not encounter any permanent postoperative neurological worsening. We believe that only further experience with different therapeutic alternatives and long term studies will give us the optimal guidelines for the management of this complex anomaly.

References

1. Auer LM, Gallhofer B, Ladurner G, Sager WD, Heppner F, Lechner H (1981) Diagnosis and treatment of middle fossa arachnoid cysts and subdural haematomas. J Neurosurg 54: 366–369

2. Bhandari YS (1972) Non-communicating supratentorial subarachnoid cysts. J Neurol Neurosurg Psychiatry 35: 763–770

3. Dyck P, Gruskin P (1977) Supratentorial arachnoid cysts in adults. Arch Neurol 34: 276–279

4. Galassi E, Piazza G, Gaist G, Frank F (1980) Arachnoid cysts of the middle cranial fossa: a clinical and radiological study of 25 cases treated surgically. Surg Neurol 14: 211–219

5. Galassi E, Tognetti F, Gaist G, Fagioli L, Frank F, Frank G (1982) CT scan and metrizamide CT cisternography in arachnoid cysts of the middle cranial fossa: classification and pathophysiological aspects. Surg Neurol 17: 363–369

6. Galassi E, Tognetti F, Frank F, Fagioli L, Nasi MT, Gaist G (1985) Infratentorial arachnoid cysts. J Neurosurg 63: 210–217

7. Galassi E, Tognetti F, Pozzati E, Frank F (1986) Extradural haematoma complicating middle fossa arachnoid cyst. Child's Nerv Syst 2: 306–308

8. Geissinger JD, Kohler WC, Robinson BW, Davis FM (1978) Arachnoid cysts of the middle cranial fossa. Surgical considerations. Surg Neurol 10: 27–33

9. Ghatak NR, Mushrush GJ (1971) Supratentorial intra-arachnoid cysts. Case report. J Neurosurg 35: 477–482

10. Go KG, Houthoff HJ, Blaauw EH, Havinga P, Hartsuiker J (1984) Arachnoid cysts of the Sylvian fissure. Evidence of fluid secretion. J Neurosurg 60: 803–813

11. Harsh GR, Edwards MSB, Wilson CB (1986) Intracranial arachnoid cysts in children. J Neurosurg 64: 835–842

12. Kaplan BJ, Mickle JP, Parkhurst R (1984) Cystoperitoneal shunting for congenital arachnoid cysts. Child's Brain 11: 304–311

13. Kato M, Nakada Y, Ariga N, Kokubo Y, Makino H (1980) Prognosis of primary middle fossa arachnoid cysts in children. Child's Brain 7: 195–204

14. La Cour F, Trevor R, Carey R (1978) Arachnoid cyst and associated subdural haematoma. Arch Neurol 35: 84–89

15. Naidich TP, Mc Lone DG, Radkowski MA (1986) Intracranial arachnoid cysts. Pediat Neurosci 12: 112–122

16. Rengachary SS, Watanabe I (1981) Ultrastructure and pathogenesis of intracranial arachnoid cysts. J Neuropathol Exp Neurol 40: 61–83

17. Robinson RG (1964) The temporal lobe agenesis syndrome. Brain 87: 87–106

18. Robinson RG (1971) Congenital cysts of the brain: arachnoid malformations. Progr Neurol Surg 4: 133–174

19. Seur NH, Kooman A (1976) Arachnoid cyst of the middle fossa with paradoxical changes of the bony structures. Neuroradiology 12: 177–183

20. Smith RA, Smith WA (1976) Arachnoid cysts of the middle cranial fossa. Surg Neurol 5: 246–252

21. Sprung CH, Mauersberger W (1979) Value of Computed Tomography for the diagnosis of arachnoid cysts and assessment of surgical treatment. Acta Neurochir (Wien) [Suppl] 28: 619–626

22. Starkman SP, Brown TC, Linell EA (1958) Cerebral arachnoid cysts. J Neuropathol Exp Neurol 17: 484–500

23. Stein SC (1981) Intracranial developmental cysts in children: treatment by cystoperitoneal shunting. Neurosurgery 8: 647–650

Address for correspondence: Dr. E. Galassi, Divisione di Neurochirurgia, Ospedale Bellaria, I-40139 Bologna, Italy.

Acta Neurochirurgica, Suppl. 42, 205–209 (1988)
© by Springer-Verlag 1988

Intracranial Arachnoid Cysts in Adults

M. Garcia-Bach, F. Isamat, and **F. Vila**

Service of Neurosurgery, Hospital "Princeps d'Espanya", University of Barcelona, Barcelona, Spain

Summary

Twenty-two cases of intracranial arachnoid cysts in adult patients have been studied and treated surgically. The authors analysed the correlation between the size and the location of the cysts and their clinical manifestations. It seems clear that the symptomatology and the problems of intracranial arachnoid cysts in the adult are quite different from those occurring with infants. The majority of our cases have been diagnosed by CT and NMR, and have been treated by open excision of the cystic membranes with the establishment of a wide communication with the basal cisterns or the subarachnoid space. With this technique the results have been very gratifying, while in cases managed with shunts the morbidity has been high.

Keywords: Arachnoid cyst; intracranial space-occupying lesion; CT scan; NMR; cysto-peritoneal shunt.

Introduction

The first description of an arachnoid cyst was made by Bright[2] in 1831. Arachnoid cysts represent approximately 1% of all atraumatic expansive lesions[20]. Robinson[19] suggested the agenesis of the temporal lobe with a passive dilatation of the subarachnoid space as an aetiological factor. More recently Starkman *et al.*[22] concluded that arachnoid cysts evolve from a splitting in two layers of the arachnoid membrane[18]. Nevertheless the true mechanism for their enlargement is still unknown, although three possibilities have been postulated[4]: a ball-valve mechanism, a fluid filtration through the cystic wall due to an osmotic gradient, and a secretion of fluid by the lining of the cyst[10]. Most of the arachnoid cysts produce clinical symptoms in infancy, particularly during the first six months of life[3]. As a result large series of intracranial arachnoid cysts are available only in the paediatric group[3, 12, 17, 24]. The differences in adults concerning clinical behaviour and surgical management is the reason of this paper.

Clinical Material

Our series consists of a group of twenty-two patients with intracranial arachnoid cysts aged between 11 and 70 years. Seventeen cases (77%)

Table 1

Age (years)	No. patients
0–10	0
11–20	5
21–30	4
31–40	6
41–50	2
51–60	2
61–70	3

were older than 20 years, with a mean age of 35.5 years (Table 1). Fourteen patients were male and eight female.

Symptoms and Findings

Epilepsy (6 with generalized seizures and 2 with focal fits) has been the commonest first symptom. Headache was the main complaint in 7 cases, and motor deficit in 4, two of them with a sudden onset. Finally, one patient had gait ataxia, and two were asymptomatic. These asymptomatic cases were diagnosed following CT studies after a head injury in one, and after a subarachnoid haemorrhage due to an aneurysm of the carotid bifurcation in the other. Other clinical symptoms less frequently encountered have been lower cranial nerve involvement, intracranial hypertension, vertigo and behavioural disturbances. The clinical data is summarized in Table 2.

In 19 out of 22 patients the diagnosis was made by CT scans, which is at present the elective method[7, 16, 26]. The other three cases, seen before 1974, were diagnosed by angiographic studies. Magnetic Resonance (NMR) images were obtained in two patients with infratentorial arachnoid cysts, which added further information in the study of the cyst itself and its relations to the surrounding structures. NMR images are extremely valuable and precise for diagnostic purposes (Fig. 1).

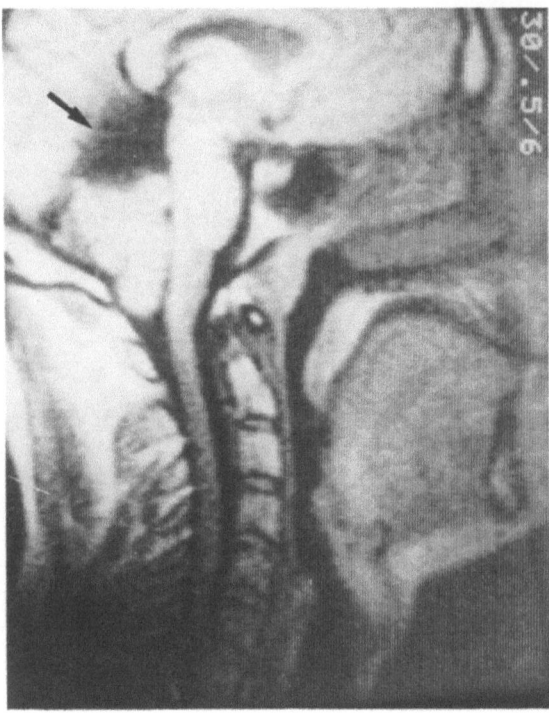

Fig. 1. NMR image of a supracerebellar arachnoid cyst

On CT the cysts have been shown to have two different appearances. Some have a triangular or quadrangular shape with a straight inner margin (Fig. 2); while others are seen as a round well-defined low attenuation area pushing back the brain (Fig. 3). After contrast injection the cortex surrounding the cyst may show a fair enhancement. One case showed a spontaneous intracystic haematoma without subdural blood[1, 15]. Two patients underwent radionuclide cisternography[5] that revealed a communication between the cyst and the subarachnoid space, but presenting a clear delayed contamination.

Location

Seventeen cysts were located in the supratentorial compartment, and all of them were superficial[9]. Ten were on the left hemisphere and seven on the right one. Arachnoid cysts of the Sylvian fissure were the most common, with 7 patients in this group. Four cysts were in the frontal lobe and another four in the temporal lobe. Only 5 arachnoid cysts were infratentorial[6, 21, 25]: 2 retrocerebellar, 1 supracerebellar, 1 in the cerebellopontine angle and 1 retro-laterocerebellar.

The analysis of location, size and symptomatology reveals that Sylvian fissure cysts have shown motor deficits (progressive or sudden ictal type hemiparesis), while convexity cysts are more apt to present with epilepsy. Posterior fossa cysts have caused gait ataxia and lower cranial nerve dysfunctions. Headache has been presented in almost every case regardless of the location of the cyst (Table 2).

Classification

Regarding size, we have classified them into three types according to the largest diameter. Three cases were smaller than 3 cm; seven between 3 and 6 cm; and twelve patients had a cyst larger than 6 cm. Therefore, in our series large cysts have been predominant most of them being located in the Sylvian fissure. There has been a natural correlation between the size of the cyst and the mass effect they produce. In 14 of our cases this mass effect was present. In 3 this effect was limited to the ventricular horn, but in 11 there was displacement of the midline structures (Table 2).

Treatment and Results

Twenty patients have been treated by craniotomy and wide excision of the cyst walls, creating a large communication between the cyst and the subarachnoid space. One of these cases, located in the paracentral medial convexity, required an insertion of cysto-atrial shunt 3 years later due to recurrence. This shunt became later infected, and had to be replaced. Two patients with Sylvian fissure cysts were treated with cysto-atrial shunts, and one of them had to be replaced into the peritoneum 4 months later due to shunt malfunction. There was no surgical mortality in relation to surgery. As postoperative morbidity we had one patient with a transient diplopia due to third nerve paresis, and serious problems (malfunction and infection) in 2 of the 3 patients treated with shunts.

The follow-up period varies from 4 months to 9 years, with a mean of 2.5 years. 13 patients had a complete regression of their symptomatology, 6 were improved and 2 have remained unchanged. The last patient showed neurological improvement following surgery, but he died 3 months later from bronchopulmonary infection (Table 3).

In 14 cases serial postoperative CT controls could be evaluated. In 11 patients a reduction of the cyst was seen, this reduction was considered very important in 7 and only moderate in 4. In 3 cases the situation remained unchanged (Table 4).

Histological studies of the excised membranes were performed in 20 out of 22 patients; in 2 cases treated with shunts there was no histological verification. In all of them the microscopic studies showed normal

Table 2

Case	Symptoms and signs	Location	Size (cm)	Mass effect
1	Headache	sylvian	6.1	+
2	Headache, vertigo	sylvian	12.3	+ + +
3	Headache, generalized epilepsy	temporal	1.5	—
4	Asymptomatic head injury, hemiparesis and aphasia	sylvian	8.2	+ + +
5	Generalized epilepsy	frontal	1.5	—
6	Headache, generalized epilepsy	retrocerebellar	12	+ + +
7	Asymptomatic subarachnoid haemorrhagy	temporal	4.5	—
8	Headache, left hemiparesis, focal epilepsy	Frontal	7.1	+ + +
9	Headache, generalized epilepsy	Frontal	6.1	+
10	Headache, vertigo	Temporal	3.6	—
11	Headache, monoparesis, dysphagia	Supracerebellar	4.5	+ + +
12	Generalized epilepsy	frontal	6	—
13	Headache, gait ataxia, vertigo, IX–X paresis	PC angle	4.1	+ + +
14	Focal epilepsy	temporal	2.4	—
15	Generalized epilepsy	parietal	4.6	+
16	Headache, gait ataxia, intracranial hypertension	retrocerebelar	5.1	+ + +
17	Right hemiparesis, aphasia (ictal onset)	sylvian	13.3	+ + +
18	Generalized epilepsy, left hemiparesis, and intracranial hypertension	occipital	8	—
19	Gait ataxia, vertigo, bilateral dysmetria, nystagmus, lower cranial nerves paresis	retro-lateral cerebellar	7.6	+ + +
20	Right hemiparesis	sylvian	10.7	+ + +
21	Right hemiparesis, and aphasia (ictal onset)	sylvian	10.7	+ + +
22	Headache, generalized epilepsy	sylvian	8	—

— no mass effect, + mass effect upon ventricular horn, + + + mass effect upon midline structures.

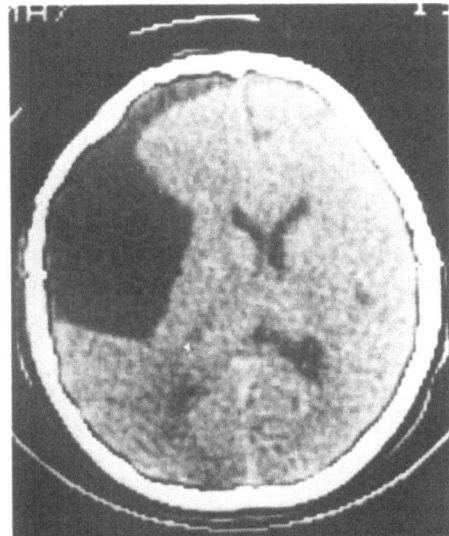

Fig. 2. Arachnoid cyst of the Sylvian fissure with a straight inner margin

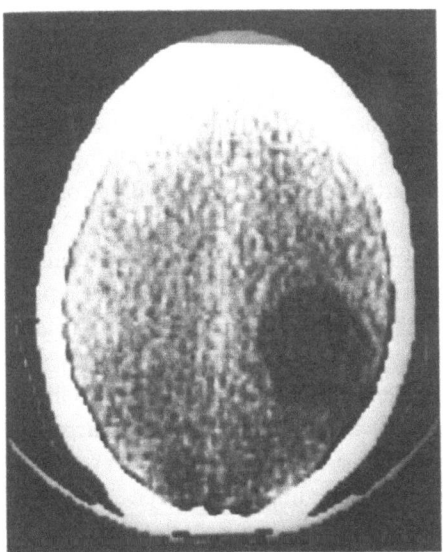

Fig. 3. Round parietal arachnoid cyst

Table 3. *Postoperative Follow-up*

Size (cm)	No.	Asymptomatic	Improved	Unchanged
<3	3	3		
3–6	7	4	3	
>6	12	6	4*	2
	22	13	7	2

* One died 3 months later due a pneumonia.

Table 4. *Postoperative CT Scan Evolution*

Size (cm)	no.	Marked reduction	Moderate reduction	Unchanged
<3	2		2	
3–6	4	2		2
>6	8	5	2	1
	14	7	4	3

arachnoidal membrane without inflammatory changes[14].

Discussion

Intracranial arachnoid cysts are lesions generally observed *in children*, and particularly during the first 6 months of life. The large published series belong to this age group[3, 12, 17, 24]. It is at this early age when the most dramatic clinical symptomatology is found. Intracranial hypertension can be acute, and therefore treatment has to be established as an emergency in many cases. Cyst drainage into the peritoneum or the atrium with valved shunts is the present treatment of choice for this group of patients[8, 9].

Intracranial arachnoid cysts *in the adult population* have not received enough attention in the medical literature. No large series is available to depict their clinical behaviour and to recommend the best form of treatment. From our series of 22 patients, with 17 cases over 20 years, some conclusions may be obtained. Acute intracranial hypertension seems not to be a prevalent problem in this group, even in cases with large cysts. Focal symptomatology is far more common, with epilepsy or a motor deficit usually marking the clinical onset. The majority of the cysts have been large and symptomatic, and most of them have shown some degree of intracranial expanding mass effect. As pointed out by Galassi[7] there seems to be a relation, with the middle cranial fossa cysts, between the existence and the degree of communication and the size and the degree of cerebral displacement. In fact we have observed a relation between the clinical improvement and the decrease in size of the Sylvian fissure cysts. CT scans[9, 13, 23] and NMR studies are correct means for diagnosis. Radionuclide cisternography seems not to be a prominent factor in relation to prognosis.

Small and medium size cysts have had an excellent outcome following surgery in our series. The large cysts over 6 cms in diameter have shown a postoperative clinical evolution parallel to their reduction in size[7].

Our results indicate the value of open surgery, craniotomy and membrane excision, versus shunt procedures. In this age group this more radical approach has offered a better solution with less morbidity. We believe that cystoperitoneal drainage is indicated only when a surgical wide communication between the cystic cavity and the basal cisterns is not feasable.

References

1. Auer LM, Gallhofer B, Ladurner G, Sager WD, Heppner F, Lechner H (1981) Diagnosis and treatment of middle fossa arachnoid cysts and subdural haematomas. J Neurosurg 54: 366–369
2. Bright R (1831) Serious cysts in the arachnoid. In: Reports of medical cases selected with a view of illustrating the symtpoms and cure of diseases by reference to morbid anatomy, vol II. Diseases of the brain and nervous system. Part I. Longman, Rees, Orme, Brown and Green, Paternoster-Row and S. Highley, London, pp 437–439
3. Choux M, Raybard C, Pinsard N (1978) Intracranial supratentorial cyst in children excluding tumour and parasitic cysts. Childs Brain 4: 15–32
4. Dyck P, Gruskin P (1977) Supratentorial arachnoid cysts in adults. Arch Neurol 34: 276–279
5. Ferreira S, Jhingran SG, Johnson PC (1980) Radionuclide cisternography for the study of arachnoid cysts: a case report. Neuroradiol 19: 167–169
6. Galassi E (1985) Infratentorial arachnoid cysts. J Neurosurg 63: 210–217
7. Galassi E, Tognetti F, Gaist G, Fagioli L, Frank F, Frank G (1982) CT scan and metrizamide, CT cisternography in arachnoid cysts of the middle cranial fossa. Classification and pathophysiological aspects. Surg Neurol 17: 363–369
8. Geissinger J, Kohler W, Robinson B, Devis F (1978) Arachnoid cysts of the middle cranial fossa: surgical considerations. Surg Neurol 10: 27–33
9. Guidicelli G, Hasson J, Choux M, Tonon C (1982) Les kystes "aracnoïdiens" supra-tentoriels. J Neuroradiol 9: 179–201
10. Go KG, Hendrik JM, Blaauw EH, Havinga P, Hartsuiker J (1984) Arachnoid cysts of the Sylvian fissure. J Neurosurg 60: 803–813
11. Harrison MJG (1971) Cerebral arachnoid cysts in children. J Neurol Neurosurg Psychiatry 34: 316–323
12. Hoffman HJ, Bruce E, Humphreys R, Armstrong E (1982) Investigation and management of suprasellar arachnoid cysts. J Neurosurg 67: 597–602

13. Jakubiak P, Dunsmore RH, Beckett RS (1968) Supratentorial brain cysts. J Neurosurg 28: 129–136

14. Krawchenko J, Collins GH (1979) Pathology of an arachnoid cyst. Case report. J Neurosurg 50: 224–228

15. La Cour F, Trevor R, Carey M (1978) Arachnoid cysts and associated subdural haematoma. Observations on conventional roentgenographic and computerized tomographic diagnosis. Arch Neurol 35: 84–89

16. Leo JS, Pinto RS, Hulvat GF, Epstein F, Krichell I (1979) Computed tomography of arachnoid cysts. Radiology 130: 675–680

17. Menezes A, Bell W, Perret G (1980) Arachnoid cysts in children. Arch Neurol 37: 168–172

18. Rengachary S, Watanabe I, Brackett Ch (1978) Pathogenesis of intracranial arachnoid cysts. Surg Neurol 9: 139–144

19. Robinson RG (1964) The temporal lobe agenesis syndrome. Brain 87: 87–106

20. Robinson RG (1971) Congenital cysts of the brain: arachnoid malformations. In: Krayenbühl H (ed) Progress in neurological surgery, vol 4. S Karger, Basel

21. Rousseaux M, Dhellemmes P, Clarisse J, Devos P (1982) Les kystes et les pseudokystes arachnoïdiens sous, retro, et supracéré-belleux de l'adulte. Neurochirurgie 28: 245–253

22. Starkman SP, Brown TC, Linell EA (1958) J Neur Exp Neurol 17: 484–500

23. Smith RA, Smith WA (1976) Arachnoid cysts of the middle cranial fossa. Surg Neurol 5: 246–252

24. Stein SC (1981) Intracranial developmental cysts in children: treatment by cysto-peritoneal shunting. Neurosurg 8: 647–650

25. Vaquero J, Carrillo R, Cabezudo J, Nombela L, Bravo G (1981) Arachnoid cysts of the posterior fossa. Surg Neurol 16: 117–121

26. Vizioli L, Cerillo A, Falivene R, Mottolese C, Tedeschi G (1983) Les kystes aracnoïdiens supratentoriels. Neurochirurgie 20: 339–347

Address for correspondence: Dr. M. Garcia-Bach, La Pau n° 44 1° 1ª Manresa, SP-08240 Barcelona, Spain.

Acta Neurochirurgica, Suppl. 42, 210–215 (1988)

Arachnoid Cysts of the Middle Fossa Predispose to Subdural Haematoma Formation Fact or Fiction?

A. C. Page, D. Mohan[1], and **R. M. Paxton**

Department of Radiodiagnosis and [1] Neurosurgery, Freedom Fields Hospital, Plymouth, U.K.

Summary

An association between arachnoid cysts of the middle fossa (ACMF) and complicating subdural haematoma (SDH) has been previously noted. More recently it has been hypothesized that ACMF may predispose to SDH formation.

The Plymouth Neurosurgical Unit has treated twenty patients with ACMF between 1976–1985, seven of these being complicated by SDH. There was an age range of 11–56 years in those with SDH. Six of the seven patients with ACMF and SDH gave no significant trauma history. Four of these were males aged 11 to 20 years. The presenting histories, clinical findings and subsequent management of these patients were compared with the age-matched males with SDH alone (twelve patients). In the SDH alone group 100% suffered major skull trauma and 80% had demonstrable skull fractures.

In addition the patients with ACMF were compared with patients presenting with other supratentorial arachnoid cysts in Plymouth. Only ACMF were associated with the development of SDH in our study. Three patients demonstrated total masking of the ACMF by isodense intracystic haematoma on computed tomography. In two of these patients the presence of an ACMF was suspected due to plain radiographic and CT enlargement of the middle fossa.

It is advocated that with special reference to young males, in the absence of major skull trauma an ACMF should be suspected as a predisposing factor to SDH. Postoperative CT scans for at least one year are advisable in young patients with SDH as demonstration of the presence of an ACMF and SDH affects future management.

Keywords: Arachnoid cyst; middle cranial fossa; subdural haematoma.

Introduction

Arachnoid cysts of the middle fossa (ACMF) have become increasingly recognised, along with other intracalvarial cystic lesions since the advent of CT head scanning in the mid 1970's. The last major review was by Robinson[14] in 1971, although since that time a plethora of case reports have been published in the neurosciences literature.

ACMF are thought to be congenital lesions[6, 14, 15, 17, 19, 22]. There is a male preponderance and ACMF usually present below 20 years, either with skull asymmetry (especially in infants and young children) or with evidence of decompensation, either due to haematoma formation, intracalvarial space occupation or raised intracranial pressure.

An association between ACMF and the formation of subdural haematoma has been noted[1, 2, 3, 4, 7, 11, 13, 15, 16, 22]. Robinson[14] hypothesized that ACMF may predispose to subdural haematoma formation, this having been reiterated by other authors[8, 15, 20, 22]. There may be a history of mild trauma but subdural haematoma may arise idiopathically. Ispilateral, contralateral or bilateral haematoma have been demonstrated.

The Plymouth Neurosciences Department is a subregional referral centre serving a population of 1.2 million. Scanning in this study was conducted during a 9 year period using an EMI 1010 head scanner. Twenty cases of ACMF have been treated from 1976 to 1985. Seven of these patients had associated subdural haematoma. In addition six other cases of supratentorial cysts at other sites have been treated.

Patients and Results

The twenty patients with arachnoid cysts of the middle fossa (ACMF) had an age range of 3 to 66 years, those seven with associated subdural haematoma (SDH) being aged 11 to 57, six being below 24 years. The full case histories of these latter seven patients are discussed in detail in Page *et al.* 1987[12] but a summary of the clinical and radiological findings and treatment is shown in Table 1. During the same study period 166 males and 88 females presented with SDH alone. The age distribution of these patients and those with an ACMF in addition are shown in Figs. 1 and 2.

The seven patients with ACMF and SDH all presented with evidence of raised intracranial pressure (see Table 1) while those patients with ACMF and no SDH had a variety of presentation, the majority demonstrating evidence of space occupation within the calvarium on CT head scanning (Table 2). The six non-middle fossa supratentorial arachnoid cysts presenting either had headache with

Table 1. *Treatment Regime all Recovery of Patients with Arachnoid Cyst of the Fossa and Subdural Haematoma*

Patient	Age and sex	Presentation	Skull radiographs	ACMF	SDH	Treatment	Recovery
1	15 male	raised intracranial pressure parasthesia left arm mild head injury 4 weeks previously	expanded right middle fossa	right	right	craniotomy right membranectomy with communication to basal cistern	good
2	17 male	occipital headache vomiting no papilloedema no head injury	expanded right middle fossa	right	right	craniotomy right membranectomy with communication to basal cistern	good
3	11 male	raised intracranial pressure mild head injury 2 weeks previously	normal	left (masked)	left	burr hole drainage left cysto-peritoneal shunt left craniectomy and membranectomy with communication to basal cistern	satisfactory asymptomatic after 4 operations
4	25 female	raised intracranial pressure mild head injury 3 weeks previously	expanded right middle fossa	right	right + left	burr hole drain left craniotomy right membranectomy cysto-peritoneal shunt	good
5	57 male	headaches nausea no papilloedema 30 seconds amnesia—post mild trauma	normal	left (masked)	right + left	craniectomy left membranectomy and communication to basal cisterns	satisfactory continuing headaches no residual midline shift
6	17 male	raised intracranial pressure no trauma	expanded left middle fossa	left (masked)	left	burr hole drainage only	good
7	12 female	raised intracranial pressure no trauma	expanded left middle fossa	left	left	craniotomy membranectomy and communication to basal cisterns	good

Table 2. *Comparison of Findings in Patient with Arachnoid Cysts of the Middle Fossa and Other Supratentorial Arachnoid Cysts Treated in the Plymouth Neurosciences Department 1976—1985*

		Symptoms and signs in association with space occupation				Symptoms and signs without space occupation	
		Subdural haematoma	Epilepsy	Headache with localising signs	Headache alone	Headache alone	Asymptomatic
ACMF	male	5	1	2**	3	1	2*
	female	2	1	2	0	1	0
Parietal frontal and occipital cysts	Male	0	0	0	1	0	0
	female	0	0	2	1	1	1

* One patient presented post road traffic accident with an extradural haematoma contralateral to an arachnoid cyst of the middle fossa. ACMF was not directly treated. Extradural evacuated. Now fit and well.
** Infants presented with macrocrania.

Table 3. *Case History Summary of Males Aged 10–20 Years with Subdural Haematoma Compared with Males Aged 10–20 Years with Subdural Haematoma and Arachnoid Cysts of the Middle Fossa*

	Group 1	Group 2
	subdural haematoma	subdural haematoma and arachnoid cysts of the middle fossa
Number of patients	12	4
Characteristics	100% major trauma (RTA or sporting)	50% minor trauma 50% no trauma
	80% skull fracture	no skull fractures
Skull asymmetry	none	3 enlarged ipsilateral middle fossa 1 no asymmetry

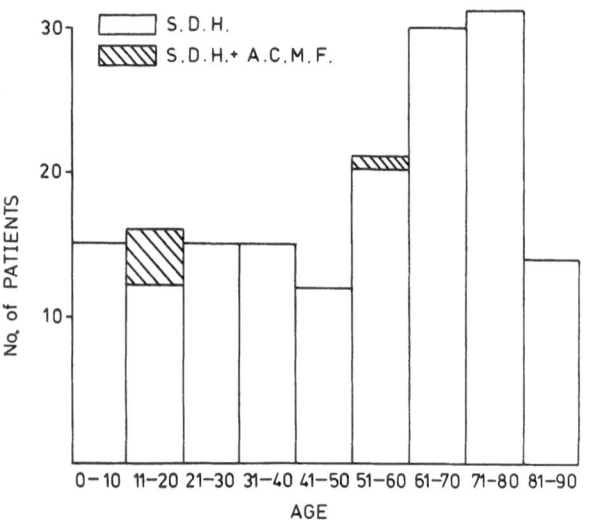

Fig. 1. Histogram demonstrating the age at presentation in male patients with subdural haematoma alone and patients with subdural haematoma and accompanying arachnoid cyst of the middle fossa presenting to the Plymouth Neurosciences Department, 1975–1986

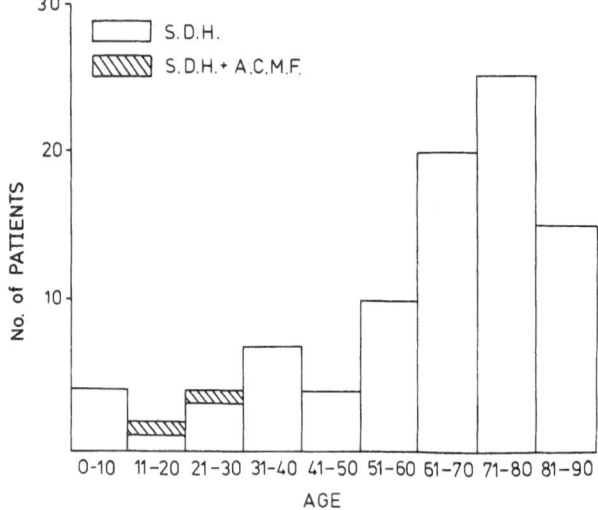

Fig. 2. Histogram demonstrating the age at presentation of female patients with subdural haematoma alone and subdural haematoma in association with arachnoid cyst of the middle fossa to the Plymouth Neurosciences Department from 1976–1985

localising neurological signs, headache alone or were asymptomatic. None were associated with subdural or intracystic haematoma (see Table 2).

Three of the patients with ACMF and SDH demonstrated intracystic haematoma masking the presence of the ACMF. Two of these showed cranial asymmetry with middle fossa expansion on the side of the cyst. One patient, case number 5, demonstrated intracystic haemorrhage in evolution (Figs. 3a to c). This followed mild head injury in a riding accident, the intracystic haemorrhage was not demonstrated at initial presentation but was seen six weeks later. The cyst became visible again again following craniectomy and drainage procedure. Five patients had evidence of intracystic haemorrhage at craniotomy.

Figs. 1 and 2 demonstrate an increasing incidence of subdural haematoma formation with increasing age in both sexes overall, although a higher base line is seen in males in the first four decades. The patients with ACMF and SDH are a younger age group, definitive clustering seen in males aged 10 to 20, thus agreeing with other authors[2, 9, 14]. The case histories of males aged 10 to 20 with ACMF and associated SDH and those with SDH alone were compared and contrasted (Table 3). It is apparent there are two completely different populations, those with SDH alone all suffered major skull trauma, while of those four patients in the ACMF and

SDH group, two gave a history of minor trauma and the other two gave no history of trauma. Skull asymmetry was noted in three of the four patients with ACMF and SDH and in none of the twelve with SDH alone.

Discussion

The haemorrhagic complications occurring in patients with ACMF are demonstrated in seven cases in this paper and by many other authors[1, 2, 3, 4, 9, 10, 11, 15, 16, 17, 22, 23]. Isodense haematoma may be difficult to identify[3] but with modern fourth generation CT scanners the grey/white matter interface position in relation to the inner table of the skull vault gives an indication of

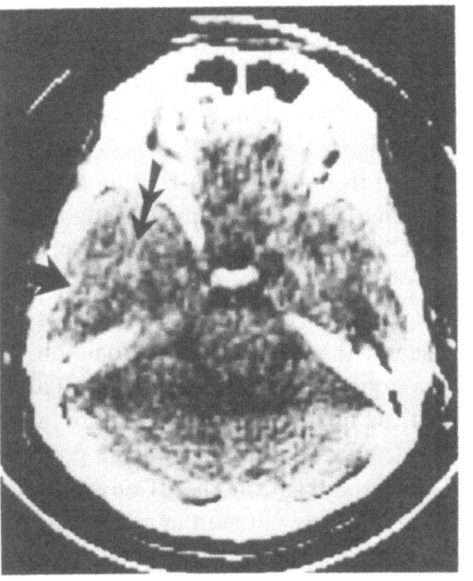

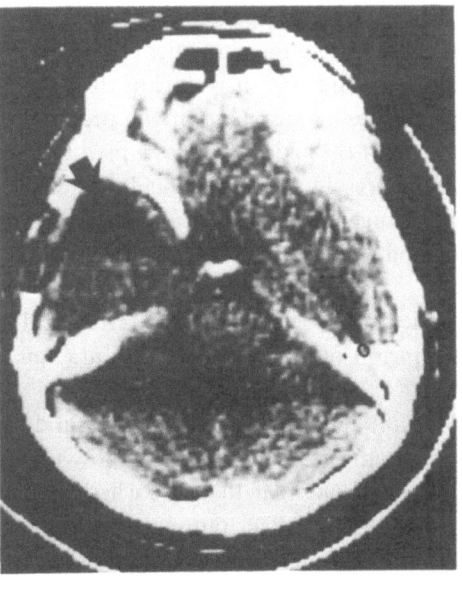

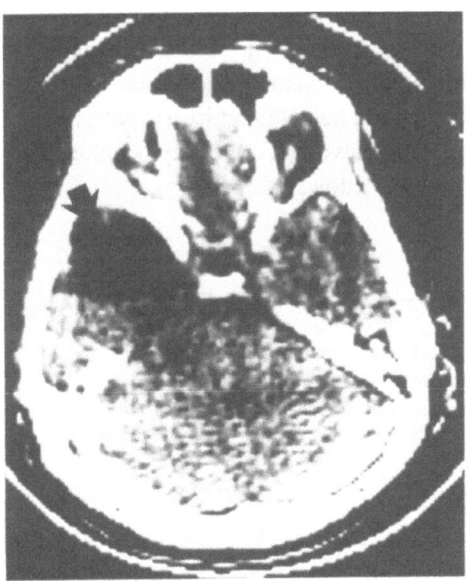

Fig. 3. Sequential CT scans of case number 5 (L + 44, W 75). (a) At the level of the middle fossa and greater wing of sphenoid demonstrated a low attenuation area (arrow) in the anterior part of a non-enlarged left middle fossa. This was conducted one week post initial riding accident in which the patient the patient suffered 30 seconds posttraumatic amnaesia. (b) 6 weeks post (a). CT head scan at same level as (a). An isodense subdural haematoma (single arrow), and isodense intracystic haemorrhage (double arrow) mask the arachnoid cyst previously seen in the left middle fossa. (c) CT head scan [level as (a) and (b)] six weeks post craniotomy, membranectomy and communication to basal cisterns demonstrating re-appearance of low attenuation area in the left middle fossa (arrow). Histological confirmation of arachnoid lining the cystic area was obtained at operation

their presence. In equivocal cases the use of further invasive investigations *e.g.* angiography has been advocated[9].

Intracystic haemorrhage has been demonstrated on CT[4, 8, 18] and was found in three cases in this study (Table 1). La Cour[9] hypothesized the potential masking of an ACMF by intracystic haemorrhage on CT. One case in the paper by Auer[3] and in cases 3, 5, and 6 in this study demonstrated this. Fig. 3 elegantly illustrated intracystic haemorrhage and cyst masking in evolution.

The occurrence of intracystic haemorrhage may be more common than at first realised, five out of seven patients in this study having demonstrable haematoma within the cyst at operation. This concurs with work by other authors[4, 5, 21]. A review of patients with SDH alone (see Figs. 1 and 2) and specific comparison of males aged 10 to 20 (Table 3) allowed us to confirm the hypothesis[8, 14, 15, 20, 22] that ACMF predisposed to subdural haematoma formation. Whether one can extend this to other age groups in both sexes cannot be accurately assessed from this study. To date no other study has compared the two groups, although most patients with ACMF and complicating SDH present under 20 years[2, 9, 14].

La Cour[9] suggests that the presence of skull asymmetry in a patient with a subdural haematoma on CT and in addition a greater than expected contralateral displacement of the middle cerebral arteries on angiography should lead to the suspicion of the presence of an ACMF. This study confirms the CT observations, however, angiography was not used in any of our patients. Referall to Table 3 shows that not only the presence of middle fossa expansion but also the history itself *i.e.* no major trauma, should lead to the suspicion of the presence of an ACMF in these patients.

We therefore advocate that in patients with a subdural haematoma and an insignificant history of trauma, a high index of suspicion of the presence of an arachnoid cyst of the middle fossa as a predisposing factor should be present. This is particularly relevant for those patients aged under 20. If there is skull asymmetry in addition, the diagnosis becomes almost certain.

In those patients who satisfy the above criteria, thus raising the suspicion of the presence of an arachnoid cyst either:

1. Following initial burr-hole drainage of a subdural haematoma, the patients should receive follow-up CT scans for a period of one year to assess whether there is a predisposing ACMF. The rationale for this approach is that although burr-hole drainage alone may

be satisfactory for subdural haematoma in isolation, in the patient with an arachnoid cyst in addition the most successful and comprehensive treatment is craniectomy, membranectomy and communication to the basal cisterns[1, 4, 21]. This, as discussed above is particularly relevant in those patients who presend under the age of 20.

2. Craniectomy, membranectomy and cyst communication to the basal cisterns should be planned on the basis of clinical and CT or MRI alone, or

3. Further investigation preoperatively such as angiography should be undertaken.

Conclusions

1. Arachnoid cysts of the middle fossa do predispose to subdural haemtoma formation. (Evidence definitive for males aged 10 to 20 years.)

2. Intracystic haemorrhage may mask the presence of an arachnoid cyst. (In the presence of minimal or no trauma history in a patient with a subdural haematoma, an arachnoid cyst of the middle fossa should be suspected as a predisposing factor.)

3. Follow-up for at least one year is mandatory in those patients under 20 years who present with subdural haematoma and no trauma history. Alternatively further preoperative investigations may be undertaken if the clinical status allows.

Acknowledgements

We wish to acknowledge Mr. E. Strachan and Mr. H. Gossman for allowing us to use some of their patients in this study. We wish to thank Mrs. H. R. Blakely for typing the manuscript and Mr. T. McCausland for preparing the illustrations.

Fig. 3 reproduced with permission, Journal of Neurology, Neurosurgery and Psychiatry.

References

1. Anderson FM, Segall HD, Caton WL (1979) Use of CT scanning in supra-temporal arachnoid cysts. J Neurosurg 56: 333–338
2. Akardi J, Baumann F (1975) Supratentorial extracerebral cysts in infants and children. J Neurol Neurosurg Psychiatry 38: 57–68
3. Auer LM, Gallhoper B, Ladurner G, Sager WD, Heppner F, Lechler H (1981) Diagnosis and treatment of middle fossa arachnoid cysts and subdural haematomas. J Neurosurg 54: 366–369
4. Galassi E, Piazza G, Gaist G, Frank F (1980) Arachnoid cysts of middle cranial fossa: a clinical and radiological study of 25 cases treated surgically. Surg Neurol 14: 211–219
5. Geissinger JD, Kohler WC, Robinson BW, Davis FM (1978) Arachnoid cysts of middle fossa; surgical considerations. Surg Neurol 10: 27–33
6. Guidicella G, Hassoun J, Choux M, Tonon C (1982) Les kystes "arachnoidiens" supratentorials. J Neuroradiol 9: 179–201

7. Kadowaki H, Ide M, Takara E, Yamamoto M, Imanaga W, Jimbo M (1983) A case of arachnoid cyst associated with chronic subdural haematoma. No Shinkei Geka 11: 431–436

8. Kusuno K, Yoshida Y, Takahashi A, Isaii S (1984) Chronic subdural hygroma caused by rupture of arachnoid cyst. Neurol Med Chir Tokyo 24: 349–354

9. La Cour F, Trevor R, Carey M (1978) Arachnoid cysts and associated subdural haematomas. Arch Neurol 35: 84–89

10. Lesoin F, Dhellemmes P, Rousseux M, Jomin M (1983) Arachnoid cysts and head injury. Acta Neurochir (Wien) 69: 43–51

11. Oliver LC (1958) Primary arachnoid cysts. BMJ 1147–1149

12. Page AC, Paxton RM, Mohan D (1987) A reappraisal of the relationship between arachnoid cysts of the middle fossa and subdural haematomas. J Neurol Neurosurg Psychiatry 50: 1001–1007

13. Robinson RG (1964) The temporal lobe agenesis syndrome. Brain 87: 87–106

14. Robinson RG (1971) Congential cysts of the brain: arachnoid malformations. Progr Neurosurg 4: 133–174

15. Smith RA, Smith WA (1976) Arachnoid cysts of the middle fossa. Surg Neurol 5: 246–252

16. Sprung CH, Mauersberger W (1979) Value of CT for the diagnosis of arachnoid cysts and assessment of surgical treatment. Acta Neurochir (Wien) 28: 619–626

17. Starkman SP, Brown TC, Linell EA (1958) Cerebral arachnoid cysts. J Neuropath Exp Neurol 17: 484–500

18. Stein SC (1981) Intracranial developmental cysts in children: treatment by cystoperitoneal shunting. Neurosurgery 8: 647–650

19. Tiberin P, Gruszkiewicz J (1969) Chronic arachnoidal cysts of the middle fossa and their relation to trauma. J Neurosurg Psychiatry 24: 86–91

20. Urich H (1976) In: Bachwood W, Corsellis JAN (eds) Greenfields neuropathology. Edward Arnold, London, pp 394

21. Van der Meche EAA, Braakman R (1983) Arachnoid cysts in the middle cranial fossa: care and treatment of progressive and non-progressive symptoms. J Neurol Neurosurg Psychiatry 46: 1102–1107

22. Varma TRK, Sedzimir CB, Miles JB (1981) Posttraumatic complications of arachnoid cysts and temporal lobe agenesis. J Neurol Neurosurg Psychiatry 44: 29–34

23. Weinberg PE, Flom RA (1973) Intracranial subarachnoid cysts. Radiology 106: 329–333

Address for correspondence: A. C. Page, M.D., Plymouth General Hospital, Department of Radiodiagnosis, Longfield House, Longfield Place, Plymouth PL1 1BR, U.K.

Acta Neurochirurgica, Suppl. 42, 216–220 (1988)

Clinical and Neuropsychological Results After Operative and Conservative Treatment of Arachnoidal Cysts of the Perisylvian Region

U. Kunz, N. Rückert, J. Tägert, and **H. Dietz**

Neurosurgical Department of Hannover Medical School, Federal Republic of Germany

Summary

Most supratentorial arachnoid cysts are found in the perisylvian region. They are thin extracerebral fluid filled pouches and occasionally symptoms are caused by space occupation.

We have studied 28 cases with these cysts, with follow-up time varying from one to ten years. 8 of the cases presented with a sudden onset of seizures, while 5 cases had a history of seizures dating from childhood. 14 cases underwent surgery, of whom most had histological evidence of previous haemorrhage and some had evidence of subdural haematoma. Surgery was uneventful in all cases regardless of the method used and all had satisfactory outcome.

All 28 cases were followed up systematically including intelligence testing and specific testing of memory and other temporal lobe functions. The group scored normal in standardized tests with a distribution similar to the normal population. Results did not show any deficit of function related to the localization of the cyst.

Despite of the severity of their initial symptoms, postoperative cases performed just as well at psychological testing. Those cases who continued to experience seizures showed a memory disturbance and seemed to lack initiative. Most cases studied had normal error scores in the dichotic listening examinations. There was no connection between the side of the cyst and a unilateral deficit. The results are discussed.

Keywords: Arachnoidal cyst; supratentorial; findings; psychological testing; treatment; results.

Introduction

Arachnoid cysts are thin fluid filled pouches, extracerebrally located. Some have a space occupying activity. 50 to 68% are found in the perisylvian region. Robinson[22] put forward the view that a developmental anomaly of mainly the temporal lobe was the basis of these cysts. He suggested the term "temporal lobe agenesis syndrome". Other authors used a different nomenclature according to different pathogenetic concepts. Raised intracranial pressure after trivial head injuries, seizures, and skull deformity have been listed as symptoms. In former time the cysts were made visible by operation or autopsy; angiography revealed only in direct signs (avascular area, displacement of vessels)[15]. Since the introduction of CAT the cysts are made visible without operation[8, 10]. The now often used diagnostic tool of CAT shows up intracranial cysts with only trivial or uncharacteristic symptoms[5] – in contrast to former times when invasive diagnostic procedure and operation had to be justified by relatively severe symptoms[13]. So it seems necessary to have a new look at arachnoid cysts of the perisylvian region including unoperated cases in which the cyst was made visible by CT scan. This study presents a follow-up of operated and unoperated cases. History, neurological, neurophysiological and radiological data have been collected. Conclusions will be drawn concerning pathogenesis, management and prognosis of patients with intracranial cysts.

Material and Methods

The sample consists of all accessible patients of our department in whom a cyst of the perisylvian region was found by CT scan, covering the period from 1976 to 1986. There are 28 patients (22 male, 6 female). Surgery was performed in one half of the cases. Age at follow-up was between 5 and 59, mean 28 years (Fig. 1). Follow-up examination included CAT scan, history, neurological and neuropsychological examination. Cysts are classified[9] according to size (3 types) and location. 10 of the 14 unoperated patients had a positron emission tomographic scan (PET) to obtain a valid diagnosis. 18 had an EEG. The psychopathological assessment was done by two examiners, using the AMDP-system[1]. For neuropsychological examination the following tests were used: From the Wechsler Adult Intelligence Scale the subtests information, similarities, picture completion and block design for general intelligence[6], the subtest digit span[6] and Benton's visual retention test[3] for short-term memory, the Rey-Osterrieth-Figure[20] and the memory scale from WIT[16] resp. for children from HSET[11] for long-term memory testing. Lateral dominance was assessed by a hand dominance examination[14] and dichotic listening test[23] (Fig. 3).

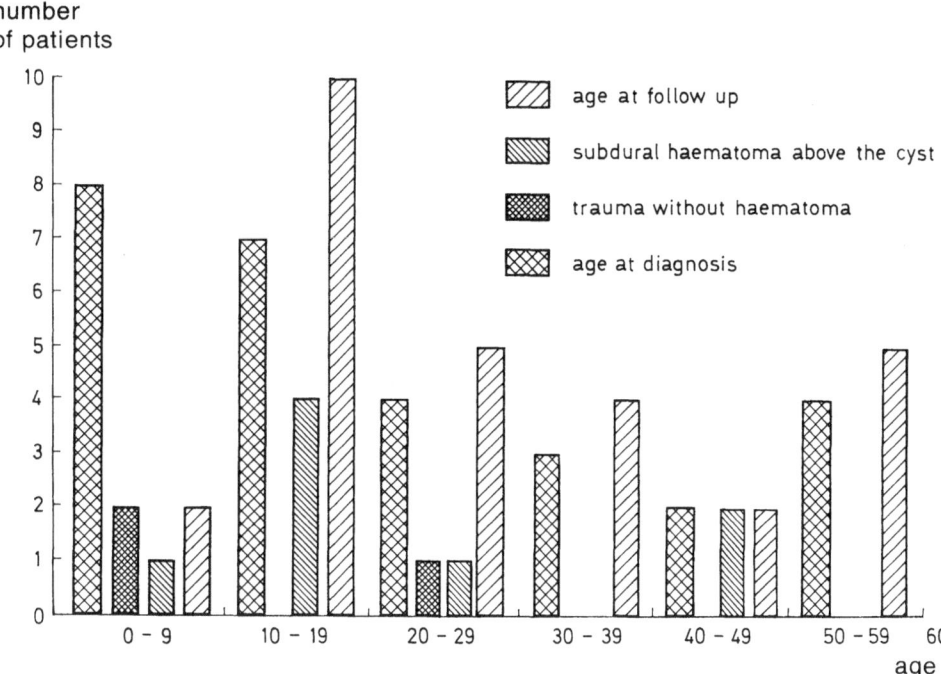

Fig. 1. Age-distribution of the groups

Results

Half of the patients showed symptoms during childhood and adolescence (Fig. 1). More cysts were on the left than on the right side (18 left – 10 right). In 22 cases the sylvian fissure formed part of the cyst. The other 6 cysts reached this lateral fissure only in parts. Three were in the "three-territory border" area, the other 3 were seen in the frontal lobe. Two babies with large frontal cysts developed cyanotic signs soon after birth. In another case of a 38-year-old man, his mother had had German measles during pregnancy; his left hemisphere is atrophied. Because of raised ICP he has been operated on three times during the past 30 years. The last operation was in our department. It showed a cyst with basal multicystic gliosis. The parents of 9 patients told us that there had been a case of head injury in the first 4 years of their children's life. In all cases they had not called a doctor. A 54-year-old patient had had a subarachnoid haemorrhage, a source was not found. CAT scan showed a temporal arachnoid cyst, which had not caused any symptoms. A 17-year-old boy was operated on ten years ago because of a hemiparesis which was considered to be due to a temporal arachnoid cyst. There is a family history of Recklinghausen's disease; the boy has neurinoma of the optic nerve with loss of the medial sphenoid wing.

Psychomotor and cognitive development during childhood was normal in most patients. Three were stutterers, an other one was classified at school as dyslexic. One had a delayed speech development. Taking into account the parents' social background the children were normal during their school years and later careers. Only the patient with hemiatrophy of the brain worked at a special workshop for the disabled.

The symptoms at first contact are seen in Fig. 2. It is natural that the symptoms of the operated group were heavier. At follow-up 3 had a hemiparesis, 4 had occasional headache, 5 continued with seizures. There were seizures in the history of 13, 6 of the complex partial type, 2 of the simple partial type, 4 other patients had partial seizures which secondarily generalized. Primary generalized seizures were only found in one patient who was an alcoholic and had no seizures after achieving abstinence. (We suppose that the seizures were caused by his alcoholism, not by his cyst.) 8 cases were diagnosed because of seizures; all were without seizures at follow-up. Of the 5 cases with a long term history of seizures, 4 were operated and only one was without at follow-up.

Facial asymmetry was only seen in the two cases shown by Markakis *et al.*[19]. One is the patient with hemispheric atrophy and the other the one with Recklinghausen's disease. 2 further patients show skull excavation above the cyst.

Psychopathological assessment showed abnormal results in 25% of the patients. Lack of initiative and dysphoria were the main symptoms. They were char-

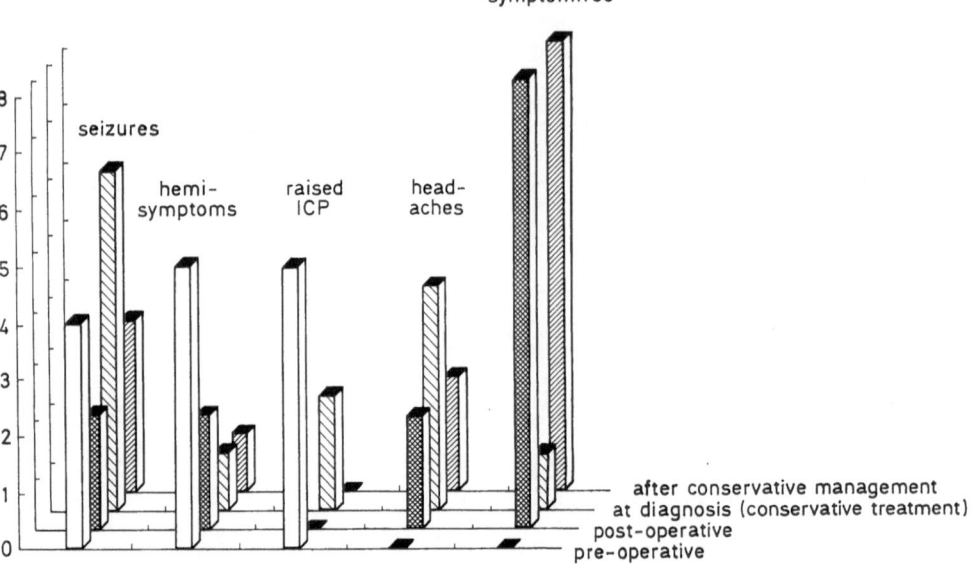

Fig. 2. Symptoms and course of patients with perisylvian arachnoidal cyst after operation and conservative treatment (n = 28)

acteristically found in patients with seizures. Symptoms were more distinct if there was a long history of seizures. In the dichotic listening test the majority of patients had a normal number of errors without pathological preference of one side. There was a minority of pathological unilateral deficits: 3 patients had right ear deficit, one a left ear deficit, three of these cases have the deficit opposite to the cyst; one on the same side. All these patients showed symptoms of the cyst in early childhood. Markakis *et al.*[19] also used this dichotic listening test. The 4 patients, who were in both groups showed the same result in both procedures. 25 have a right hand dominance – 2 a left. This is distributed as in normal population.

The intelligence test scores were normal with a distribution corresponding to the population at large. Results did not show any deficit of function related to the location of the cyst. The group of patients with seizures had a deficit of memory function (T = 2.27, p < 0.05). Despite the severity of their initial symptoms, postoperative cases performed just as well at psychological testing as patients with conservative management (Fig. 3).

The EEG was normal in 10 cases. 8 showed a focus above the cyst. Cerebral angiography was performed in a considerable part of the sample. X-rays were accessible in 4 cases: There was the avascular space occupation with displacement of middle cerebral artery. The middle cerebral (sylvian) vein was visible in 3 cases bordering the cyst.

In ten cases a PET scan was performed by the use of C^{11}marked Methionin and H_2O^{15}. The cystic area showed a reduced activity. The extent was the same as in CAT scan, but in two cases of type II cysts the lesion appeared smaller when H_2O^{15} was used. We found no reliable connection between clinical signs and the size[9] of the cysts.

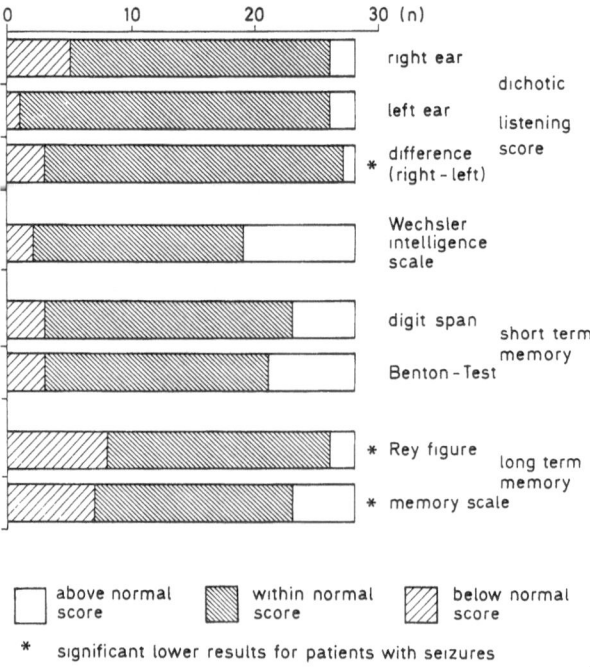

Fig. 3. Neuropsychological test results

We have operated 14 cases. The symptoms are listed in Fig. 2. Only one required a shunt after the standard procedure with craniotomy and basal opening of cyst membranes. In 8 of the patients a chronic subdural haematoma above the cyst was confirmed during surgery. In this sample there were no cases with postoperative complications. A negligible decrease in the size of the cyst was visible in 5 operated cases and in 2 type I cysts without treatment.

Discussion

There are different hypotheses concerning the pathogenesis of cysts in the perisylvian region[8, 12, 13, 15, 17, 18, 19, 22]. Robinson[22] put forward the view that there is a clinical entity which is based on agenesis of the frontotemporal region, probably occurring during the three last months of foetal life. He introduced the term "temporal lobe agenesis syndrome". According to his hypothesis the subarachnoid cyst fills the remaining space. Robinson[22] was impressed by the similarity in adult autopsies with sylvian depression during foetal development before conversion into the insula, opercula and the sylvian fissure. But this similarity holds true only for those cysts located inside the sylvian fissure. There are other locations, *e.g.* temporobasal, which cannot be explained by an arrest of development during embryogenesis.

The infarction hypothesis[17, 24] seems capable of explaining different cyst locations. According to this pathogenetic theory cysts fill the space remaining after infarction during the foetal period, birth, or early childhood. The temporopolar and presylvian regions are supplied by the middle cerebral artery, the temporobasal region by the posterior cerebral artery and the median part of the temporal lobe by the choroidal therapy. Cysts located in these different regions may attributed to an infarction of these arteries. The predilection for the perisylvian region may be explained by the course of arteries of the temporal lobe across the tentorial notch[7]; other vascular disease also seem possible[12]. We expect more information from ultrasonic examination and follow-up during pregnancy and after birth.

The findings of other authors concerning predilections of laterality and sex are confirmed. The ratio of left to right-sided cysts is 18 to 10. The ratio of male to female is 22 to 6. The same predilections are known for infarctions during childhood[12]. Up to now there is no satisfactory explanation for this predilection for males and for the left side. It remains a matter of speculation if there is a common cause. Huber's[15] explanation describing the loss of the sylvian vein cannot be held, because we have found it in 3 of 4 cases at angiography. There must be a sensitivity for subdural haematoma in trauma because of remaining veins.

The clinical signs of temporal cysts are hemisymptoms (pareses), seizures or raised ICP. If there are seizures they are partial ones, in most cases complex partial seizures. If there are primary generalized seizures the lesion cannot be a simple perisylvian cyst. The usually reported skull deformity is in our experience not a necessary symptom.

A disturbed development in childhood is rare in our cases. Because of the preference for the left side it is not surprising that there was a disorder of speech in 5 cases. In contrast to the results of Markakis *et al.*[19] there was a normal distribution of handedness in our cases. In dichotic listening there was a majority of normal error scores without pathological preference of one side, and a minority of four cases with pathological one-sided deficit. Only a deficit opposite to the cyst may be interpreted as a direct sign of perisylvian dysfunction, such a contralateral deficit was found in three cases.

We found that the subgroup of patients with seizures suffered from long term memory disturbance. In contrast to a former report[18] we found no correlation between the disturbance and the location of a cyst. It seems plausible that seizures and memory disturbance result from a common cause – namely lesion of the special brain regions like hippocampus, temporal stem and temporal pole which participate in the cause of seizures and memory deficit[4]. Side-effects of anticonvulsant medication must also be taken into account[21]. Because the psychosocial development of a patient with a cyst is impaired by the experience of seizures they must be treated early. An operative procedure seems to be successful only if the history of seizures is short.

Since the introduction of CAT reflections upon the diagnostic procedure have become unnecessary. There are also new possibilities in MR tomography and PET. But PET is necessary only for scientific information such as fluid connections and neuropsychological presentation on the brain surface. Obviously it is possible to use this method to differentiate hypodense low grade gliomas from cysts because of their metabolism. The MR tomography also brings information about the fluid movement and is more available.

The management of cysts depends on their clinical symptoms, not on their size. Patients with hemiparesis, papilloedema and a history of seizures were operated on. 8 of them had a subdural haematoma above the cyst which was evacuated. Symptoms decreased or disappeared. Only in cases with a long history of seizures was there no impressive improvement. The size of the cysts was not significantly influenced by surgery — contrary to Beltramello's[2] result after conservative management. Patients with only trivial symptoms and non-progressive headaches were not operated, independent of the cyst size; their subsequent course was also favourable. According to our findings there is no need to change the criteria of management. However, our results do not include sufficient information about shunt procedure, it is possible that this is the simpler operation with less physical strain for adult patients.

References

1. Arbeitsgemeinschaft für Methodik und Dokumentation in der Psychiatrie (Hrsg) (1983) Testmanual zum AMDP-System. Empirische Studien zur Psychopathologie. Springer, Berlin Heidelberg New York
2. Beltramello A, Mazza C (1985) Spontaneous disappearance of a large middle fossa arachnoid cyst. Surg Neurol 24: 181–183
3. Benton AL (1961) Der Benton-Test. Huber, Bern Stuttgart Wien
4. Birri R, Perret E, Wieser HG (1982) Der Einfluß verschiedener Temporallappenoperationen auf das Gedächtnis bei Epileptikern. Nervenarzt 53: 144–149
5. Colameco S, Ditomasso R (1982) Arachnoid cysts associated with psychological disturbances. J Med Soc NY 79: 209–210
6. Dahl G (1986) WIP. Handbuch zum reduzierten Wechsler-Intelligenztest. 2. Aufl., Königstein/Ts, Hain
7. Earle KM, Baldwin M, Penfield W (1953) Incisural sclerosis and temporal lobe seizures produced by hippocampal herniation at birth. Arch Neurol 69: 27–42
8. Galassi E, Piazza G, Gaist G, Frank F (1980) Arachnoid cysts of the middle cranial fossa: a clinical and radiological study of 25 cases treated surgically. Surg Neurol 14: 211–219
9. Galassi E, Toguetti F, Gaist G, Fagioli L, Frank F, Frank G (1982) CT scan and metrizamide CT-cisternography in arachnoid cysts of the middle cranial fossa: classification and pathophysiological aspects. Surg Neurol 17: 363–369
10. Galassi R, Clardulli C, Ferrara R, Lorusso S, Galassi E, Lugaresi E (1985) Asymptomatic large arachnoid cyst of the middle cranial fossa. Eur Neurol 24: 140–144
11. Grimm H, Schöler H (1987) Heidelberger Sprachentwicklungstest. Westermann, Braunschweig
12. Gross H, Jellinger K (1969) Morphologische Aspekte zerebraler Mißbildungen. Wien Z Nervenheilkunde 27: 9–37
13. Hardman J (1939) Asymmetry of the skull in relation to subdural collection of fluid. Br J Radiol 12: 455–461
14. Harris A (1974) Harris tests of lateral dominance. The psychological corporation (ed) 3rd Edition, New York
15. Huber P (1961) Die temporale Arachnoidalzyste im angiographischen Bild. Fortschr Röntgenstrahl 94: 755–768
16. Jäger AO, Althoff K (1983) Der Wilde-Intelligenz-Test. Hogrefe Verlag, Göttingen
17. Kolberg T (1978) Die incisurale Porencephalie des Temporallappens und das "temporal lobe angenesis syndrom". Neurochirurgia (Stuttg) 21: 14–20
18. Lang W, Lang M, Kornhuber A, Gallwitz A, Kriebel J (1985) Neuropsychological and neuroendocrinological disturbance associated with extracerebral cysts of the anterior and middle cranial fossa. Eur Arch Psychiatry Neurol Sci 235: 38–41
19. Markakis E, Heyer R, Stoeppler L, Werry H (1979) Die Aplasie der perisylvischen Region. Neurochirurgia (Stuttg) 22: 211–220
20. Rey A (1959) Test de cople et de reproduction de mémoire de figures géométriques complexes. Paris
21. Reynolds EH (1983) Mental effects of antiepileptic medication. Epilepsia 24 [Suppl] 2: 85–95
22. Robinson RG (1971) Congenital cysts of the brain: arachnoid malformations. Progr Neurol Surg Vol 4. Karger, Basel
23. Sipos J, Tägert J (1976) Ein neuer dichotischer Hörtest als neuropsychologisches Untersuchungsverfahren. Nervenarzt 47: 329–332
24. Zuelch KJ (1985) The cerebral infarct, pathology, pathogenesis and computed tomography. Springer, Berlin Heidelberg New York

Address for correspondence: Dr. U. Kunz, Neurosurgical Department of Hannover Medical School, Konstanty Gutschow Strasse 8, D-3000 Hannover 61, Federal Republic of Germany.

Acta Neurochirurgica, Suppl. 42, 221–224 (1988)
© by Springer-Verlag 1988

Cerebral Scintigraphy with [111]Indium Oxine-Labelled Leukocytes in the Differential Diagnosis of Intracerebral Cystic Lesions

C. Bellotti, M. Medina, G. Oliveri, F. Ettorre, S. Barrale, C. Sturiale, G. Camuzzini, M. G. Aragno, and A. L. Viglietti[1]

Neurosurgery Division and [1] Nuclear Medicine Service, Santa Croce General Hospital, Cuneo, Italy

Summary

CT scanning and scintigraphy with [111]Indium-oxide-labelled white blood cells were used to study 32 cases of intracerebral cystic lesions. The results and the criteria of positivity used to lower the false positive rate are discussed. A new criterion, designed to assess the time course of the scintiscan and so reduce still further the frequency of false positives is put forward.

Keywords: Intracranial abscess; ^{111}In-oxine-labelled leukocytes; intracranial cystic lesion; scintigraphy; computerized tomography.

Introduction

The differential diagnosis of CT hypodense intracerebral cystic lesions is a major problem of clinical and instrumental diagnosis in neurosurgery, still solved only in part despite great technical advances and the increased diagnostic accuracy that have marked the development of Computed Tomography (CT) in the past decade[3, 8, 10, 17].

Of particular importance within this field is the diagnosis of cerebral abscesses. Especially in the differential diagnosis from malignant cystic tumours, it is of key importance both in defining the criteria of operability (one need only think of sites like the thalamus or brain stem) and in the planning of the surgical approach, since we, in common with many other workers, are convinced that if the lesion is an abscess, the best course is simple evacuation by puncture, repeated, if necessary, maybe with stereotaxic procedures, and that in any case the operation should be planned and conducted in such a way as to prevent the dissemination of purulent material[14, 16].

The persistence of these diagnostic difficulties have recently stimulated keen interest in the search for complementary methods of investigation to supplement traditional neuroradiological methods of differential diagnosis of intracerebral cystic lesions with a view to certain detection of abscesses or abscessed lesions.

An almost constant response of the organism to any bacterial attack is the migration of leukocytes to the site of infection. On this assumption scintiscanning with tagged leukocytes has long been in use in the detection of occult abscesses, especially abdominal[2, 4, 5, 6]. Its use in the detection of cerebral abscesses has recently been reported[1, 9, 12, 13].

We set out to assess this method in a personal series of neurosurgical abscesses using a method of image reading that we have devised in the hope of lowering the false positive rate, reported by other authors for other regions[2, 10], related to the migration of leukocytes to tumours that have not yet formed abscesses.

Material and Method

The method used for white blood cell (WBC) separation and labelling was that proposed by Thakur in 1977[15] and modified by Hardeman in 1979[7]. Scintigraphy is performed 4 to 6 hours and again 24 hours after administration of the labelled cells using a large-field gamma camera. Anterior, posterior, lateral and, if necessary, axial views are obtained.

Results

We studied 32 patients, 21 men and 11 women, between the ages of 19 and 74 years who had consulted us for intracranial hypertension, focal neurological deficits or epileptic seizures. Immediately before undergoing scintigraphy all the patients presented an intracerebral cystic space-occupying lesion on CT scanning. In 21 cases the CT scan neither excluded nor confirmed that the lesion was an abscess while in the other 11 cases it practically established the diagnosis. Scintigraphy was

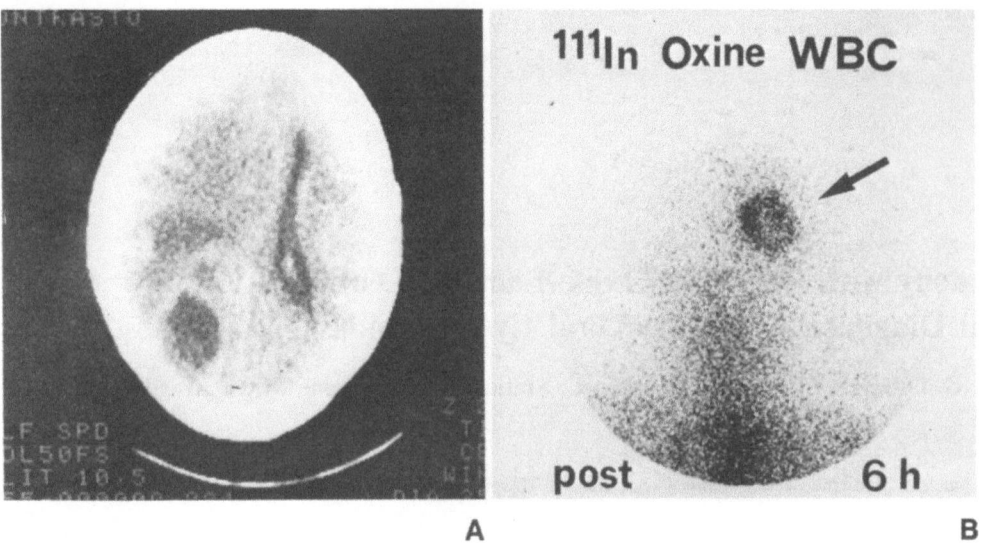

Fig. 1. Cerebral abscess. (A) CT scan. (B) Scintigram. Posterior view at 6th hour: intense focal uptake of leukocytes in the right parietal lobe

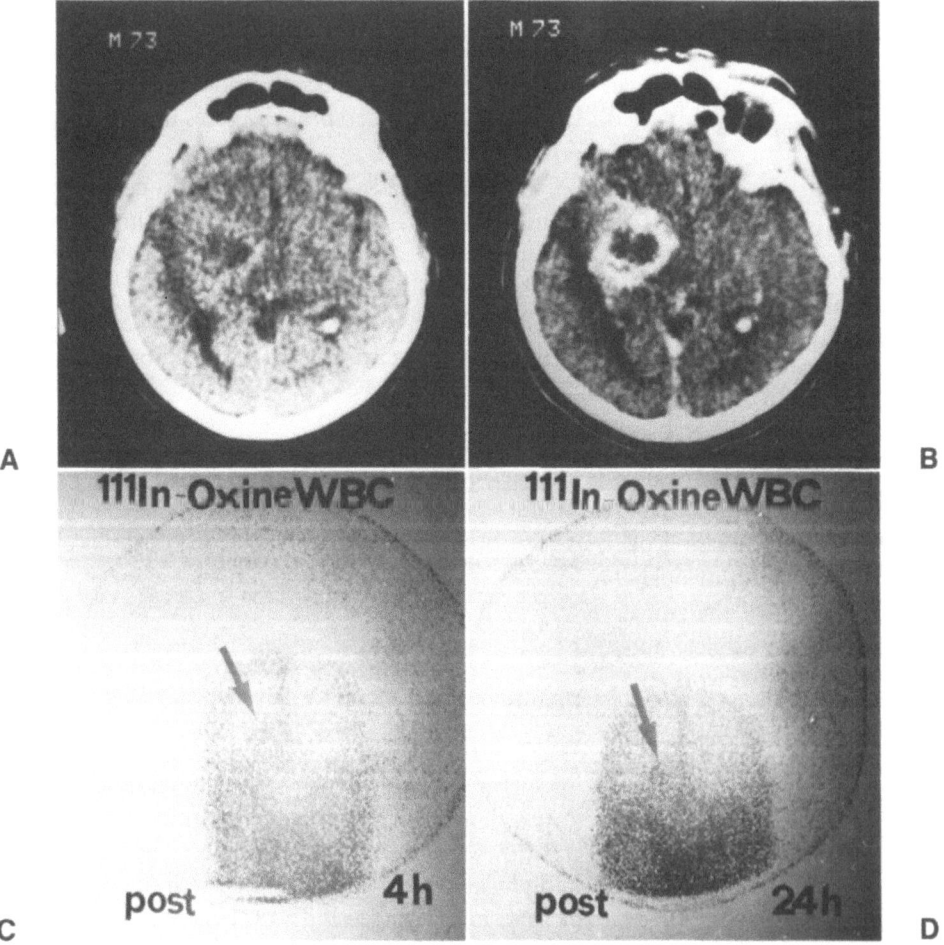

Fig. 2. Glioblastoma (case of false positive). (A–B) CT scan. (C–D) Scintigrams. Posterior view at 4 hour showing a roundish patch of focal uptake which is more intense 4 hours after injection of labeled WBC's cells than after 24 hours

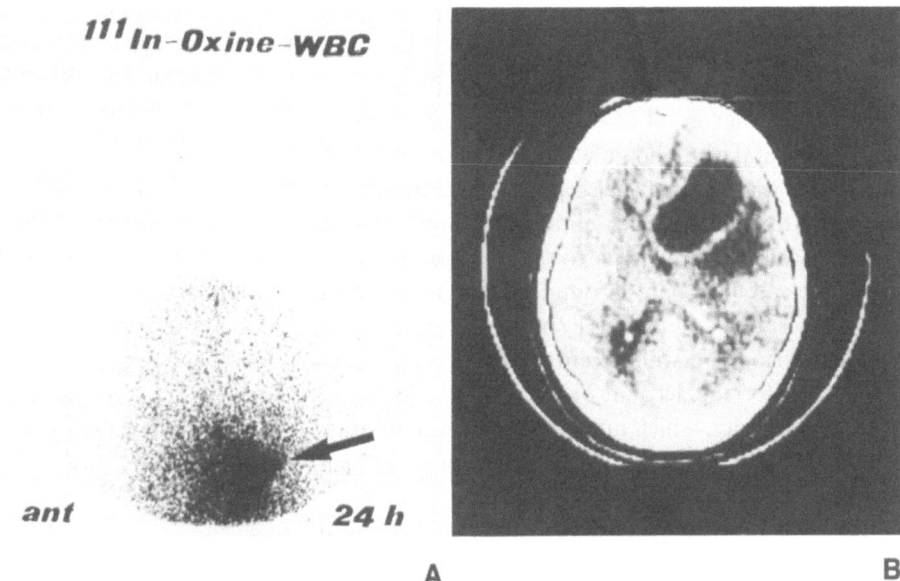

Fig. 3. Glioblastoma. (B) CT scan. (A) Scintigram. Anterior view at 24th hours. The scan shows sinusitis in the left maxillary sinus (arrow) but no intracerebral focal uptake

nonetheless done in all cases to test the procedure for sensitivity and specificty.

All the diagnosis of tumour and abscess were verified by histological examination. There were no side effects whatever after administration of the labelled cell suspension.

The findings were assessed retrospectively using three criteria postulated as capable of distinguishing uptake attributable to an abscess from the chance expression of modest WBC infiltration in a tumour. This occurrence is, as already reported for other organs, a possible source of false positives.

Criterion A: any uptake in excess of the physiological uptake of the parietal region of the skull.

Criterion B: any focal uptake.

Criterion C: any focal uptake equal to or greater than the physiological uptake of the base of the skull.

On critical review of our series we found criterion C the best for the purpose.

On criterion C ten scans were positive and four negative. Eight scans were positive 4–6 hours after administration of the labelled cells, although the uptake was less in the earlier scans than in scans taken after 24 hours. The operative findings were abscess in six cases (Fig. 1) and abscessed metastasis (from a primary tumour in the lung in two). In one of these cases the study had to be stopped after the first series of scans at 6 hours because of the need for emergency surgery. The CT image showed cavitation without a well defined capsule; necropsy showed confluent foci of menin-

goencephalitis. In the 10th patient, the only false positive (Fig. 2) on criterion C, the scintigraphic findings at 4 and 24 hours were both positive but the uptake was definitely greater at 4 hours. The histological diagnosis was glioblastoma and at operation large venous lakes were found at the tumour periphery.

The scintigraphic and CT findings were at the same site in every case.

The three lesions with focal uptake, insufficient to rank as positive on criterion C, were diagnosed histologically, two as glial tumours and one as a rare cystic meningioma.

Nine glial tumours, three metastatic tumours (one malignant melanoma and two of pulmonary origin) and two arachnoid cysts showed no uptake on the scintiscan (Fig. 3). In five other cases rated negative the clinical course and subsequent CT scans established the diagnosis of ischaemic cerebrovascular disease.

Conclusions

We formerly thought that criterion C with its classification of faint but focal uptake attributable to leukocyte infiltration within a non abscessed tumour as negative, and therefore greatly increasing the specificity and diagnostic accuracy of the investigation, would alone be sufficient for the certain identification of an abscess or the reverse[1]. In the light of more recent experience the hypothesis that abundant vascular spaces at the tumour periphery may yield an image of intense focal uptake and so give rise to false positives has proved just as well founded. Although our experience of this is confined to one case, it has to be remembered that in all cases of abscess or abscessed lesion the uptake at 6 hours was always less than at 24 hours and that the only false positive on criterion C showed a distinctly greater uptake at 4 hours than at the next scan. It is reasonable to suppose that the washing of the vascular spaces around the tumour will consistently result in a similar time course of the scintiscan.

To criterion C it will therefore be necessary to add a corollary to take account of the dynamics of scintigraphic positivity, which must be stronger at 24 hours than at 6 hours after the administration of tagged leukocytes.

It is still not possible by this method to distinguish between a cerebral abscess and an abscessed tumour. However, by combining routine neuroradiological investigations with this procedure even this doubt can be cleared.

References

1. Bellotti C, Aragno MG, Medina M *et al* (1986) Differential diagnosis of CT hypodense cranial lesions with indium-111-oxine-labeled leukocytes. J Neurosurg 64: 750–753
2. Camuzzini GF, Aragno MG, Viglietti AL *et al* (1984) Autologous WBC In[111] oxine scanning: results on cell separation and labeling; sensibility and specificity of the technique. Eur J Nucl Med 9: A 81 (Abstract)
3. Coulam CM, Seshul M, Donaldson J (1980) Intracranial ring lesions: can we differentiate by computed tomography? Invest Radiol 15: 103–112
4. Fawcett HD, Goodwin DA, Lantieri RL (1981) In[111] leukocyte scanning in inflammatory renal disease. Clin Nucl Med 6: 237–241
5. Froelich JW, Swanson D (1984) Imaging of inflammatory processes with labeled cells. Sem Nucl Med 14: 128–140
6. Goodwin DA (1983) Clinical use of In[111] leukocyte imaging. Clin Nucl Med 8: 36–38
7. Hardeman MR (1979) Possible use of radioactive-labeled blood cells in rheumatic disease evaluation. In: Feltkamp TEW, van der Korst JK (eds) Disease evaluation and patient assessment in rheumatoid arthritis: Proceedings of the 5th I.S.R.A. Symposium. Alphen an den Rijn: Stafleu Scientific, pp 93–98
8. Kandalaft N, Diehl J, Neuwelt EA (1982) Non neoplastic intracranial lesions simulating neoplasms on computed tomographic scan. Excellent sensitivity with limited specificty. JAMA 248: 2166–2168
9. Kock-Jensen C, Bøgelund Andersen B, Søgard L (1986) Leucocyte scanning – a valuable tool in diagnosing cerebral abscess – a survey. Acta Neurochir (Wien) 83: 121–124
10. Krayenbuhl H, Yaşargil MG (1965) Die zerebrale Angiographie. Thieme, Stuttgart, pp 322–324
11. McAfee JG, Samin A (1985) In[111] labeled leukocytes: a review of problems in image interpretation. Radiology 155: 221–229
12. Peters AM, Lavender JP, Macdermot J (1980) Diagnosing cerebral abscess with In[111] labeled leukocytes. Lancet 2: 309–310 (Letter)
13. Rehncrona S, Brismar J, Holtas S (1985) Diagnosis of brain abscesses with In[111] labeled leukocytes. Neurosurgery 16: 23–26
14. Stephanov S, Joubert MJ (1982) Large brain abscesses treated by aspiration alone. Surg Neurol 17: 338–340
15. Thakur ML, Lavender JP, Arnot RN *et al* (1977) In[111] labeled autologous leukocytes in man. J Nucl Med 18: 1014–1019
16. Walsh PR, Larson SJ, Rytel MW *et al* (1980) Stereotactic aspiration of deep cerebral abscesses after CT-directed labeling. Appl Neurophysiol 43: 205–209
17. Whelan MA, Hilal SK (1980) Computed tomography as a guide in the diagnosis and follow-up of brain abscesses. Radiology 135: 663–671

Address for correspondence: Dr. C. Bellotti, Neurosurgery Division, Santa Croce General Hospital, Cuneo, Italy.

Acta Neurochirurgica, Suppl. 42, 225–229 (1988)
© by Springer-Verlag 1988

RISA Cisternography in the Option of Ventriculocisternal Shunt for Infantile Nontumoural Aqueductal Stenosis

L. Palma, A. Mariottini, R. D'Addetta[1], and L. Mastronardi[1]

Chair of Neurosurgery, University of Siena, Siena, Italy and [1] Neurosurgical Clinic, University "La Sapienza" of Rome, Rome, Italy

Summary

Twenty cases of infantile triventricular hydrocephalus from nontumoural aqueductal stenosis were treated by ventriculocisternal shunt following RISA cisternography. In 11 cases RISA cisternography showed a normal pattern of CSF circulation. One patient was lost to follow-up. Two had their intrathecal shunt converted into an extrathecal one because of postoperative meningeal infection. Of the remaining 8 patients, 7 had good and 1 fair long term results. In 9 cases RISA cisternography presented an abnormal pattern without indicating a definite impairment of CSF absorption. Slow flow of the tracer leading to its complete disappearance from 36 to 48 hours and 48 to 72 hours was observed respectively in 7 and 2 patients. In both the latter as well in two of the other seven an extrathecal shunt had to be employed (44%). A retrospective analysis to assess the predictive value of CSF absorption test by RISA in the selection of this kind of intrathecal shunt is made.

Keywords: Hydrocephalus; aqueductal stenosis; treatment; ventriculocisternal shunt; RISA cisternography; CSF absorption.

Introduction

After the initial enthusiasm for extrathecal shunt in the treatment of nontumoural hydrocephalus in children we, like others[2, 5, 6, 15], became more and more alarmed at the high percentage of complications arising from this type of operation and reverted to the intrathecal shunt whenever possible. This approach was also encouraged by the widespread use of radio-iodinated serum albumin (RISA) cisternography by which it seemed possible to evaluate beforehand the competence of the CSF pathways and to select appropriately candidates for an intrathecal shunt[4].

In our clinic since 1970, an increasing number of children over the age of one year with triventricular hydrocephalus from nontumoural aqueductal stenosis have been selected by RISA cisternography for Torkildsen's ventriculocisternostomy.

Whenever blocked CSF pathways were found, we inserted an extrathecal shunt.

The present study examines the reliability of RISA cisternography, especially in borderline cases, in a sample of Torkildsen's ventriculocisternostomy on the basis of this technique.

Clinical Materials and Methods

Ten male and 10 female patients, ranging from 4 to 20 years (mean age: 11.5 years) with triventricular hydrocephalus from nontumoural aqueductal stenosis and investigated by RISA cisternography, were surgically treated by a Torkildsen's procedure during the period 1972–1976 at the Neurosurgical Clinic of the University "La Sapienza" in Rome.

The main clinical data of the patients on admission is given in Tables 1, 2, and 3. Stenosis of the aqueduct of Sylvius was demonstrated by air encephalography supplemented by iodine ventriculography until 1976 and by CT scan thereafter. All patients were also investigated by RISA cisternography in order to evaluate preoperatively the competence of the subarachnoid spaces and the feasibility of an intrathecal shunt. Isotope cisternography was performed by lumbar injection of 2–5 μCi/kg of fresh, high-specific-activity radio-iodinate serum albumin following previous administration of a saturated solution of potassium iodide to block thyroid uptake. The subjects were examined 30 minutes, 3, 6, 24, 36, 48, and 72 hours after the isotope injection. A few patients suffered a slight rise in temperature or headache but no cases of aseptic meningitis were observed.

In eleven patients the picture was that of normal CSF circulation

Table 1. *Age at Diagnosis*

Age	< 5	5–9	10–14	15–20
Number	1	6	7	6
	(5%)	(30%)	(35%)	(30%)

Table 2. *Duration of Symptoms*

Duration (mos)	< 1	1 11	12 24	> 24
Number	1	10	8	1
	(5%)	(50%)	(40%)	(5%)

Table 3

Symptoms and signs	No.	%
Headache	15	75
Gait disturbances	12	60
Papilloedema	11	55
VI cranial nerve deficit	8	40
Spasticity	8	40
Cerebellar deficit	6	30
Nausea, vomiting	6	30
Visual loss	6	30
Epilepsy	3	15
Mental retardation	3	15
Secondary optic atrophy	3	15
Stupor	1	5

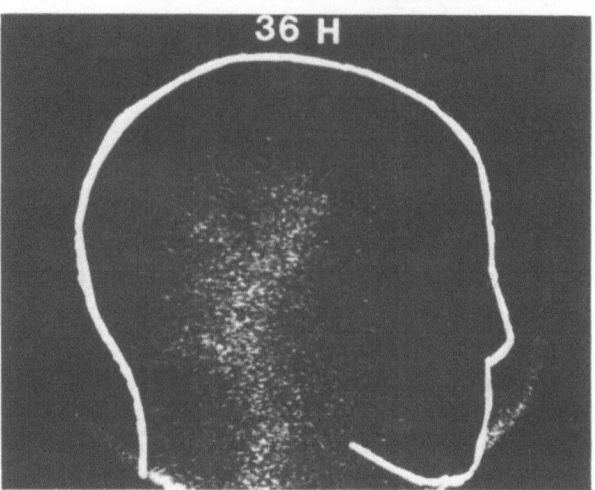

Fig. 2. RISA cisternography. Lateral view: delayed circulation of CSF. The 36-hour scan shows an atypical distribution of the tracer over the hemispheres which appears to have been moving along the posterior (circumpeduncular?) cisterns

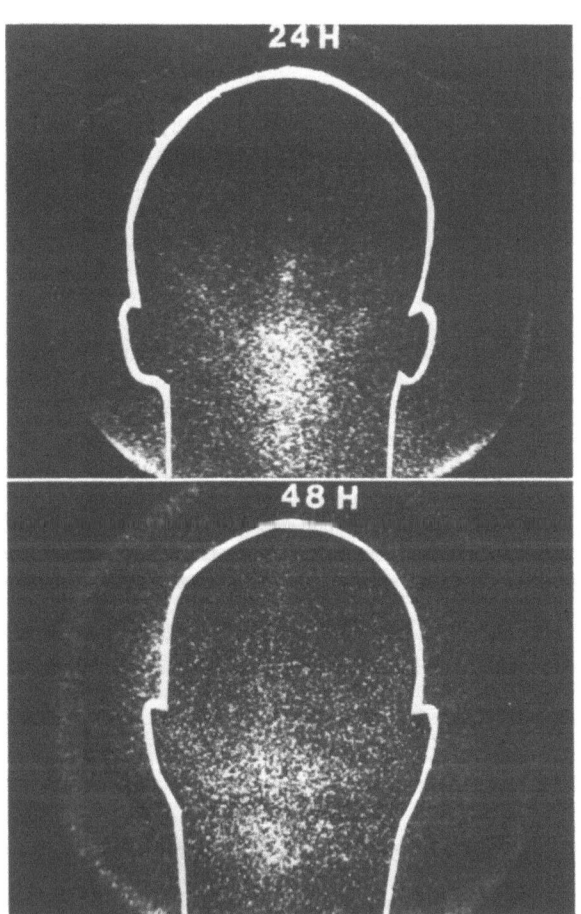

Fig. 1. RISA cisternography. Posterior view: extremely delayed circulation of CSF. Above: 24-hour scan. 131 I-SA is still confined to the basal cisterns. Below: the bulk of the tracer remains in the basal cisterns 48 hours after isotopic lumbar injection

and in the other nine RISA cisternography showed an abnormal pattern without indicating definite impairment of CSF absorption. In five of the latter there was a slow flow of the tracer through the normal pathways leading to its complete disappearance after 36 to 48 hours in three and after 48 to 72 hours in two (Fig. 1). In the other four cases slow flow towards the convexity was associated with a variability in the pattern of spread of the tracer over the hemispheres which appeared to be moving along the posterior (circumpeduncular?) cisterns (Fig. 2).

All patients underwent ventriculocisternostomy. In most cases the ventriculo-cervical variant of Torkildsen's procedure as described by Matson[9] was used. A reservoir was routinely inserted only in patients with an abnormal RISA pattern in order to facilitate the eventual conversion of the intrathecal into an extrathecal shunt. Lumbar puncture was routinely performed in the first 3–4 days after the operation in order to foster reexpansion of the compressed subarachnoid spaces, to check shunt patency, and to clear eventually clouded cerebrospinal fluid.

Results

Follow-up ranged from 1 to 14 years. Check-up included neurological and X-ray examination, subsequently supplemented by psychometric tests, EEG, and CT scan.

The overall rating of each patient was summed up as "Good" or "Fair". Good meant that the patient could be regarded as neuropsychically normal with performance at school or work consonant with age and social background. Fair meant that there were certain neurological deficits such as epilepsy, truncal ataxia, slight mental retardation, or grave visual loss due to secondary optic atrophy.

Of the eleven patients with normal RISA patterns one was lost to follow-up and two had their intrathecal

Table 4. *Summary of Data on Patients with Normal Risa Findings*

Cases	Sex age (years)	Diagnostic studies	Surgery	Postoperative course	Outcome follow-up
1) G.G. (1972)	m 20	P-V-R	T	infection V-P shunt (after 15 days)	good 14 years
2) B.D. (1975)	f 15	P-V-R	TM	satisfactory	fair 11 years
3) P.F. (1975)	m 7	P-V-R	TM	satisfactory	lost
4) Z.M. (1976)	m 7	P-V-R	TM	infection V-P shunt (after 20 days)	good 10 years
5) R.M. (1977)	m 15	CT-R	TM	satisfactory	good 9 years
6) C.F. (1978)	m 4	CT-R	TM	satisfactory	good 8 years
7) S.F. (1978)	m 5	CT-R	T	satisfactory	good 8 years
8) N.M. (1979)	f 6	CT-R	T	satisfactory	good 7 years
9) M.M. (1979)	f 12	CT-R	T	satisfactory	good 7 years
10) D.S. (1983)	f 16	CT-R	TM	satisfactory	good 3 years
11) I.G. (1985)	m 19	CT-R	T	satisfactory	good 1 year

P = Pneumoencephalography. V = Iodine ventriculography. R = Risa cisternography. CT = Computed tomography. T = Torkildsen's ventriculo-cisternostomy. TM = Torkildsen's ventriculo-cisternostomy by Matson. V-A = Ventriculo-atrial shunt. V-P = Ventriculo-peritoneal shunt.

shunt converted into an extrathecal one because of meningeal infection in the immediate postoperative period. Of the remaining eight patients, seven had a good rating and only one a fair rating because of residual epilepsy (which responded well to anticonvulsants) and slight hemilateral spasticity. One patient was rated as good result despite residual unilateral optic atrophy secondary to papilloedema (Table 4).

The fate of the nine patients with borderline RISA is summarized in Table 5. Ventriculocisternostomy worked well in five cases (55%) with good long-term functional results, but it had to be converted into an extrathecal shunt in four other cases because of the persistence of the preoperative symptomatology. Conversion was necessary within a few days from the operation in two patients and after 12 and 24 months respectively in the remaining two. It should be noted, however, in the last patient that failure of the ventriculocisternostomy became evident after a mild head trauma. The outcome of these four cases was good in three patients and fair in the fourth because of marked visual loss from secondary optic atrophy.

Discussion

Before the introduction of RISA cisternography our failure rate with Torkildsen's procedure was 40%[3]. In the present series of 20 patients the selection of candidates for ventriculocisternal shunt was mainly based on the findings in preoperative cisternography and the intrathecal shunt was not successful in only the four cases who exhibited a borderline pattern. In retrospect the success rate could have been higher had we eliminated two of the four patients in whom the CSF circulation appeared extremely slow as demonstrated by the complete disappearance of the isotope only after 48 hours (Fig. 1). Both of these patients were observed early in the series. Like many subsequent subjects, they had marked papilloedema and therefore their slow RISA pattern was attributed to intracranial hypertension, as suggested by others[8]. Later experience demonstrated that most patients[7, 11] with intracranial hypertension had a normal cisternography pattern. Such evenience has already been noted by other authors[13]. Of the other four patients with papilloedema, two were

Table 5. *Summary of Data on Patients with Abnormal Risa Findings*

Cases	Sex age (years)	Diagnostic studies	RISA	Surgery	Postoperative course	Outcome follow-up
1) C.P. (1975)	f 12	P-V	atypical 36–48 h	TM	satisfactory	good 11 years
2) P.B. (1976)	f 13	P-V	36–48 h	TM	satisfactory	good 10 years
3) G.F. (1976)	f 10	P-V	48–72 h	TM	V-A shunt (after 15 days)	good 10 years
4) R.R. (1978)	f 12	CT	36–48 h	TM	V-P shunt (after 24 months)	good 8 years
5) D.G. (1978)	f 12	CT	48–72 h	TM	V-P shunt (after 1 month)	fair 8 years
6) B.G. (1980)	m 7	CT	atypical 36–48 h	TM	satisfactory	good 6 years
7) Q.R. (1981)	m 13	CT	atypical 36–48 h	TM	V-A shunt (after 12 months)	good 5 years
8) G.S. (1981)	f 8	CT	atypical 36–48 h	TM	satisfactory	good 5 years
9) B.A. (1983)	m 17	CT	36–48 h	TM	satisfactory	good 3 years

P = Pneumoencephalography. V = Iodine ventriculography. CT = Computed tomography. TM = Torkildsen's ventriculo-cisternostomy by Matson. V-A = Ventriculo-atrial shunt. V-P = Ventriculo-peritoneal shunt.

the subjects mentioned above, in the other two the Torkildsen's shunt worked well despite slow reabsorption of the tracer which took 36–48 hours. Considering the factor of retarded RISA flow, we found that the time limit of 36–48 hours was compatible with a successfully functioning shunt in 5/7 cases. On the other hand, RISA absorption delayed from 48 to 72 hours heralded failure in 100% of cases (2/2) (Table 5).

The value of RISA cisternography in determining the appropriateness of an intrathecal shunt in obstructive hydrocephalus appears from this study to be similar to that recently outlined by Jaksche and Loew[6], who used burr hole third ventriculostomy to treat tumoural and nontumoural obstructive hydrocephalus in 79 patients. Thirty-seven were cases of nontumoural aqueductal stenosis. The authors stressed that better patient selection following preoperative CSF scintigraphy improved the success rate from 65 to 90%.

Many authors insist on the importance of evaluating the CSF pathways before performing an extrathecal shunt operation in obstructive hydrocephalus[1, 2, 5, 6, 10, 12, 15]. However opinions differ over the value of RISA cisternography in predicting the success or failure of an internal shunt. Certain authors[1, 6, 8] consider this type of examination useful and others either expressly

deny its value[14] or else omit to list it among the criteria for third ventriculostomy preoperative selection[5]. Still other authors[2, 7, 11] express doubts about the value of isotope cisternography in cases of this kind. This disparity is partly due to different approaches to the problem. Whereas normal RISA cisternography findings give a 90–100% likelihood of success of an intrathecal shunt (1, 6, 14, present series), abnormal patterns do not predict failures with precision. Pierre-Kahn *et al.*[14] found that 4/7 patients, preoperatively exhibiting a cisternographic time greater than 36 hours, had a successful third ventriculostomy. In our series nine patients showed such an abnormal time pattern of RISA cisternography and five of these had a successful ventriculo-cervical shunt insertion. It seems then that the failure rate of an intrathecal shunt after an abnormal or borderline RISA cisternography is similar to that observed without the preoperative use of this examination[5]. Nevertheless a more careful analysis of our results suggests the possibility of a reasoned selection of patients for intrathecal shunt even with abnormal RISA findings. As a matter of fact a 71% success rate in cases with isotopic clearance ranging between 36 and 48 hours and a 100% failure rate when reabsorption took from 48 to 72 hours was observed. Although the number of cases was small, these observations lead us to consider 36–48 hours as the maximum delay in isotope reabsorption which will lead to an acceptable success rate.

References

1. Akerman M, De Tovar G, Guiot G (1972) Radio-isotope cisternography and ventriculography in non-communicating hydrocephalus. In: Cisternography and hydrocephalus. A symposium. CC Thomas, Springfield, Illinois, USA, pp 483–501
2. Backlund E-O, Grepe A, Lunsford D (1981) Stereotaxic reconstruction of the aqueduct of Sylvius. J Neurosurg 55: 800–810
3. Guidetti B, Giuffré R, Palma L *et al* (1976) Hydrocephalus in infancy and childhood; our experience of CSF shunting. Child's Brain 2: 209–225
4. Harbert JC, McCullough DC, Lussenhop AJ *et al* (1972) Cisternography and hydrocephalus. A symposium. CC Thomas, Springfield, Illinois, USA, pp 1–559
5. Hoffman HJ, Harwood-Nash D, Gilday DL (1980) Percutaneous third ventriculostomy in the management of noncommunicating hydrocephalus. Neurosurgery 7: 313–321
6. Jaksche H, Loew F (1986) Burr hole third ventriculostomy. An unpopular but effective procedure for treatment of certain forms of occlusive hydrocephalus. Acta Neurochir (Wien) 79: 48–51
7. Lapras C, Bret Ph (1980) Les sténoses de l'acqueduc de Sylvius. Neurochirurgie vol 26 [Suppl], p 131
8. McCullough DC, Harbert JC, Lussenhop AJ (1972) Pediatric hydrocephalus: contribution of radioisotope cisternography to diagnosis and management. In: Cisternography and hydrocephalus. A symposium. CC Thomas, Springfield, Illinois, USA, pp 375–383
9. Matson DD (1969) Neurosurgery of infancy and childhood, 2nd ed. CC Thomas, Springfield, Illinois USA
10. McMillan JJ, Williams B (1977) Aqueduct stenosis. Case review and discussion. J Neurol Neurosurg Psychiatry 40: 521–532
11. Milhorat TN, Hammock MK, Di Chiro G (1971) The subarachnoid space in congenital obstructive hydrocephalus. Part 1: Cisternographic findings. J Neurosurg 35: 1–6
12. Patterson RH Jr, Bergland RM (1968) The selection of patients for third ventriculostomy based on experience with 33 operations. J Neurosurg 29: 252–254
13. Paraicz E, Simkovics M, Kutas V (1972) Some peculiarities of the passage and absorption in the subarachnoid space at increased ICP investigated by radioisotope methods. In: Intracranial pressure. Experimental and clinical aspects. Springer, Berlin Heidelberg New York, pp 33–36
14. Pierre-Kahn A, Renier D, Bombois B *et al* (1975) Place de la ventriculo-cisternostomie dans le traitement des hydrocéphalies non communicantes. Neurochirurgie 21: 557–569
15. Vries JK (1978) An endoscopic technique for third ventriculostomy. Surg Neurol 9: 165–168

Address for correspondence: Dr. L. Palma, Chair of Neurosurgery, University of Siena, Siena, Italy.

Acta Neurochirurgica, Suppl. 42, 230–235 (1988)

Intracranial CSF Volumes: Natural Variations and Physiological Changes Measured by MRI

G. M. Teasdale, R. Grant, B. Condon, J. Patterson, A. Lawrence, D. M. Hadley, and **D. Wyper**

Institute of Neurological Sciences, Southern General Hospital, Glasgow, U.K.

Summary

Cranial CSF volumes, for the first time including CSF in the subarachnoid space, can be measured by Magnetic Resonance Imaging (MRI). The MRI sequence causes signal from the grey matter and white matter to cancel producing a contrast of 200 : 1 between a unit of CSF and a unit of brain. We have assessed the variations between normal individuals and investigated some of the physiological factors that might influence cranial CSF volumes.

Total CSF volumes were measured in 64 normal subjects, aged from 18–64 years (mean 38 years). Ventricular, cortical sulcal and posterior fossa volumes were also calculated separately. In 20 females with a normal menstrual cycle, CSF volumes were measured mid cycle and premenstrually; 10 post menopausal females and 10 males were rescanned after an interval of 2 weeks. Total cranial CSF volume were calculated before and during inhalation of 7% CO_2 and before and during hyperventilation while breathing 60% O_2, in 12 normal subjects.

Total intracranial CSF volume ranged from 57.1–286.5 ml. Total intracranial and cortical sulcal CSF volumes increased more steeply with age than ventricular or posterior fossa CSF volumes. Males had more cranial CSF than females. Total CSF volume increased premenstrually in 19 females. Males and post-menopausal females did not have a significant change in CSF volume, on repeat examination. CO_2 inhalation produced a mean increase of $paCO_2$ of 17.2 mmHg and CSF volume decreased in all subjects (mean 9.4 ml). Cranial CSF volume increased in 11 subjects during O_2 inhalation (range −0.5 to + 26.7 ml mean 10.9 ml).

These observations show that total cranial CSF volume increases with age mainly due to an increase in cortical CSF volume. Significant changes occur during the menstrual cycle. The changes with 7% CO_2 and 60% O_2 inhalation are in keeping with the modified Monro-Kellie hypothesis and probably reflect reciprocal changes in cerebral blood volume and cerebro-spinal fluid volume. MRI measurements of cranial CSF volume are likely to increase knowledge and advance management of many intracranial disorders.

Keywords: CSF volume; intracranial CSF volume; MRI; normal values.

Introduction

In 1846, George Burrows modified the Monro-Kellie Doctrine to account for the fundamental role played by cranial CSF volume[1]. Methods of estimating CSF volume have either been highly invasive or subject to large errors[21]. Condon has described an accurate method of measuring total cranial and ventricular CSF volumes using Magnetic Resonance Imaging (MRI)[3]. We have used this method to measure the total, ventricular, posterior fossa and supra-tentorial cortical sulcal CSF volumes in healthy subjects and to establish physiological factors that might influence intracranial dynamics[8, 9].

Measurement of CSF Volume

Total cranial CSF volume was measured in a 0.15 T resistive imager using an IRCP 300/400/5,000 pulse sequence. This produces a contrast of greater than 200 : 1 between a unit volume of CSF and a unit volume of mixed grey and white matter. The result is a image of CSF only. A reference phial containing a known volume of water at 37 °C was strapped to the head. Two sagittal CSF volume scans are performed. The first has a slice width of 240 mm and includes signals from all CSF in the head. The second scan has a narrower slice width which includes all ventricular CSF and posterior fossa CSF but excludes the overlying and underlying cortical sulcal CSF. The signal and area of a region of interest is then compared with that of the reference phial and total or regional cranial CSF volume can be calculated.

Study Groups and Subjects

(a) Age and Sex Relationships

CSF volumes were measured in 64 healthy volunteers. There were 25 males and 39 females. Ages ranged from 18 to 64 years, and were similarly distributed in both sexes with a mean age of 38.4 years for males and 37.1 year for females. Subjects did not complain of neurological symptoms at the time of MRI or any previous history of neurological or cardiovascular disorder. Age, sex, and skull circumference were recorded. Total cranial, ventricular, cortical sulcal and posterior fossa CSF volumes were measured.

(b) Repeat Studies After Two Weeks and Menstrual Cycle Effects

Thirty females, aged from 18–75 years and 10 males aged from 19 61 years were studied. Ten of the females had a normal menstrual cycle and were not taking oral contraceptives (mean age 29.8 years), 10 had a normal menstrual cycle and were taking combined low oestrogen dosage oral contraceptive (mean age 26.2 years) and 10 women postmenopausal (mean age 55.3 years). The males were aged from 19–61 with a mean age of 37.4 years. Females with a regular menstrual cycle were imaged at mid-cycle and premenstrually. Post-menopausal females and also the males had MRI studies repeated after 2 weeks.

(c) Cerebral Blood Volume Changes

Total cranial CSF volumes were measured in 12 healthy volunteers before and during 7% carbon dioxide inhalation. There were 9 males and 3 females (age range 19–41, mean 28.8 years). Another 12 healthy subjects were studied before and during hyperventilation with high flow O_2 ($> 60\%$ O_2). There were 8 males and 4 females aged from 20–38 years (mean 29.1 years).

A rubber mouthpiece was placed in the subject's mounth and then connected to a two way valve. Three metres of tubing extended from the inspiratory limb of the two way valve to outside the MRI radiofrequency shield. The expiratory limb from the valve was connected to 1 metre of tubing which was open to the air.

During the assessment of "resting" CSF volume, the inspiratory limb of the tubing was left open so that the subject was able to breathe fresh air. A repeat scan was taken after a cylinder containing 7% CO_2 was attached to a Douglas bag and the outlet from the Douglas bag was attached to the inspiratory limb of the tubing to the subject.

Mean arterial blood pressure (MABP) was recorded using a Critikon Dinamp 1846 P Version 028, before entering the imager and during scanning. Cerebral blood flow (CBF) response to 7% CO_2 inhalation and end expiratory CO_2 were measured after MRI by the Xenon-133 inhalation technique[17] and using a capnograph (Godart Statham 17070).

The method used to induce hypocapnia was similar to that of CO_2 inhalation but high flow O_2 was delivered at a flow rate of 10 litres/min via the inspiratory limb directly. During the "resting" scan the inspiratory limb of the system was open to air and at the start of the second scan the high flow O_2 was connected to the inspiratory limb and the subject asked to hyperventilate at a rate of approximately 30 breaths/min. The average O_2 concentration delivered at the mask was 60–65%.

Results

(a) Age and Sex Relationships

Total cranial CSF volume ranged from 57.1–286.5 mls. The mean total CSF volume for males (146 mls) was significantly greater than that of females (114.5 mls) and for this reason the results in males and females were considered separetely. Head circumference was significantly larger in males ($p < 0.001$), but neither in males nor in females was there a significant relationship between an individual's head circumference and total cranial CSF volume. Total cranial CSF volume in-

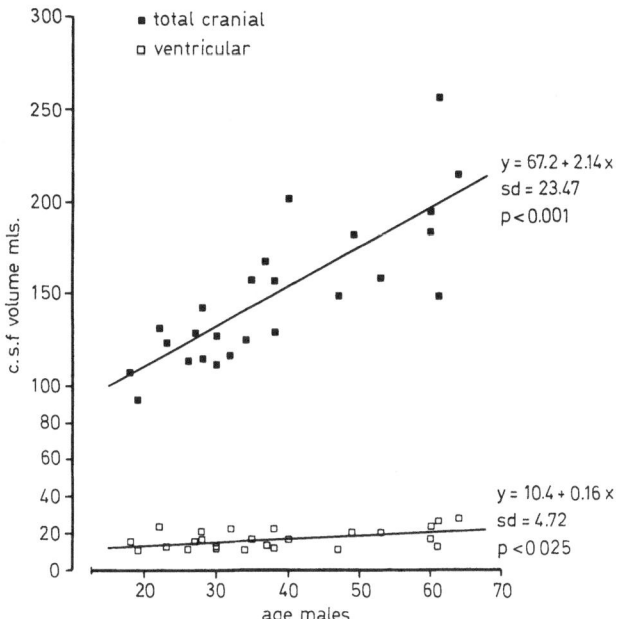

Fig. 1. Relationship between total cranial and ventricular CSF volumes and age in males

creased significantly with age in both sexes ($p < 0.001$). The slopes of the lines were not significantly different and the mean change per year in males was 1.9% and in females by 1.6%. Ventricular CSF volumes ranged from 6.8–30 mls. A difference in ventricular volume between males and females was not demonstrated. There was a small but significant increase in ventricular volume with age in males but not in females. Cortical sulcal volume ranged from 45–256.5 mls and increased steeply with age in both males and females ($p < 0.001$). Posterior fossa CSF volume ranged from 4.8–26.9 mls, was greater in males than in females and increased significantly with age ($p < 0.025$).

(b) Menstrual Cycle Studies

Total cranial CSF volume in women with normal menstrual cycle ranged from 57.1 mls to 150.9 mls (mean 107 mls). Post-menopausal women and males had total cranial CSF volumes ranging from 95.1–170.5 mls (mean 126.8 mls) and 90.2–181.4 mls (mean 134.9 mls), respectively.

The total cranial CSF volume increased from midcycle to premenstrually in all but one subject. The mean increase premenstrually was by 11.5 mls (s.d. 10.2) (paired T test: $p < 0.0001$). In women taking the oral contraceptive pill the change in CSF volume (9.6 mls [s.d. 10.4]) was less than that of women not taking oral contraceptives (13.4 mls [s.d. 10.1]), but the difference

Table 1. *Premenstrual and Mid-cycle CSF Volumes and 2 Week Repeat in Males and Post-menopausal Females*

		Females: not taking OCP						Females: taking OCP			
Subj.	Age (years)	Total ventric		Total ventric		Subj.	Age (years)	Total ventric		Total ventric	
		CSF (mls)	CSF (mls)	CSF (mls)	CSF (mls)			CSF (mls)	CSF (mls)	CSF (mls)	CSF (mls)
		Mid-cycle		Pre-menstrual				Mid-cycle		Pre-menstrual	
1.	18	83.2	10.2	101.1	13.4	1.	22	97.6	6.7	101.2	9.5
2.	20	85.7	11.5	101.7	14.5	2.	23	73.4	11.2	97.3	14.0
3.	24	98.3	14.0	131.3	17.9	3.	23	105.2	13.9	115.6	13.2
4.	26	123.2	15.9	126.3	13.2	4.	24	130.1	18.0	130.5	15.1
5.	26	91.8	16.3	95.9	16.8	5.	25	60.8	7.2	57.1	6.8
6.	28	97.8	10.6	122.9	11.1	6.	25	133.2	10.4	135.8	10.7
7.	33	95.9	30.0	104.0	33.9	7.	26	101.6	8.1	110.4	10.9
8.	35	81.6	4.0	90.8	7.8	8.	29	105.0	13.1	136.1	17.5
9.	41	102.5	9.2	117.8	9.7	9.	32	95.0	7.4	113.0	9.3
10.	46	148.6	14.0	150.9	13.5	10.	33	115.4	13.8	133.7	20.2
Mean	29.8	100.9	13.6	114.3	15.2	Mean	26.2	101.7	11.0	111.3	12.7

Table 2. *Results of Initial and 2 Week Repeat CSF Volume Measurements in Post-menopausal Females and Males*

		Females: not taking OCP						Females: taking OCP			
Subj.	Age (years)	Total ventric		Total ventric		Subj.	Age (years)	Total ventric		Total ventric	
		CSF (mls)	CSF (mls)	CSF (mls)	CSF (mls)			CSF (mls)	CSF (mls)	CSF (mls)	CSF (mls)
		Mid-cycle		Pre-menstrual				Mid-cycle		Pre-menstrual	
1.	48	108.7	12.0	102.4	10.5	1	19	91.2	10.4	90.2	11.9
2.	50	97.8	9.7	147.1	11.0	2.	23	123.0	11.6	107.6	10.9
3.	51	135.8	8.0	104.9	6.9	3.	26	107.6	10.4	111.6	11.9
4.	51	136.8	20.3	144.6	21.2	4.	28	114.0	16.7	113.5	17.1
5.	53	144.2	11.9	170.5	104.9	5.	30	126.5	11.9	122.3	11.5
6.	55	117.2	21.3	141.4	22.0	6.	38	124.5	9.7	128.2	11.9
7.	56	107.1	10.8	95.1	9.8	7.	47	147.5	10.8	151.4	10.2
8.	57	116.5	11.3	114.1	11.5	8.	49	173.0	20.3	181.4	20.5
9.	57	142.7	14.8	136.7	14.5	9.	53	179.4	21.7	157.8	19.7
10.	57	135.7	17.7	135.0	16.7	10.	61	173.7	14.0	173.9	16.9
Mean	55.3	124-3	13-8	129.2	13.3	Mean	37.4	136.0	13.8	133.8	14.3

was not significant. Ventricular CSF volume increased premenstrually to an extent similar to the increase in total CSF volume (ventricular mean 13.4%, total mean 11.3%).

Studies to assess the repeatability of the measurements after two weeks were performed on 30 females, aged from 18–75 years. Total cranial CSF volumes did not change significantly in post-menopausal females or in males when measured two weeks apart; the mean change was + 4.9 mls (s.d. 22.8 mls) and − 2.2 mls (s.d. 9.35 mls) for post-menopausal females and males, respectively (paired t test).

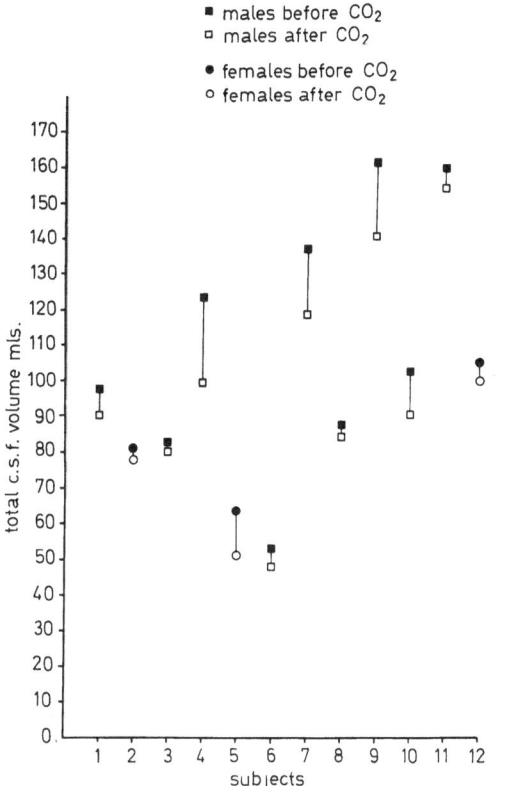

Fig. 2. Effect of CO_2 inhalation on total cranial CSF volume

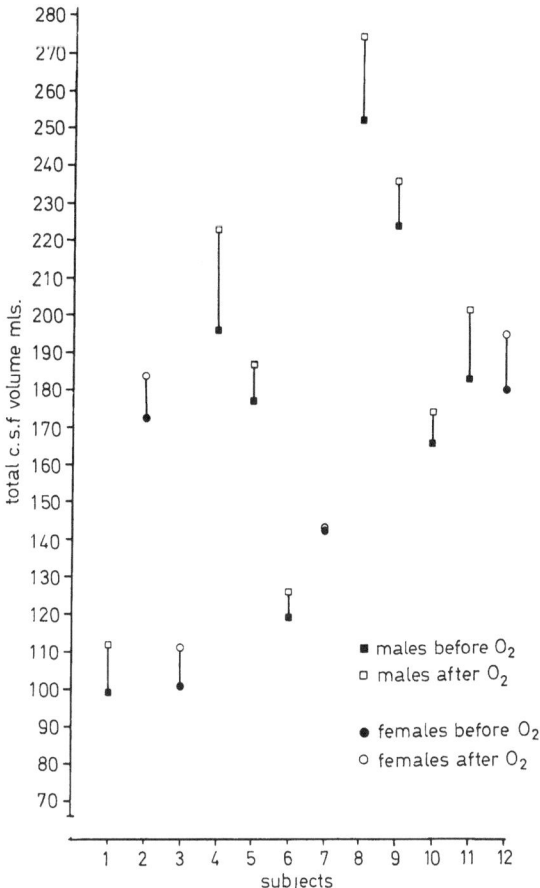

Fig. 3. Effect of 60% O_2 hyperventilation on total cranial CSF volume

c) Cerebral Blood Volume Changes

A reduction in total CSF volume was recorded following inhalation of 7% CO_2 in all cases. This ranged from −0.7 to −23.7 (mean −9.36 mls, SD 7.67). MABP increased during hypercapnia in 11 of the 12 subjects and did not alter in one case. The mean rise in MABP was 9.25 mmHg (SD 6.19). An increase in CBF was recorded in all cases after 7% CO_2 inhalation. The percentage increase in CBF ranged from 10–69% (mean 40.8%, SD 17.2) of the resting flow. The end expiratory CO_2 increased by a mean of 13.25 mmHg (SD 4.18) after 7% CO_2 inhalation. There was not a correlation between individual subject's change in MABP, CBF or end expiratory CO_2 and the reduction in CSF volume.

During hyperventilation there was an increased total CSF volume in all subjects. The change in CSF volume ranged from + 0.7 ml to + 26.7 mls (mean 12.7 mls, SD). Hyperventilation decreased pCO_2 from a mean of 40.05 mmHg (SD 0.95) to a mean value of 29.85 mmHg (SD 1.15).

Discussion

Some studies suggest an increase in the incidence of cortical atrophy and ventricular dilatation with increasing age[6, 11, 13] while other authors[12, 14] report that 86% of normal elderly individuals show little or no cortical atrophy and Zatz *et al.*[22] suggest that ventricular and cortical sulcal volumes do not change significantly below the age of sixty. Our findings, using MRI, demonstrate a significant difference in total cranial CSF volume between males and females and that there is a marked increase in total CSF volume and cortical sulcal CSF between the ages 18–64 years, in both sexes. The increase in CSF volume will reflect the amount of cerebral atrophy[1].

It has been postulated that neurological premenstrual symptoms are due to cerebral oedema, occurring as part of premenstrual fluid retention[19], but evidence to indicate the occurrence of brain swelling is lacking. The increase in total cranial CSF volume indicates that, rather than being swollen premenstrually, brain volume is reduced. An increased cranial CSF volume is consistent with other premenstrual changes related to sodium and water retention observed premenstrually. These are thought to result from changes in serum progesterone level stimulating the renin-angiotensin-aldosterone mechanism[15].

The mechanism causing the increase in CSF volume can only be speculated upon and could involve either increased rate of secretion or decreased absorption. Secretion of the CSF by the choroid plexus is predominantly under the control of the autonomic nervous system[18]. It has been suggested however, that increased CSF production due to oestrone stimulus is responsible for Benign Intracranial Hypertension[5]. Oestrogen administration has also been shown to increase brain water content in some animals[23]. A reduction in cerebral blood volume would also be reflected in increased cranial CSF volumes. CSF and arterial pCO_2 are significantly lower (mean 3.7 mmHg) during the luteal phase of the menstrual cycle, with an inverse relation to plasma progesterone level[16]. It is unlikely, however, that hypocapnic vasoconstriction could account for the observed changes in CSF volume premenstrually, as the cerebral circulation adapts to changes in $paCO_2$ if these are sustained for several hours. Moreover, a much greater change in $paCO_2$ would be needed to change CSF volume by the amount observed.

Our findings of a consistent reduction in CSF volume with hypercapnia and of increases in CSF volume during hyperventilation provide a measure of the extent of the reciprocal interactions between CSF volume and cerebral blood volume in human subjects and are a direct confirmation of the modified Monro-Kellie doctrine.

Cerebral blood volume (CBV) measurement in man is technically difficult to perform, and will depend on individual differences in brain size, vascular compliance, intracranial pressure and vascular responsiveness to changes in arterial CO_2 tension. Greenberg *et al.*[10], using emission tomography and ^{99m}Tc-labelled red blood cells, demonstrated a change in blood volume of 0.0495 ml/100 g brain/torr $paCO_2$. For a brain weighing 1,400 g, the change in CBV resulting from an increase in pCO_2 of 13.25 mmHg, would be estimated to be 9.2 mls. This is similar to the mean decrease in CSF volume (9.36 mls) that we have observed.

The reduction in CSF in response to increase in CBV may reflect CSF being displaced into the distensible spinal subarachnoid space, which is known to play an important role in intracranial spatial compensation[17]. Hypercapnia might also affect the production or absorption of CSF. CSF is formed at 20 mls/hr, therefore reduced CSF production is unlikely to significantly contribute to our findings, however, absorption is dependent on intracranial pressure (ICP)[4], and as ICP increasing during hypercapnia, the reduction in cranial CSF volume may have resulted partially from increased CSF absorption. Conversely, hypocapnia reduces intracranial pressure and CBV, therefore absorption of CSF may decrease and CSF may be displaced from the spinal subarachnoid space into the skull.

The CSF volume is an important factor in the study of physiological and pathological conditions. The ability to measure it directly in life in man may provide fundamental information about many intracranial conditions.

References

1. Brody H (1973) Aging of the vertebrate brain. In: Rockstein S, Sussman ML (eds) Development and aging in the nervous system. Academic Press Inc 7: 121–133
2. Burrows G (1846) On disorders of the cerebral circulation. London
3. Condon B, Patterson J, Wyper D, Hadley D, Grant R, Teasdale GM, Rowan J (1986) Intracranial CSF volumes determined using Magnetic Resonance Imaging. Lancet i: 1355–1358
4. Cutler RWP, Page L, Galicich J, Watters GV (1968) Formation and absorption of cerebrospinal fluid in man. Brain 91: 707–720
5. Donaldson JO, Horak E (1982) Cerebrospinal fluid oestrone in pseudotumour cerebri. J Neurol Neurosurg Psychiatry 45: 734–736
6. Earnest MP, Heaton RK, Wilkinson WE, Manke WF (1979) Cortical atrophy, ventricular enlargement and intellectual impairment in the aged. Neurology 29: 1138–1143
7. Grant R, Condon B, Lawrence A, Hadley DM, Patterson J, Bone I, Teasdale GM (1987) Human cranial CSF volumes measured by MRI: sex and age influences. Magnetic Resonance Imaging 5: 465–468
8. Grant R, Condon B, Lawrence A, Hadley DM, Patterson J, Bone I, Teasdale GM (1988) Is cranial CSF volume under hormonal influence? An MR study. J Comp Ass Tomog 12 (1) 36: Jan/Feb
9. Grant R, Condon B, Patterson J, Wyper DJ, Hadley DM, Teasdale GM (in press) Changes in cranial CSF volume during hypercapnia and hypocapnia. J Neurol Neurosurg Psychiatry
10. Greenberg JH, Alavi A, Reivich M, Kuhl D, Uzzell B (1978) Local cerebral blood volume response to carbon dioxide in man. Circ Res 43(2): 324–331
11. Gyldensted C (1977) Measurement of the normal ventricular system and hemispheric sulci in 100 adults with computed tomography. Neuroradiology 14: 183–192
12. Huckman MS, Fox J, Topel J (1975) The validity of criteria for evaluation of cerebral atrophy by computed tomography. Radiol 116: 85–92
13. Kaszniak AW, Garron DC, Fox JH, Bergen D, Hackman M (1979) Cerebral atrophy, EEG showing age, education and cognitive functioning in suspected dementia. Neurology 29: 1273–1279
14. Laffey PA, Peyster RG, Hathan R, Haskins ME, McGinlay JA

(1984) Computed tomography and ageing; Results in a normal elderly population. Neuroradiology 26: 273–278

15. Landau RL, Bergenstal DM, Lugibihl K, Kascht ME (1955) The metabolic effects of progesterone in man. J Clin Endocrin 15: 1194–1215

16. Machida H (1981) Influence of progesterone on arterial blood and CSF acid-base balance in women. J Appl Physiol 15 (6): 1433–1436

17. Martins AN, Wiley JK, Myers PW (1972) Dynamics of the cerebrospinal fluid and the spinal dura mater. J Neurol Neurosurg Psychiatry 35: 468–473

18. McComb JG (1983) Recent research into the nature of cerebrospinal fluid production and absorption. J Neurosurg 59: 369–383

19. McQuarrie I (1932) Some recent observations regarding the nature of epilepsy. Ann Intern Med 6: 497–505

20. Wyper DJ, Lennox GA, Rowan J (1976) Two minute slope technique for CBF measurement in man. J Neurol Neurosurg Psychiat 39: 141–146

21. Wyper DJ, Pickard JD, Matheson M (1979) Accuracy of ventricular volume estimation. J Neurol Neurosurg Psychiatry 42 (4): 345–350

22. Zatz LM, Jernigan TL, Ahumada AJ (1982) Changes on computed cranial tomography with ageing; Intracranial fluid volume. AJNR 6: 1–11

23. Zuckerman S, Palmer A, Hanson DA (1950) The effect of steroid hormones on the water content of tissues. J Endocrinol 6: 261–276

Address for correspondence: G. M. Teasdale, MRCP, FRCS, Professor of Neurosurgery, Institute of Neurological Sciences, Southern General Hospital, Glasgow, U.K.

V. Central Pain Syndromes

Acta Neurochirurgica, Suppl. 42, 239–242 (1988)

Thalamic Pain and Stereotactic Mesencephalotomy

C. Shieff and B. S. Nashold, Jr

Department of Surgery, Division of Neurosurgery, Duke University Medical Center, Durham, North Carolina, U.S.A.

Summary

The severe pain that can be experienced by stroke patients is refractory both to drugs and to non-medical therapies. Various surgical procedures are widely advocated for its relief, stereotactic mesencephalic tractotomy in particular providing good results.

Twenty seven patients with pain of central origin following stroke underwent stereotactic mesencephalic tractotomy by thermocoagulation at one of two alternative sites. Fourteen had lesions created at the original target adjacent to the superior colliculus, 75% reporting long term relief of their pain. Of this group, 83% had residual postoperative ocular dysfunction (50% symptomatic) and two died soon after surgery. Thirteen patients had surgery at the revised target at the level of the inferior colliculus: 58% had long term pain relief, 23% had ocular problems (none symptomatic) and mortality was nil.

Keywords: Thalamic pain; stereotactic mesencephalic tractotomy.

Introduction

The pain syndromes occuring after cerebrovascular accidents are well recognised although their incidence is unknown[11, 14, 21]. Sensory disturbance with pain and paresis with abnormal movements was first described by Dejerine and Roussy[4]. The pain results from lesions of the spinothalamic pathways and it is often, but not exclusively, associated with altered sensation; this is a phenomenon of deafferentation[3, 11, 19]. Patients may complain of a continuous background pain with exacerbations occurring either spontaneously or following "normal" stimuli. In some cases it has been shown that this is of an epileptic nature[24, 25]. In many cases there is also a strong emotional component either as a precipitating factor or as a secondary feature.

Surgical procedures designed to relieve thalamic pain are either ablative or stimulatory. The earliest successes followed destruction of spinothalamic pathways or excision of sensory cortex[5, 7, 23, 24]. Stereotactic ablative lesions within the sensory thalamus or adjacent structures are commonplace and often effective in providing pain relief. Unfortunately these procedures may result in further postoperative deafferentation symptoms[19]. Chronic stimulation of deep brain structures is effective but it is expensive and requires both patient comprehension and compliance[9, 12, 20]. Interruption of the limbic connections alleviates "suffering" but it does not abolish the continuing experience of pain itself[6, 8, 13]. The production of discrete lesions within the upper brain stem has been refined since Spiegel and Wycis' original case[26] and can now be shown to relieve both the pain and emotional factors without further problems of deafferentation.

Patients and Methods

Twenty-seven patients aged between 37 and 73 years (mean 56 years) underwent stereotactic mesencephalotomy at Duke University Medical Center between 1963 and 1985 for the relief of unremitting chronic pain following cerebrovascular accidents. Twenty (thirteen men and seven women) had supratentorial infarcts and seven (six men, one woman) infarcts of the brain stem. Ten patients had onset of their pain at intervals of up to two years following their stroke. The period between the onset of pain and surgery ranged from eight weeks to seventeen years.

Six patients had severe motor disability limiting daily activities and ten had movement disorders. Seven had speech disorders. Four had visual field defects, one disordered conjugate eye movements and two anisocoria. Twenty were found to have hypoaesthesia alone on sensory examination and three exhibited allodynia without any clinical sensory deficit. Four patients had epilepsy and four had major psychiatric problems.

Eight patients had previously undergone surgery in attempts to abolish their pain, one having had stereotactic mesencephalic spinothalamic tractotomy at a different target site which had produced dysaesthesiae and gait disturbance without pain relief.

The techniques employed are described elsewhere[15, 16, 18]. All procedures were performed under local anaesthesia alone or with adjuvant neuroleptanaesthesia. A frontal approach from the coronal suture was made and contrast ventriculography used to determine the target coordinates. The target was contralateral to the side of the greatest pain in all except one patient whose third procedure was ipsilateral. The initial target (lesion I, superior colliculus) was used for the first fourteen patients. It was level with, and three millimetres behind, the posterior commissure, its laterality (up to 7 millimetres) being determined from the effects of stimulation. This was revised in the last thirteen cases (lesion II, inferior colliculus) to five millimetres behind, below and lateral to the posterior commissure[2]. Physiological corroboration was by electrical stimulation. Contralateral sensory phenomena were obtained in twenty-one patients and anxiety or agitation observed together with complaints of visceral or corporeal sensations in twelve. Horizontal gaze paresis was noted in sixteen patients, vertical gaze paresis in eleven and impaired ocular convergence in three. Nystagmus developed in four patients, pupillary changes in seventeen and eyelid fluttering in six. Ablative lesions were then produced by thermocoagulation and this was repeated until pain or hyperpathia had been abolished. Four patients had surgery repeated early because of unsustained effects and six were readmitted for further surgery after late return of their symptoms. One patient was pain free only after four procedures.

Results

These were assessed in terms of pain relief, residual dysaesthesiae, demand for analgesics and the ability to participate in daily activities. They were graded good, fair, poor and bad. Complete relief of pain was a good result. A fair result was recorded when there was minimal residual pain, physical activity was enhanced and when non-narcotic analgesics only were required. Poor results were obtained in those with significant residual pain and dysaesthesiae but with less spontaneous complaints of their symptoms, and whose spouses could often report a reduction in apparent suffering together with increased participation in social activities. Patients were assessed initially at the time of discharge from hospital and at later review (up to five years); this was by letter or by telephone (to patient or physician) and in some cases by out patient attendance.

Pain Relief at Discharge

Twenty one of the original twenty seven patients (78%) were pain free or substantially improved, lesion I improving eleven patients (79%) and lesion II ten (77%).

Long Term Relief

Follow-up was from three months to five years. One patient could not be traced and two had died shortly after discharge. Six patients had improved and nine had deteriorated. Improvement occurred in three patients from each group (one early patient from "bad" to "good"). Deterioration by two grades took place in two of the earlier patients and in four from the later group. Sixteen of the twenty four patients available were in good or fair categories (67%); of twelve patients from each group, nine (75%) and seven (58%) respectively reported relief.

Table 1. *Lesion I—Superior Colliculus*

	Good	Fair	Poor	Bad
Early (14 patients)	5	6	0	3
	11/14 = 79%			
Late (12)	7	2	1	2
	9/12 = 75%	(1 died, 1 lost to follow up)		

Table 2. *Lesion II—Inferior Colliculus*

	Good	Fair	Poor	Bad
Early (13 patients)	5	5	1	1
	10/12 = 83%	(1 moribund)		
Late (12)	4	3	2	3
	7/12 = 58%			

The results from each target site are presented in Tables 1 and 2.

Complications

Nine patients had temporary alteration in their conscious level – four becoming drowsy and five agitated. Two had transient worsening of a preexisting hemiparesis, one of dysphasia and one of dysarthria.

All those undergoing surgery at the original site, and seven at the revised site, had abnormal postoperative orthoptic examination, ten and two respectively of these persisting through follow-up. The commonest problem was a vertical or horizontal gaze palsy but in only one patient was this symptomatic[22].

Two patients developed aspiration pneumonia, one recovering and the other requiring tracheostomy before his death (both had lesions at superior collicular level). Another died from pulmonary embolus three days after discharge. A third death occurred in a patient who was still pain free one year after his discharge, following infection of a ventriculoperitoneal shunt inserted elsewhere.

Discussion

Lesions of the neospinothalamic pathways, whether pathological or surgically produced, result in diminished or altered sensation. Older pathways within the periaqueductal and periventricular grey matter also contribute to the delicate balance of signals that are interpreted either as normal epicritic sensations or as pain. Deafferentation arises when an imbalance in these messages has occurred. Solitary lesions are seldom dis-

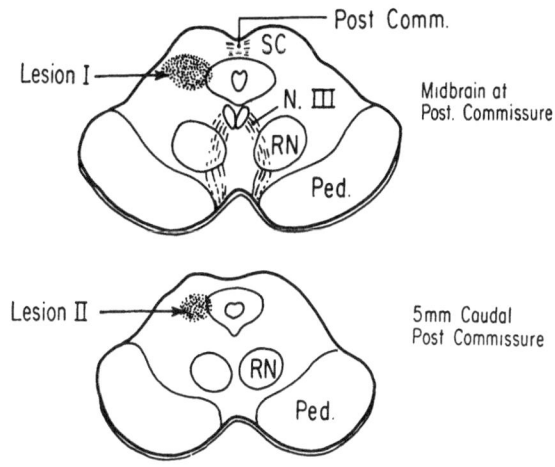

Fig. 1. Diagrammatic transverse sections of mesencephalon showing lesions

crete enough to diminish discriminative sensation alone and "pain" arises when signals are transmitted by the polysynaptic pathways which cannot then be integrated with those that would be carried by more highly evolved pathways[3, 10, 19].

Ablation of portions of the neospinothalamic system or the thalamus relieves pain when this has a nociceptive origin, but pain of central origin is often exacerbated. Therapeutic lesions within the atavistic network comprising the central grey matter cannot be localised adequately to alter pain perception alone and in consequence the practice of creating lesions which include part of both neo- and palaeo-spinothalamic pathways has developed (Fig. 1). We have observed that, while the subjective experience of pain from peripheral stimulation is abolished immediately, background pain resolves over several days or weeks together with general reduction in anxiety and emotional lability following, we believe, secondary degeneration of less specific neural circuits. Similarly, late recurrence may be due to the emergence of new connections by regeneration or by sprouting.

The change in target site has reduced ocular morbidity[17, 22] while any reduction in the number obtaining pain relief is probably due to a greater reluctance to create multiple lesions either in single or multiple operations. This is almost certainly also responsible for a general reduction in morbidity.

We believe that our results, confirmed by others[1], indicate that stereotactic mesencephalotomy offers the neurosurgeon a valuable technique for use in the treatment of this distressing condition which defies most other therapies.

References

1. Amano K, Kawamura H, Tanikawa T, Kawabatake H, Notani M, Iseki H, Shiwaku T, Nagao T, Iwata Y, Taira T, Umezawa Y, Simizu T, Kitamura K (1986) Long term follow-up study of rostral mesencephalic reticulotomy for pain relief – report of 34 cases. Appl Neurophysiol 49: 105–111
2. Amano K, Kitamura K, Sano K, Sedino H (1976) Relief of intractable pain from neurosurgical point of view with reference to present limits and clinical indications. A review of 100 consecutive cases. Neurologia Medico-chirurgica 16: 141–153
3. Cassinari V, Pagni CA (1969) Central pain, a neurosurgical survey. Harvard University Press, Cambridge, p 192
4. Dejerine J, Roussy G (1906) Le syndrome thalamique. Rev Neurol (Paris) 14: 521–535
5. Erickson TC, Bleckwenn WJ, Woolsey CN (1952) Observations on the postcentral gyrus in relation to pain. Trans Am Neurol Assoc 77: 57–58
6. Foltz EL, White LE (1962) Pain "relief" by frontal cingulotomy. J Neurosurg 19: 89–100
7. Frazier CH, Lewy FH, Rowe SN (1937) The origin and mechanism of paroxysmal neuralgic pain and the surgical treatment of central pain. Brain 60: 44–51
8. Freeman W, Watts JW (1948) Pain mechanisms and the frontal lobes: a study of prefrontal lobotomy for intractable pain. Ann Intern Med 28: 747–754
9. Gybels JM (1979) Electrical stimulation of the central grey for pain relief in humans: a critical review. In: Bonica JJ, Liebeskind JC, Albe-Fessard DG (eds) Advances in pain research and therapy, vol 3. Raven, New York, pp 499–508
10. Hassler R (1975) Central interactions of the systems of rapidly and slowly conducted pain. In: Penzholz H, Brock M, Hamer J, et al (eds) Advances in neurosurgery, vol 3. Springer, Berlin Heidelberg New York, pp 143–156
11. Head H, Holmes G (1911) Sensory disturbances from cerebral lesions. Brain 34: 102–254
12. Hosobuchi Y, Adams JE, Rutkin B (1973) Chronic thalamic stimulation for the control of facial anaesthesia dolorosa. Arch Neurol 29: 158–162
13. Jelasic F (1966) Relations of the lateral part of the amygdala to pain. Confin Neurol 27: 53–55
14. Martin JJ (1969) Thalamic syndromes. In: Vinken PJ, Bruyn GW (eds) Handbook of clinical neurology, vol 2. North-Holland, Amsterdam, pp 469–496
15. Nashold BS (1975) Mesencephalotomy. A current appraisal. In: Voris HC, Whisler WW (eds) Treatment of pain. Thomas, Springfield, pp 121–131
16. Nashold BS (1982) Brain stem stereotaxic procedures. In: Schaltenbrand G, Walker AE (eds) Stereotaxy of the human brain, 2nd ed. Thieme, Stuttgart, pp 475–483
17. Nashold BS, Seaber JH (1972) Defects of ocular morbidity after stereotactic midbrain lesions in man. Arch Ophthal 88: 245–248
18. Nashold BS, Wilson WP, Slaughter DG (1969) Stereotaxic midbrain lesions for central dysaesthesia and phantom pain. Prelimary report. J Neurosurg 30: 116–126
19. Pagni CA (1979) Pain due to central nervous system lesions: physio-pathological considerations and therapeutical implications. In: Bonica JJ (ed) Advances in neurology, vol 4. Springer, Berlin Heidelberg New York, pp 339–348
20. Richardson DE, Akil H (1977) Pain reduction by electrical brain stimulation in man. J Neurosurg 47: 178–194

21. Riddoch G (1938) The clinical features of central pain. Lancet 2234 (1): 1093–1098, 1150–1156, 1205–1209
22. Seaber JH, Nashold BS (1980) Comparison of ocular effects of unilateral stereotactic midbrain lesions in man. Neuro-ophthalmology 1: 95–99
23. White JC (1963) Anterolateral chordotomy – its effectiveness in relieving pain of nonmalignant disease. Neurochirurgie 6: 83–102
24. Wilson SAK (1927) Dysaesthesiae and their neural correlates. Brain 50: 428–462
25. Wilson WP, Nashold BS (1968) Epileptic discharges occurring in the mesencephalon and thalamus. Epilepsia 9: 265–273
26. Wycis HT, Spiegel EA (1962) Long range results in the treatment of intractable pain by stereotaxic midbrain surgery. J Neurosurg 19: 101–107

Address for correspondence: C. Shieff, FRCS, Midland Centre for Neurosurgery and Neurology, Holly Lane, Smethwick, Warley, West Midlands B67 7JX, U.K.

Acta Neurochirurgica, Suppl. 42, 243–247 (1988)
© by Springer-Verlag 1988

Stimulation of Internal Capsule, Thalamic Sensory Nucleus (VPM) and Cerebral Cortex Inhibited Deafferentation Hyperactivity Provoked After Gasserian Ganglionectomy in Cat

S. Namba and A. Nishimoto

Department of Neurological Surgery, Okayama University Medical School, Okayama, Japan

Summary

Facilitation of the opiate-mediated system provides relief of excess pain but not of deafferentation pain. Influence on neuronal hyperactivity, which was provoked after the deafferentation of the peripheral trigeminal nerve (deafferentation hyperactivity; DH), by stimulation of the internal capsule (IC), VPM nucleus of the thalamus and by the cerebral sensorimotor cortex, was examined. This experiment demonstrates that a suppressive effect is exerted on the sustained neuronal hyperactivity, which was proved to have a close relationship with deafferentation pain.

In the preliminary experiment, it was confirmed that DH was provoked by coagulation of unilateral (left side) Gasserian ganglion in 29 adults cats, who were allowed to survive up to 72 days. DH, sustained high amplitude discharge without any stimuli on the face or neck, was detected in the subnucleus caudalis of the spinal trigeminal nucleus (STNcd) of the denervated side. DH was never suppressed by facilitation of the opiate-mediated system. 87 of 113 neurones identified in the STNcd of the denervated side showed DH in another 18 cats.

In 30 of 55 neurones examined, DH was conspicuously suppressed by the stimulation of contralateral (right) side, 11 of 29 neurones by ispilateral IC. Five of 13 neurones were suppressed by contralateral VPM stimulation, 2 of 3 neurones by ipsilateral stimulation. Stimulation of cortical area, SM_1 of MS_1, in other 15 cats, considerably suppressed DH in STNcd.

Microinjection of wheat-germ-agglutinin in the cerebral cortex revealed in direct projection of the descending fibre from MS_1 to STNcd and brain stem reticular formation via IC.

From these experimental results, it was considered that descending fibre inhibiting deafferentation acitvity exists, diverging bilaterally from cerebral cortex, and stimulation of IC or VPM probably facilitates this descending fibre at each stimulating site.

Keywords: Deafferentation pain; neuronal hyperactivity; trigeminal nerve; stimulation; internal capsule; VPM nucleus of the thalamus; cerebral sensorimotor cortex; pain inhibition.

Introduction

Facilitation of the opiate-mediated pain inhibiting system provides reliev of excess pain but not of deaffer-

entation pain. This experiment demonstrates the suppressive effect of a stimulation of the internal capsule, thalamic sensory nucleus (VPM) and cerebral cortex on the sustained neuronal hyperactivity, which is postulated to have a close relationship with the so-called "deafferentation pain".

Methods and Material

Part I: Setting up an Experimental Model of Deafferentation Pain

As a preliminary experiment, the anatomical location of the subnucleus caudalis of the spinal trigeminal nucleus (STNcd) was determined by Klüver-Barrera stain; also by physiological studies by provoking pain with a pinprick of pinch on the facial skin in 16 adult cats and placing the extracellular recording electrode in STNcd (Fig. 1). These studies made it clear that the STNcd is located from obex to 2 mm caudally, and an electrophysiological study revealed nociceptive neurones distributed 3–4 mm lateral to midline.

The Gasserian ganglion on the left side was totally coagulated through a subtemporal extradural approach in 6 of these 16 cats. It was found that a definitive neuronal hyperactivity could be observed from about 10 days after the denervation.

The main experiment was performed using the other 13 adult cats, which were unilaterally (left side) denervated. The longest survival period of the cats was 63 days after the denervation. Forty-three neurones were identified in the STNcd of the intact (nondenervated) side, the majority of which (40 neurones) being the wide dynamic range neurone. Twenty-two neurones were identified in the STNcd of the denervated (left) side, 12 of which (55%) showed sustained high amplitude firings in the absence of any noxious stimuli to the face (denervation hyperactivity). Such spontaneous remarkable and continuously hyperactive neurones were never detected in the neurones of the intact (right) side (Fig. 1). After the denervation, the cat usually showed abnormal behaviour in the form of ceaseless rubbing of the denervated side of the face with their paws or against the cage. Intraventricular injection of morphine (1 mg) of enkephalinamide (0.55 mg), or electrical stimulation of periaqueductal gray obviously and significantly suppressed the nociceptive neuronal

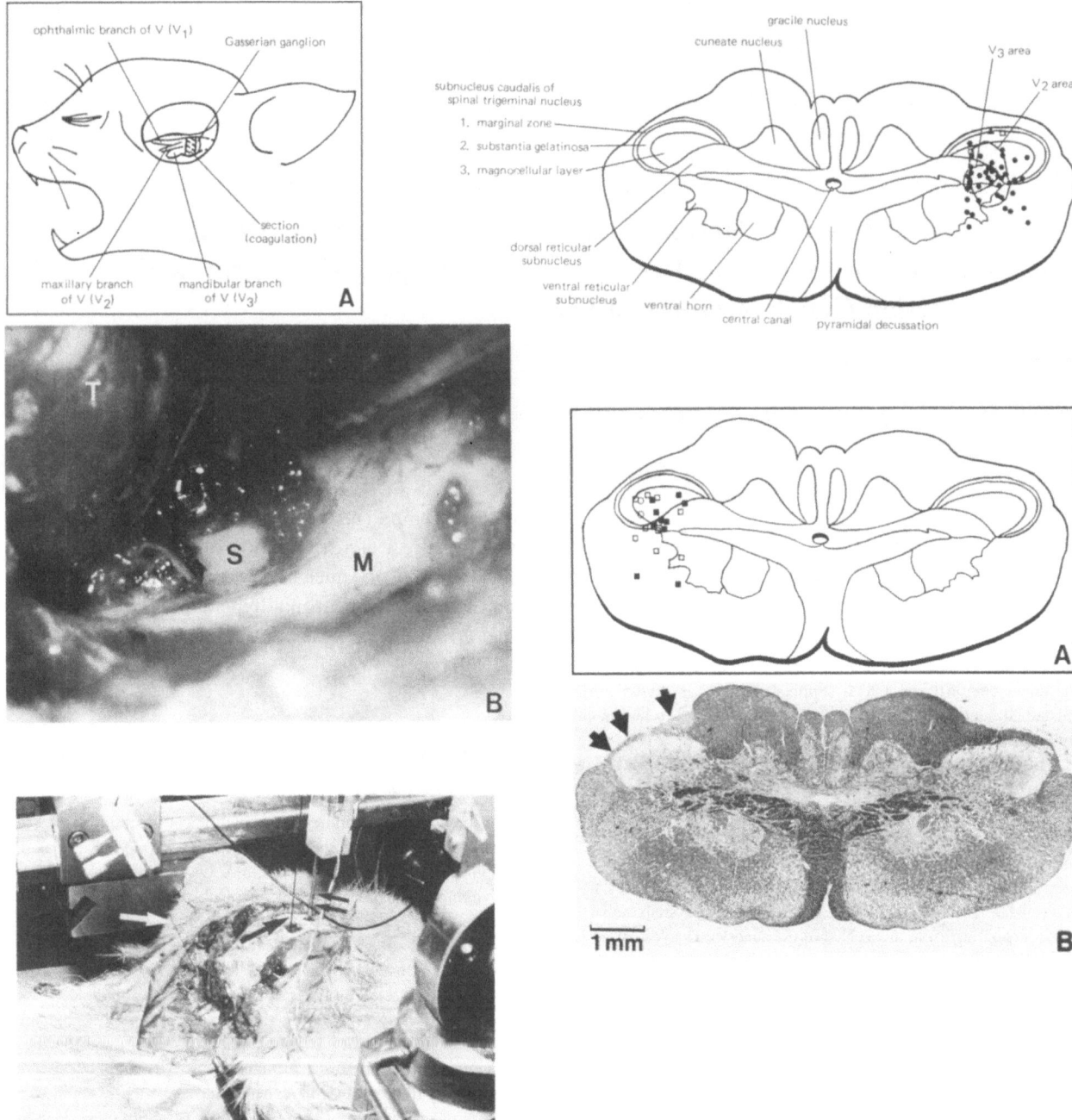

Fig. 1. Upper left: (A) Schematic drawing of the section of the left Gasserian ganglion. (B) High frequency coagulation is applied as widely as possible to attain complete denervation. *T* indicates temporal lobe, *S* section (coagulation), *M* middle cranial base. Lower left: Microelectrode (white arrow) is inserted in the right subnucleus caudalis of the spinal trigeminal nucleus (STNcd). Another electrode (black arrow) is into the periaqueductal grey. A small cannula (double arrow) is introduced into the lateral ventricle. Upper right: Distribution of the perceptive neurones in STNcd of the intact (non-denervated) side. Location of the neurones is projected on the Berman's[3] cat atlas. Solid triangle: nociceptive specific neurone, open square: low threshold mechano-receptive neurone, solid circle: wide dynamic range neurone. Lower right: (A) Distribution of the neurones identified on the denervated side (22 neurones), open squares indicate the neurones with denervation hyperactivity, two open circles are the hyperactive neurones as well as responding to noxious stimuli on the cervical skin, and solid squares are those without the hyperactivity. (B) Horizontal section of the spinal cord of a cat at 1 mm below the obex, showing degeneration (arrows) of the spinal trigeminal tract 23 days after the deafferentation (Klüver-Barrera stain)

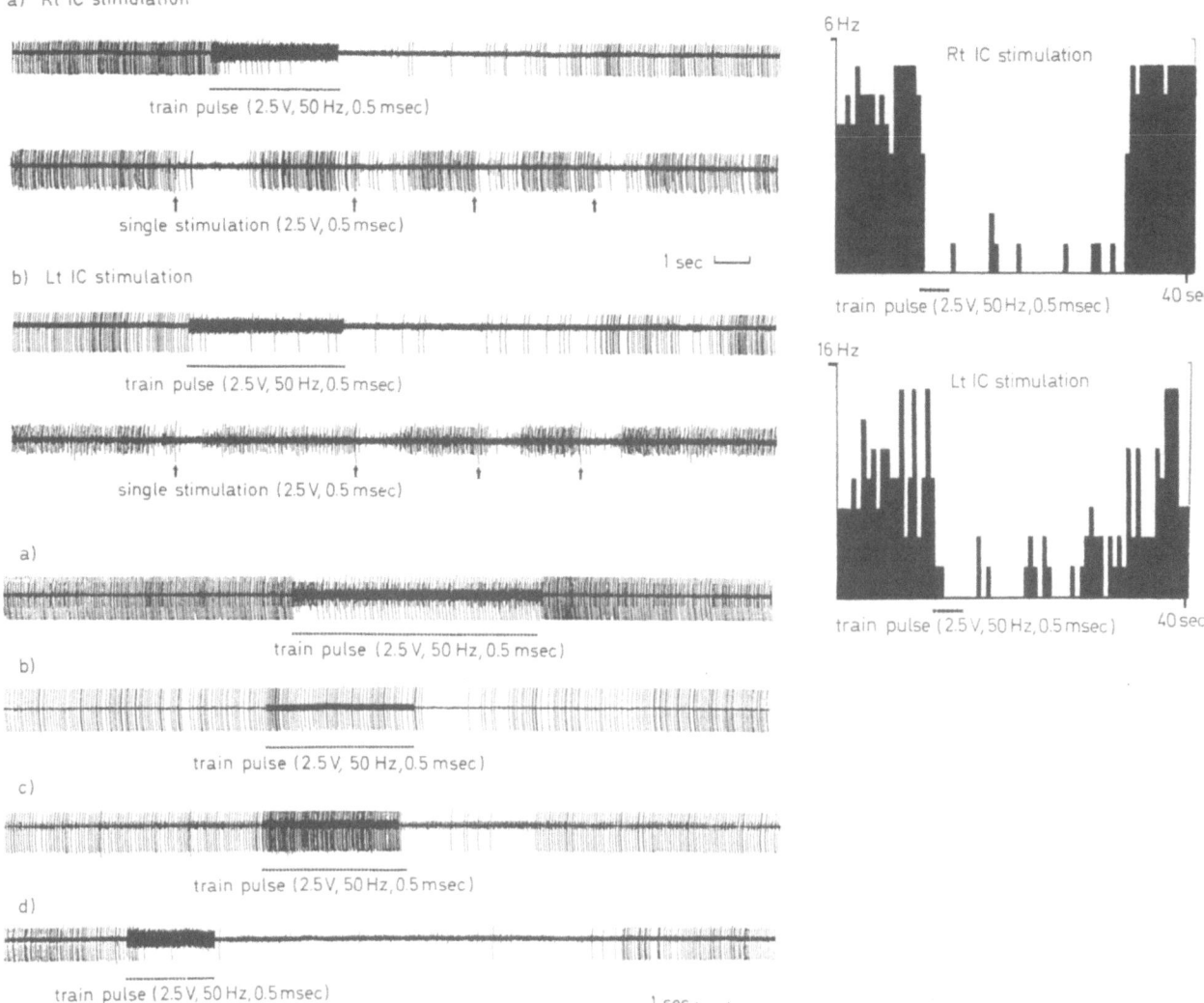

Fig. 2. Upper left: Train or single pulse stimulation of either side of the internal capsule clearly suppressed the deafferentation hyperactivity recorded in the STNcd of denervated (left) side. Upper right: Histogram showing suppression of deafferentation hyperactivity in STNcd by either side of the internal capsule stimulation. Lower left: There were different patterns in suppression of deafferentation hyperactivity by internal capsule stimulation

firing of the intact side. By contrast, denervation hyperactivity was never affected.

Part II: Inhibition of Deafferentation Hyperactivity by Electrical Stimulation of Internal Capsule or Thalamic Sensory Nucleus

In this experiment, 28 adult cats were used. This study investigated: 1. the influence on the neuronal activity in STNcd of the intact side responding to noxious stimuli on the face, to stimulation of the internal capsule (A : 17, L : 6.0, H : 1–5) or of the thalamic sensory nucleus, VPM (A : 8.5, L : 4.5–5.5, H : 1–2) ipsilaterally or contralaterally in 4 cats, 2. the influence on deafferentation hyperactivity in the STNcd, by stimulation of the internal capsule or VPM in another 18 cats, 3. the spreading of deafferentation hyperactivity up to neurone in thalamic sensory nucleus after the deafferentation in 6 cats.

Results: 1. Twenty-one neurones were identified in the STNcd of the non-denervated cats. All of these neurones were wide dynamic

range neurones. Two of 4 these neurones investigated were inhibited by stimulation of contralateral internal capsule, and 3 of 6 neurones were inhibited by stimulation of the ipsilateral internal capsule. A similar inhibitory effect of STNcd neurones was observed by stimulation of ipsilateral or contralateral VPM. 2. On one hundred and thirteen occasions, stable neuronal activity was recorded in 9 of the 18 cats, which survived and were used for the experiment in the period between 24–72 days after the denervation. Eighty seven of 113 neurones showed deafferentation hyperactivity. Contralateral (right side) internal capsule stimulation inhibited 30 of 55 neurones (55%) with deafferentation hyperactivity analyzed in 8 cats. Ipsilateral (left side) stimulation inhibited 11 of 29 neurones (38%) with deafferentation hyperactivity. The pattern of inhibition on deafferentation hyperactivity could be classified into 3 groups: 1. the suppressive effect is weak and lasts only for a short time, 2. no suppressive effect during stimulation but a pronounced suppression after ceasing stimulation, 3. suppression appearing shortly after beginning stim-

Table 1. *Effect of Cerebral Cortical Stimulation on Neuronal Activity in STNcd of Denervated Cats*

	Control (31 neurons)	DH (80 neurons)
Ms. I	15/21 (71%)	5/36 (14%)
Sm. I	5/10 (50%)	12/36 (33%)
Sm. II	0/1	0/10 (10%)
Sm. III	0/1	0/6
Orb. Cx		0/5

On the deafferented (left) side, 80 neurones with deafferentation hyperactivity were identified. Seventeen of 72 neurones (24%) were suppressed by electrical stimulation of MS_1 or SM_1.

ulation and lasting for a long time after ceasing stimulation (Fig. 2). Five of 13 neurones with deafferentation hyperactivity (38%) were inhibited by stimulation of contralateral VPM, 2 of 3 neurones by stimulation of ipsilateral VPM. Neurones with similar continuous high amplitude firing, which was not affected by any exogeneous stimuli, were found in VPM on the right side. The influence on 5 neurones with such hyperactivity in VPM by stimulation of internal capsule of either side, was examined. Hyperactivity was suppressed by stimulation of the internal capsule of either side, the more striking suppression being elicited by ipsilateral internal capsule stimulation.

Part III: Influence of Stimulation of Cerebral Cortex on Deafferentation Hyperactivity in STNcd

In this study, 15 adult cats were used. Seven cats were used as a control study, and 8 were unilaterally (left side) denervated and survived more than 30 days. Only 2 of 31 neurones identified in the STNcd of control cats were nociceptive specific neurones, and others were wide dynamic range neurones. They were remarkably suppressed by the stimulation of contralateral somatomotor (MS_1) or somatosensory (SM_1) cerebral cortical area. Eighty neurones were found in the left SNTcd showing deafferentation activity. Seventeen of 72 neurones (24%) were suppressed by the stimulation of MS_1 or SM_2 (Table 1). Furthermore, wheat germ agglutinin labelling horseradish peroxidase was injected into the cerebral cortex; MS_1 and SM_1, and the descending fibres from the cerebral cortex were pursued to their terminals. The study demonstrated that the efferent fibre passes near that portion of the internal capsule which was stimulated in the former study (Part II) which investigated the influence of stimulation of internal capsule on deafferentation hyperactivity in STNcd. It was also shown that the efferent fibre terminated in the dorsal and ventral reticular subnucleus of STNcd on the contralateral side.

Discussion and Conclusion

It is fairly easy to set up an excessive pain model[6, 7, 10]. For example, stimulation of tooth pulp is believed to induce excessive pain, by stimulation of pure C fibres[7, 10]. In this study a deafferentation pain model was created by unilateral Gasserian ganglionectomy. After a few weeks, the cat showed abnormal behaviour, which was considered to have some relation at least to uncomfortable sensations induced by the deafferentation[2, 5, 8]. Of course there is no definitive

method to determine whether such abnormal behaviour in the cat is really related to deafferentation pain or not. However, deafferentation hyperactivity in STNcd was evoked in cats which showed such abnormal behaviour. Further, the deafferentation hyperactivity in STNcd of the denervated left side was not inhibited by stimulation of the opiate-mediated system — intraventricular injection of morphine or periaqueductal grey stimulation, in sharp contrast to the remarkable inhibition of nociceptive neuronal firing. After these considerations, the Gasserian ganglionectomised cat who showed the abnormal behaviour was accepted as a deafferentation pain model.

Stimulation of internal capsule or VPM showed an inhibitory effect on deafferentation hyperactivity. Stimulation of the same sites also inhibited nociceptive neuronal firing in STNcd. The inhibitory effects was bilateral. If one considers the complexity of the reticular formation in the brain stem, it is not surprising that this inhibitory effect on the neurones in STNcd is bilateral and affects both deafferentation hyperactivity and nociceptive neuronal firing in response to facial stimuli. It is interesting that three different patterns in suppression of deafferentation hyperactivity were evoked by stimulation of internal capsule. This may result from slight differences either in the point of stimulation in the internal capsule or else in the nature of the neurone recorded in the STNcd. Alternatively, this may be the result of a difference in the descending pathway from the stimulating site of the internal capsule to the brain stem; in other words, variability may be a consequence of the fact that the inhibitory effect is conducted to brain stem via such a complex reticular formation. We certainly believe that the brain stem reticular formation plays an important role in the conducting of the inhibitory effect shown in this experiment. However, an involvement of cerebral cortex could not be excluded. There is a possibility that the inhibitory effect of internal capsule or VPM stimulation is conducted up to the cerebral cortex and then descends via an inhibitory pathway to the brain stem. There have been some studies that describe the direct connection between cerebral cortical neurones to those in the brain stem[1, 4, 9], also that stimulation of cerebral cortex inhibits neuronal firing of the brain stem[1]. In this experiment both these phenomena were confirmed. Wheat germ agglutinin injection in the cerebral somatomotor (MS_1) or somatosensory (SM_1) area revealed the direct projection of the cortical neurones to the STNcd or nucleus reticularis in the brain stem, and that the stimulation of the areas of the cortex inhibits both deafferentation hyperactivity and nociceptive neuronal firing in the brain stem.

The most interesting observation in this series of experiments is that descending fibres from cerebral cortex were found to pass through one portion of internal capsule where stimulation showed remarkable inhibition of deafferentation hyperactivity or nociceptive neuronal firing in the STNcd.

The experimental result is so complex that it may not be fully explained by these theories. However, it should be accepted as a fact that there is a pain inhibiting pathway other than the opiate-mediated system, and that this pathway plays an important role in the control of deafferented pain. It is considered that further study is needed concerning the role of brain stem reticular formation.

Acknowledgement

This study was done in cooperation with Yozi Shimizu, M.D., Takao Wani, M.D., Tetsuya Masaoka, M.D. and Shigeo Nakamura, M.D. in the Department of Neurological Surgery, Okayama University Medical School.

References

1. Anderson P, Eccles JC, Sears TA (1962) Presynaptic inhibitory action of cerebral cortex on spinal cord. Nature (London) 194: 740–741
2. Anderson LS, Block RG, Abraham J, Ward AA Jr (1971) Neuronal hyperactivity in experimental trigeminal deafferentation. J Neurosurg 35: 444–452
3. Berman AL (1968) The brain stem of the cat, a cytoarchitectonic atlas with stereotaxic coordinates. Madison, The University of Wisconsin, p 175
4. Berrevoets CE, Kuypers HGJM (1975) Pericurciate cortical neurons projecting to brain stem reticular formation, dorsal column nuclei and spinal cord in the cat. Neuroscience Letters 1: 257–262
5. Losser JD, Ward AA Jr (1967) Some effects of deafferentation on neurons of the cat spinal cord. Arch Neurol 17: 629–639
6. Nakao Y (1983) Experimental and clinical studies on pain reduction by electrical stimulation of the internal capsule. Part I: Electrical stimulation of the posterior limb of the internal capsule for treatment of thalamic pain. Acta Med Okayama 95: 1209–1226
7. Nakao Y (1983) Experimental and clinical studies on pain reduction by electrical stimulation of the internal capsule. Part II: Experimental study of pain relief by electrical stimulation of internal capsule. Acta Med Okayama 95: 1227–1241
8. Namba S, Shimizu Y, Wani T, Nakamura S (1985) An experimental model for deafferented pain in cats. Neurol Med Chir (Tokyo) 25: 715–722
9. Niimi K, Kishi S, Miki M, Fujita S (1963) An experimental study of the course and termination of the projection fibres from cortical areas 4 and 6 in the cat. Folia Psychiatr Neurol Jp 17: 167–216
10. Toda K, Iriki A (1979) Effect of electroacupuncture on thalamic evoked responses recorded from the ventrobasal complex and posterior nuclear group after pulp stimulation in rat. Exp Neurol 66: 419–422

Address for correspondence: S. Namba, M.D., Department of Neurological Surgery, Okayama University Medical School, Okayama, Japan.

Acta Neurochirurgica, Suppl. 42, 248–252 (1988)

The Tolosa-Hunt Syndrome: a Problem in Differential Diagnosis

W. E. Hunt and **R. P. Brightman**

The Division of Neurosurgery, The Ohio State University Hospitals, Columbus, Ohio, U.S.A.

Summary

The authors give a review on the Tolosa-Hunt Syndrome. They discuss its differential diagnosis, appropriate diagnostic and therapeutic measures for the evaluation of this painful ophthalmoplegia which may be combined with neurological deficit referrable to the anterior cavernous sinus.

Introduction

In 1954 Dr. Eduardo Tolosa of the Neurologic Institute of Barcelona reported a case of remittent painful ophthalmoplegia which at autopsy showed a granulomatous inflammation of the cavernous sinus that he termed "periarteritis"[1]. Hunt *et al.*, in 1961 described a group of patients with a similar clinical syndrome, having reviewed Dr. Tolosa's original slides, and postulated a similar aetiology in their 6 cases[2]. It was characterized by the following criteria[3]:

1. Retro-orbital pain which may precede the ophthalmoplegia by several days, or may not appear until later. It is not a throbbing hemicrania occurring in paroxysms, but a steady pain behind the eye that is often described as "gnawing" or "boring".

2. Neurologic involvement is not confined to the third cranial nerve, but may include the fourth, sixth and first division of the fifth cranial nerves. Periarterial sympathetic fibres and the optic nerve may be involved.

3. Symptoms last for days to weeks.

4. Spontaneous remissions occur, sometimes with residual neurological deficit.

5. Attacks may recur at intervals of months or years.

6. Exhaustive studies, including angiography and surgical exploration, have produced no evidence of involvement of structures outside the cavernous sinus. There is no systemic reaction.

The value of corticosteroid therapy was also noted.

This benign, self-limited but recurring, steroid-sen-sitive condition was called the Tolosa-Hunt Syndrome (THS) by J. Lawton-Smith in 1966[4]. Since that time numerous investigators have supported and challenged these initial observations. Although new imaging techniques have allowed more detailed examination of the cavernous sinus and its contents, the aetiology of the idiopathic inflammation remains obscure. The question of taxonomy has arisen, as it often does in eponymic syndromes where the aetiology of the process is unclear. Thus, it seems appropriate to review the differential diagnosis and the diagnostic and therapeutic techniques now available. We will aso revise the initial criteria, attempt to elucidate the taxonomy and speculate on the aetiology of this puzzling syndrome.

The symptom complex of painful ophthalmoplegia consisting of retroorbital pain associated with paralysis of ocular movements and other cranial nerve involvement has been described by many authors. These symptoms alone do not identify a specific disease but rather localize the process to the region of the superior orbital fissure and anterior cavernous sinus. Patients presenting with symptoms of pathology in this region require a thorough diagnostic investigation since the differential diagnosis includes a variety of lesions. Rochon-Duvigneaud in the late nineteenth century recognized the superior orbital fissure syndrome as an entity and attributed it to syphilitic periostitis[5]. Tuberculous periostitis as well as traumatic lesions were also described in the early literature[6, 7]. More recently, the differential diagnosis has included metastatic or primary tumours, aneurysm, diabetic ophthalmoplegia, collagen-vascular disease, carotid-cavernous fistula, cavernous sinus thrombosis and ophthalmoplegic migraine. Orbital cellulitis or myositis and pseudotumour orbiti must also be excluded. The diversity of causes of this symptom complex has been emphasized by many, as has the importance of early identification of the more malignant processes. Only 3% of patients with a parasellar syndrome of painful ophthalmoplegia were found by Thomas and Yoss to have the idiopathic steroid-sensitive process[8]. It is critical to realize that the diagnosis of Tolosa-Hunt Syndrome is one of exclusion and is but one phenomenon which produces the symptom complex of painful ophthalmoplegia.

A thorough physical examination, looking particularly for malignancy is essential. Haematological and cerebrospinal fluid evaluation and neuroradiological studies help to detect or exclude specific infectious processes, collagen-vascular disease, tumour or aneurysm. Once the diagnosis of THS is suspected and other pathology is eliminated, some specific diagnostic findings may help to confirm the diagnosis. Rarely, various haematological abnormalities suggesting an autoimmune aetiology have been identified in some patients, as have CSF abnormalities[9]. None of these, however, are specific or constant. Orbital venography has shown superior orbital vein occlusion and partial occlusion of the affected cavernous sinus, although this too, is a nonspecific finding[10]. Angiography, usually normal, may show an irregularly narrowed or flattened intracavernous carotid artery which may be due to mass effect or spasm[11]. Arterial stationary waves have also been described[12]. All these changes again are nonspecific. Angiography, however, is essential to rule out aneurysm.

Cranial computed tomography has greatly improved the evaluation of pathology of the orbit, superior orbital fissure and the cavernous sinus. Definition of the optic nerve, adjacent cranial nerves and dural boundaries of the cavernous sinus is possible. A soft tissue mass in this region seen on CT has disappeared with steroid therapy[13].

Magnetic resonance imaging is another excellent method for visualizing this region[14]. The resolution of the cranial nerves and vascular channels in this area is superior to computerized tomography. Novel information about this perplexing syndrome may yet be obtained by this device.

We reiterate that, although surgical exploration and biopsy may provide a definitive diagnosis, this is not indicated until there is failure of other diagnostic efforts and failure of clinical response to steroids. In patients with negative angiography and CT scans, venography and other invasive procedures are not required and may pose unnecessary risks. Newer imaging techniques may ultimately eliminate the need for surgical intervention entirely.

A revision of the initial criteria proposed by Hunt

et al., is in order. The characterization of the pain as constant, boring and retroorbital, lasting for days to weeks with spontaneous remission and occasional recurrence is still accurate. It has been observed by many others since the initial report. The criterion of involvement of the third, fourth, sixth, first division of the fifth, sympathetic and parasympathetic fibres and even the optic nerve should be expanded. Patients meeting the other criteria of the THS have been found to have involvement of the second and occasionally even the third division of the trigeminal nerve and the facial nerve[15, 16, 17]. Thus, the involvement of cranial nerves which lie outside of the cavernous sinus also necessitates a revision of the final criterion that "there is no evidence of involvement of structures outside of the cavernous sinus", perhaps saying that "rarely involvement of structures outside of the cavernous sinus has been reported".

Many investigators have searched for associated systemic disease but the evidence is sparse. Elevated erythrocyte sedimentation rates, eosinophilia, positive rheumatoid factor, the presence of LE cells, as well as a case of positive ANA have been noted[9, 18]. These findings however, are uncommon. The evidence for an autoimmune aetiology or associated collagen-vascular disease is interesting but not convincing. However, the initial statement that "there is no associated systemic reaction" is no longer entirely valid.

As first suspected by Hunt and coworkers and confirmed by many others, the therapeutic effect of corticosteroids is important[2, 4]. In fact, a clinical response to a steroid trial has been considered by many to be diagnostic of the THS but caution must be exercised in using steroid responsiveness as the primary diagnostic measure for this syndrome. Several authors have emphasized the clinical improvement of symptoms seen in patients with chordoma, epidermoid and other parasellar masses[3, 8, 19]. With this in mind, a seventh criterion should be added to the original 6, stating that "systemic corticosteroid administration most often results in a dramatic improvement in pain and even neurological deficit within 24–48 hours. Failure of symptoms to respond to steroid therapy necessitates further evaluation".

Clarification of the use of the eponym "Tolosa-Hunt Syndrome" to better delineate the taxonomy of the syndrome is needed. There follows, accordingly, a brief review of the syndromes of the superior orbital fissure, orbital apex and pseudotumour orbiti.

First, the diagnosis requires a regional pattern as first discussed by Rochon-Duvigneaud. Second, the

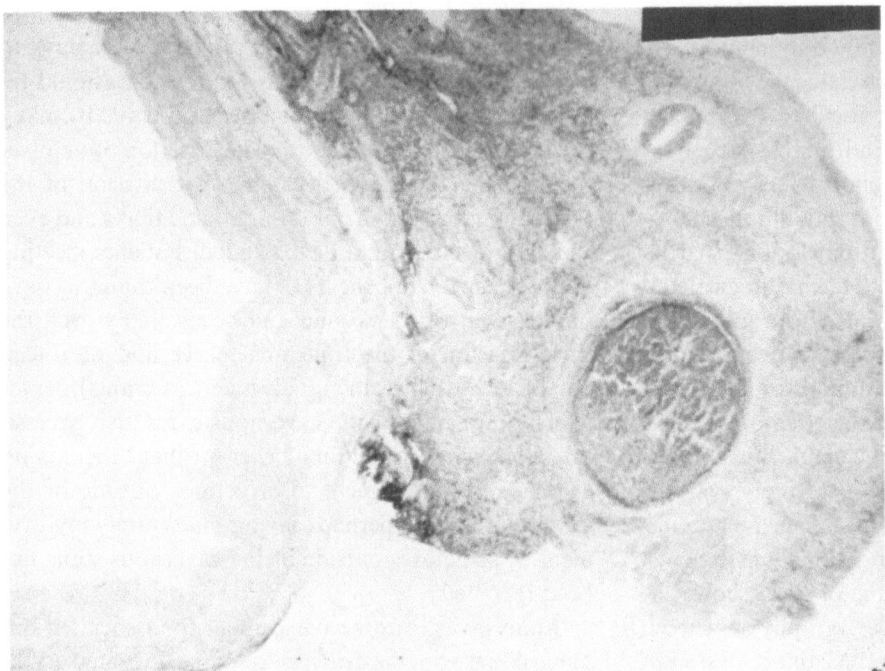

Fig. 1. Original autopsy specimen showing granulomatous process fusing the abducens nerve to the carotid artery. Note absence of involvement of the adventitia of the artery. (Slide courtesy of Dr. Tolosa)

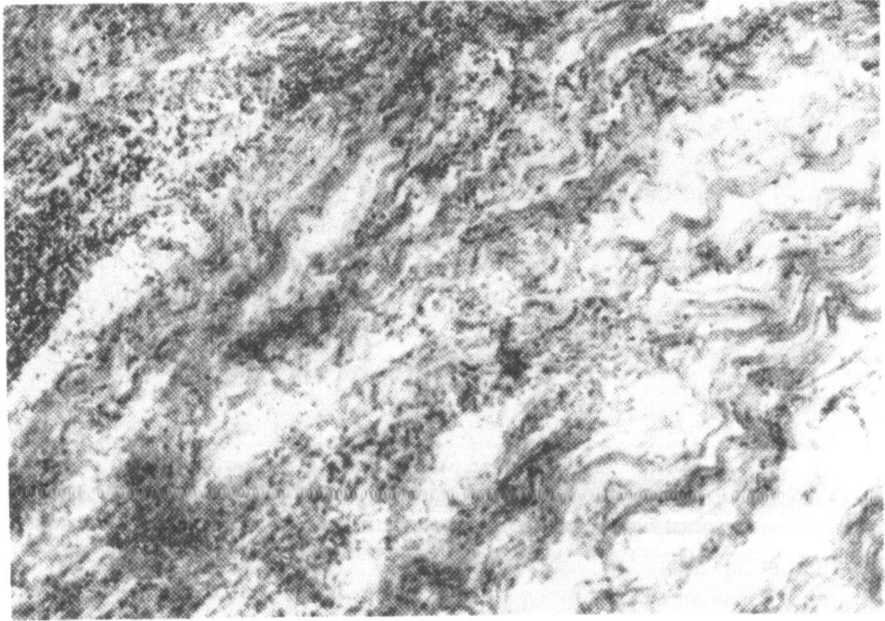

Fig. 2. The dural wall of the lateral cavernous sinus showing oedema and invasion by inflammatory cells. (Slide courtesy of Dr. Tolosa)

disease process must be identified by excluding other causes, by its self-limited character and by its response to therapy.

Collier in 1921 reported a group of patients with self-limited ophthalmoplegia and orbital pain which he attributed to an inflammatory lesion of the superior orbital fissure[20]. Tait, in 1934 described a similar case which he termed "rheumatic periostitis" of the superior orbital fissure[21]. Many years later, Lakke pathologically demonstrated a case of nonspecific pachymen-ingitis of this region in a patient with similar symptoms[22]. This so-called "fissuritis" seemed to be clinically similar to the Tolosa-Hunt Syndrome or cavernous sinus inflammation. The orbital apex syndrome with involvement of the optic nerve has also been thought to be due to a similar inflammatory process[7]. The nonspecific granlomatous inflammation described by Tolosa and Hunt, has been thought by some also to be responsible for the syndrome of orbital pseudotumour[23]. Both are exquisitely steroid-sensitive and some overalp in their clinical presentation may be seen, especially when the inflammation is particularly severe. The proptosis and orbital congestion seen with orbital pseudotumour is rare in THS, but has been observed[15]. Orbital myositis of unknown aetiology is a chronic inflammatory condition with clinical similarities to these other syndromes[24]. Collier speculated that there could be an association of THS with cranial polyneuropathy. In his study of 40 patients with ophthalmoplegia he suggested a similarity to Bell's palsy with regard to symptoms, causation and recovery[20]. Other authors have reported cases of peripheral facial palsy accompanying Tolosa-Hunt ophthalmoplegia[16, 17]. A review of idiopathic cranial polyneuropathy by Juncos and Beal states that after reviewing the original criteria for THS "the only distinguishing criterion from multiple cranial neuropathy is therefore the requirement that structures involved be confined to the cavernous sinus"[25]. There is, however, no evidence of a granulomatous process in multiple cranial neuropathy, vague as this condition is.

Pathological identification of these various syndromes is necessarily rare. Tolosa in 1954 reported his autopsy finding of nonspecific granulomatous "periarteritis" of the cavernous sinus[1]. Zeman in the initial paper disagreed, insisting that the artery was not involved (Figs. 1 and 2)[2]. Lakke, Schatz, and Framer found similar pathology in surgical biopsy specimens of the superior orbital fissure and cavernous sinus from patients with painful ophthalmoplegia[22, 26]. Since surgical exploration is rarely indicated and autopsy material virtually nonexistent, all hypotheses are speculative. Indeed, a clinicopathological spectrum of disease may exist. Variation in the signs and symptoms can be attributed to the variable extent of involvement of the structures of the orbit, superior orbital fissure and cavernous sinus. Perhaps the common denominator is involvement of the dural venous channels at the base of the skull.

As to when to apply the eponym "Tolosa-Hunt Syndrome", we emphasize again that, at present, the diagnosis is entirely one of exclusion. This diagnosis must not be made until all other aetiologies are confidently ruled out. Once this is done, if the case otherwise fits the above listed criteria, a trial of steroids is indicated. Prompt relief of pain, often accompanied by improvement of neurological symptoms, is characteristic (but not pathognomonic) of this syndrome. Spontaneous remission and exacerbation, as well as residual neurological deficit, occur. Thus, the eponym Tolosa-Hunt, if used, should apply solely to a benign self-limited syndrome of pain and neurological deficit referrable to the orbital apex, superior orbital fissure or anterior cavernous sinus which responds to steroid therapy. Some variations on this theme are to be expected.

In summary, the syndrome is still a curiosity. Many questions remain unanswered. What causes this subacute inflammtory response? Is it an allergic or autoimmune response? Why is the area of the orbit, superior orbital fissure and cavernous sinus peculiarly affected by such a process? What relationship, if any, does it have to other cranial polyneuropathies, periarteritis nodosa, ophthalmoplegic migraine, etc.? Perhaps new insights into immune mechanisms and other autoimmune diseases, aided by new imaging techniques, will shed further light on this mystery.

References

1. Tolosa E (1954) Periarteritic lesions of the carotid siphon with the clinical features of a carotid infraclinoid aneurysm. J Neurol Neurosurg Psychiatry 17: 300–302
2. Hunt WE, Meagher JN, LeFever HE, Zeman W (1962) Painful ophthalmoplegia. Its relation to indolent inflammation of the cavernous sinus. Neurology 11: 56–62
3. Hunt WE (1976) Tolosa-Hunt syndrome: one cause of painful ophthalmoplegia. J Neurosurg 44: 544–549
4. Smith JL, Taxdal DSR (1966) Painful ophthalmoplegia. The Tolosa-Hunt syndrome. Am J Ophthalmol 61: 1466–1472
5. Rochon-Duvigneaud (1896) Quelques cas de paralysie de tous les nerfs orbitaires (ophtalmoplégie totale avec amaurose et anesthésie dans le domaine de l'ophtalmique) d'origine syphilitique. Arch Ophtalmol (Paris) 16: 746–760
6. Hirschfield L (1858) Epanchement de sang dans le sinus caverneux du côte gauche diagnostiqué pendant la vie. CR Soc Biol 10: 138–140
7. Holt H, deRötth A (1940) Orbital apex and sphenoid fissure syndrome. Arch Ophthalmol 24: 731–741
8. Thomas JE, Yoss RE (1978) The parasellar syndrome, problems in determining etiology. Mayo Clin Proc 45: 617–623
9. Mathew NT, Chandy J (1970) Painful Ophthalmoplegia. J Neurol Sci 11: 243–256
10. Milstein BA, Morretin LB (1971) Report of a case of sphenoid fissure syndrome studied by orbital venography. Am J Ophthalmol 72: 600–603
11. Sondheimer FK, Knapp J (1973) Angiographic findings in the Tolosa-Hunt syndrome. Painful Ophthalmoplegia. Radiology 106: 105–112
12. Kettler HL, Martin JD (1975) Arterial stationary wave phenomenon in Tolosa-Hunt syndrome. Neurology 25: 765–770
13. Neigel JM, Rootman J, Robinson RG, Durity FA, Nugent RA (1986) The Tolosa-Hunt syndrome: computed tomographic changes and reversal after steroid therapy. Can J Ophthalmol 7: 287–290
14. Daniels DL, Pech P, Mark L, Pojunas K, Williams AL, Haughton VM (1985) Magnetic resonance imaging of the cavernous sinus. AJNR 6: 187–192

15. Kline LB (1982) The Tolosa-Hunt syndrome. Surv Ophthalmol 14: 79–95

16. Vallat JM, Vallat M, Julien J *et al* (1980) Painful ophthalmoplegia (Tolosa-Hunt) accompanied by peripheral facial paralysis. Ann Neurol 8: 645

17. Swerdlow B (1980) Tolosa-Hunt syndrome: a case associated with facial nerve palsy. Ann Neurol 8: 542–543

18. Lesser R, Jampol LM (1974) Tolosa-Hunt syndrome and antinuclear factor. Am J Ophthalmol 77: 732–734

19. Kline LB, Galbraith JG (1981) Parasellar epidermoid tumour presenting as painful ophthalmoplegia. J Neurosurg 54: 113–117

20. Collier J (1921) Discussion on ocular palsies. Proc Roy Sec Med 14: 10–11

21. Tait CBV (1934) Ophthalmoplegia associated with bony changes in the region of the sphenoidal fissure. Br J Ophthalmol 18: 532–535

22. Lakke JPWF (1962) Superior orbital fissure syndrome. Report of a case caused by local pachymeningitis. Arch Neurol 7: 289–300

23. Levy IS, Wright JE, Lloyd GAS (1975) Orbital and retro-orbital pseudotumours. Med Probl Ophthalmol 14: 364–367

24. Donin JF, Borit A (1977) Orbital myositis: its relationship to the Tolosa-Hunt syndrome. In: Smith JL (ed) Neuro-ophthalmology update. Masson Publishing Inc, New York, pp 99–103

25. Juncos JL, Beal MF (1987) Idiopathic cranial polyneuropathy. A fifteen year experience. Brain 110: 197–211

Address or correspondence: William E. Hunt, M.D., Division of Neurologic Surgery, The Ohio State University, 410 W. 10th Avenue, Room N-935 Doan Hall, Columbus, OH 43210 U.S.A.

Druck von Adolf Holzhausens Nfg., Universitätsbuchdrucker, Wien

 # Springer-Verlag

Acta Neurochirurgica

Supplementum 40

1987. 80 figures. X, 130 pages.
Cloth DM 165,–, öS 1150,–
Reduced price for subscribers to
"Acta Neurochirurgica":
Cloth DM 148,50, öS 1035,–
ISBN 3-211-82025-6

Supplementum 41

1987. 95 figures. V, 125 pages.
Cloth DM 165,–, öS 1150,–
Reduced price for subscribers to
"Acta Neurochirurgica":
Cloth DM 148,50, öS 1035,–
ISBN 3-211-82027-2

G. Csécsei, O. Hoffmann, N. Klug,
A. Laun, R. Schönmayr, J. Zierski

Primary and Secondary Brain Stem Lesions

This volume is the first to describe all clinically and experimentally relevant aspects of primary and secondary brain stem lesions important to clinicians. It contains a detailed description of the computer-tomographical and morphological changes of the cerebral cisterns in acutely and chronically increased intracranial pressure. The prognostic value of clinical parameters of primary and secondary brain stem lesions is demonstrated. The possibilities of assessing the clinical course by computer-aided evaluation are presented. In addition to that comprehensive view of morphological, radiological and clinical findings, extensive investigations concerning blink reflex (BR) and auditory evoked brain stem potentials (BAEP) supply highly relevant functional aspects of those lesions. The effects of raised intracranial pressure upon BR, BAEP as well as upon cerebral blood flow and focal flow in different brain areas were studied in animal experiments and reveal new and fascinating conclusions. Based on these investigations a mathematical model following modern concepts of system analysis was developed. The model includes the intracranial system, autoregulation of cerebral flow (cardiovascular components) and the short-time behaviour of arterial blood pressure regulation.

K. Sano, S. Ishii (Eds.)

Plasticity of the Central Nervous System

Proceedings of the Second Convention of the Academia Eurasiana Neurochirurgica, Hakone, October 5–8, 1986, Japan

The Leitmotiv of the Second Convention of the Academia Eurasiana Neurochirurgica was "Cerebrum convalescit"– literally "the brain recovers". The focus of the meeting was on plasticity of the central nervous system, one of the most decisive factors in recovery and readaption after cerebral lesions.

Distinguished experts from the fields of neurosurgery, neurology, neurophysiology, anatomy, pathology, oncology, and pharmacology discussed the following topics:

- Molecular and cellular basis of plasticity
- Regeneration and growth in the CNS
- Self-organization of neuronal network
- Brain oedema–a reparatory process?
- Growth factors and carcinogenesis

Acta Neurochirurgica/Supplementum 40   Acta Neurochirurgica/Supplementum 41

Springer-Verlag   Wien   New York
Moelkerbastei 5, Postfach 367, A-1011 Wien
Heidelberger Platz 3, D-1000 Berlin 33
175 Fifth Avenue, New York, NY 10010, USA
37-3, Hongo 3-chome, Bunkyo-ku, Tokyo 113, Japan

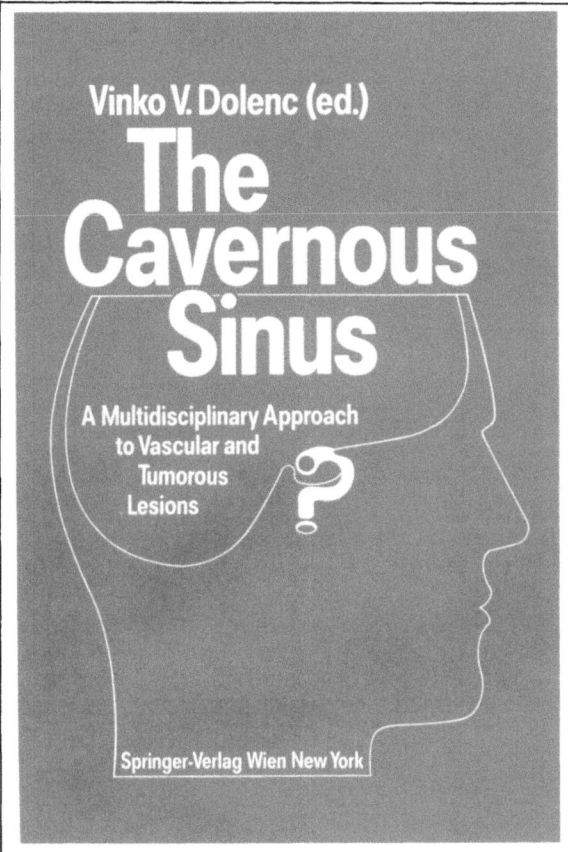